T0202684

Lecture Notes in Business Information Processing

487

LNBIP reports state-of-the-art results in areas related to business information systems and industrial application software development – timely, at a high level, and in both printed and electronic form.

The type of material published includes

- Proceedings (published in time for the respective event)
- Postproceedings (consisting of thoroughly revised and/or extended final papers)
- Other edited monographs (such as, for example, project reports or invited volumes)
- Tutorials (coherently integrated collections of lectures given at advanced courses, seminars, schools, etc.)
- Award-winning or exceptional theses

LNBIP is abstracted/indexed in DBLP, EI and Scopus. LNBIP volumes are also submitted for the inclusion in ISI Proceedings.

Joaquim Filipe · Michał Śmiałek ·
Alexander Brodsky · Slimane Hammoudi
Editors

Enterprise
Information Systems

24th International Conference, ICEIS 2022
Virtual Event, April 25–27, 2022
Revised Selected Papers

 Springer

Editors
Joaquim Filipe
Polytechnic Institute of Setúbal/ INSTICC
Setúbal, Portugal

Michał Śmiałek
Warsaw University of Technology
Warsaw, Poland

Alexander Brodsky
George Mason University
Fairfax, VA, USA

Slimane Hammoudi
ESEO, ERIS
Angers, France

ISSN 1865-1348 ISSN 1865-1356 (electronic)
Lecture Notes in Business Information Processing
ISBN 978-3-031-39385-3 ISBN 978-3-031-39386-0 (eBook)
https://doi.org/10.1007/978-3-031-39386-0

This Springer imprint is published by the registered company Springer Nature Switzerland AG
The registered company address is: Gewerbestrasse 11, 6330 Cham, Switzerland

Preface

The present book includes extended and revised versions of a set of selected papers from the 24th International Conference on Enterprise Information Systems (ICEIS 2022), which was exceptionally held as an online event, due to COVID-19, from 25–27 April 2022.

ICEIS 2022 received 197 paper submissions from 44 countries, of which 10% were included in this book.

The papers were selected by the event chairs and their selection is based on a number of criteria that include the classifications and comments provided by the program committee members, the session chairs' assessment and also the program chairs' global view of all papers included in the technical program. The authors of selected papers were then invited to submit a revised and extended version of their papers having at least 30% innovative material.

The purpose of the International Conference on Enterprise Information Systems (ICEIS) is to bring together researchers, engineers and practitioners interested in advances in and business applications of information systems. Six simultaneous tracks were held, covering different aspects of Enterprise Information Systems Applications, including Enterprise Database Technology, Systems Integration, Artificial Intelligence, Decision Support Systems, Information Systems Analysis and Specification, Internet Computing, Electronic Commerce, Human Factors and Enterprise Architecture.

The papers selected to be included in this book contribute to the understanding of relevant trends of current research on Enterprise Information Systems, including: Deep Learning, Business Modelling and Business Process Management, Digital Transformation, Industrial Applications of Artificial Intelligence, Operational Research, Software Engineering, User Experience, Applications of Expert Systems, Knowledge Management, Problem Solving, Models and Frameworks.

We would like to thank all the authors for their contributions and also the reviewers who have helped to ensure the quality of this publication.

April 2022

Joaquim Filipe
Michał Śmiałek
Alexander Brodsky
Slimane Hammoudi

Organization

Conference Co-chairs

Alexander Brodsky   George Mason University, USA
Slimane Hammoudi   ESEO, ERIS, France

Program Co-chairs

Joaquim Filipe   Polytechnic Institute of Setubal / INSTICC, Portugal
Michał Śmiałek   Warsaw University of Technology, Poland

Program Committee

Amna Abidi	INETUM-Paris, France
José Alfonso Aguilar	Universidad Autónoma de Sinaloa, Mexico
Adeel Ahmad	Université du Littoral Côte d'Opale, France
Zahid Akhtar	State University of New York Polytechnic Institute, USA
Patrick Albers	École Supérieure d'Électronique de l'Ouest, France
Javier Albusac	Universidad de Castilla-La Mancha, Spain
Julien Aligon	IRIT, France
Mohammad Al-Shamri	Ibb University, Yemen
Omar Alvarez-Xochihua	Universidad Autónoma de Baja California, Mexico
Leandro Antonelli	Universidad Nacional De La Plata, Argentina
Ada Bagozi	University of Brescia, Italy
Veena Bansal	Indian Institute of Technology Kanpur, India
Ken Barker	University of Calgary, Canada
Balbir Barn	Middlesex University London, UK
Rémi Bastide	University of Toulouse, France
Smaranda Belciug	University of Craiova, Romania
Marta Beltrán	Universidad Rey Juan Carlos, Spain
Jorge Bernardino	Polytechnic of Coimbra - ISEC, Portugal
Ilia Bider	Stockholm University, Sweden

Fotios Kokkoras	University of Thessaly, Greece
Christophe Kolski	Univ. Polytechnique Hauts-de-France, France
Rob Kusters	Open Universiteit Nederland, The Netherlands
Ramon Lawrence	University of British Columbia Okanagan, Canada
Carlos León de Mora	University of Seville, Spain
Giorgio Leonardi	University of Eastern Piedmont, Italy
Therese Libourel	University of Montpellier II, France
Kecheng Liu	University of Reading, UK
Stephane Loiseau	University of Angers, France
Maria Lopes	Universidade Portucalense Infante D. Henrique, Portugal
Pedro Lorca	University of Oviedo, Spain
Regis Magalhães	Federal University of Ceara, Brazil
Antonio Martí Campoy	Universitat Politècnica de València, Spain
Riccardo Martoglia	University of Modena and Reggio Emilia, Italy
Rafael Mayo-García	CIEMAT, Spain
Michele Melchiori	University of Brescia, Italy
Marcin Michalak	Silesian University of Technology, Poland
Michele Missikoff	ISTC-CNR, Italy
Hiroyuki Mitsuhara	Tokushima University, Japan
Norshidah Mohamed	Prince Sultan University, Saudi Arabia
Lars Mönch	FernUniversität in Hagen, Germany
Fernando Moreira	Universidade Portucalense, Portugal
Edward Moreno	Federal University of Sergipe, Brazil
Hamid Mukhtar	National University of Sciences and Technology, Pakistan
Pietro Murano	Oslo Metropolitan University, Norway
Omnia Neffati	College of Science and Arts in Sarat Abidah, Saudi Arabia
Leandro Neves	São Paulo State University, Brazil
Ovidiu Noran	Griffith University, Australia
Joshua Nwokeji	Gannon University, USA
Edson OliveiraJr	State University of Maringá, Brazil
Ermelinda Oro	National Research Council (CNR), Italy
Malgorzata Pankowska	University of Economics in Katowice, Poland
Silvia Parusheva	University of Economics, Varna, Bulgaria
Ricardo Pérez-Castillo	University of Castilla-La Mancha, Spain
Dana Petcu	West University of Timisoara, Romania
Dessislava Petrova-Antonova	Sofia University "St. Kliment Ohidski", Bulgaria
Matus Pleva	Technical University of Kosice, Slovak Republic
Filipe Portela	University of Minho, Portugal
Naveen Prakash	ICLC, India

Belen Vela Sanchez   Rey Juan Carlos University, Spain
Gualtiero Volpe   Università degli Studi di Genova, Italy
Miljan Vucetic   Vlatacom Institute of High Technologies, Serbia
Frank Wang   University of Kent, UK
Xikui Wang   University of California Irvine, USA
Janusz Wielki   Opole University of Technology, Poland
Adam Wójtowicz   Poznan University of Economics and Business,
                                      Poland
Mudasser Wyne   National University, USA
Surya Yadav   Texas Tech University, USA
Geraldo Zafalon   São Paulo State University, Brazil
Zaki Brahmi   Taibah University, Saudi Arabia
Qian Zhang   University of Technology Sydney, Australia
Yifeng Zhou   Southeast University, China
Eugenio Zimeo   University of Sannio, Italy

Additional Reviewers

Marcello Barbella   University of Salerno, Italy
Rafael Barbudo   University of Córdoba, Spain
Miguel Calvo Matalobos   Universidad Rey Juan Carlos, Spain
Andre Cordeiro   State University of Maringá, Brazil
Hércules do Prado   Catholic University of Brasília, Brazil
Guillermo Fuentes-Quijada   University of Castilla-La Mancha, Spain
Ricardo Geraldi   State University of Maringá, Brazil
Alix Vargas   University of Hertfordshire, UK

Invited Speakers

Davor Svetinovic   Vienna University of Economics and Business,
                                      Austria
Brian Fitzgerald   University of Limerick, Ireland
Jan Mendling   Humboldt-Universität zu Berlin, Germany
Birgit Penzenstadler   Chalmers University of Technology, Sweden

Contents

Databases and Information Systems Integration

Balancing Simplicity and Complexity in Modeling Mined Business Processes: A User Perspective

D. G. J. C. Maneschijn(✉), R. H. Bemthuis, J. J. Arachchige, F. A. Bukhsh, and M. E. Iacob

University of Twente, Drienerlolaan 5, 7522 NB Enschede, The Netherlands
d.g.j.c.maneschijn@student.utwente.nl

Abstract. Process mining techniques use event logs to discover process models, check conformance, and aid business analysts in improving business processes. A proper interplay between the complexity of discovered process models and the expertise of human interpreters, such as stakeholders, domain experts, and process analysts, is key for the success of process mining efforts. However, current practice often suffers from a misalignment between aligning the technical or business background of a human subject and the results obtained through process mining discovery techniques. This results in either overly complex or overly simplistic discovered process models for particular users. In this study, we propose and validate a methodology for aligning users of process models with different abstractions of discovered process models. We present two case studies to illustrate the usability of our six-phase methodology and demonstrate its potential. We hope that these findings will encourage systematic approaches for better alignment between users' willingness to employ process models and process model abstraction.

Keywords: Process mining · Methodology · Abstraction levels · Stakeholder analysis · Process models

1 Introduction

The growing availability of data in information systems offers a rich source for examination and analysis [20]. One prevalent data entity is historical execution data recorded in event logs [1]. The relatively young discipline of process mining has emerged to scrutinize those event logs, discover process models, evaluate conformance, and improve processes. Process mining techniques can offer insights into real-world data, including performance evaluations, root-cause analysis, process prediction, and optimization [1]. Another application of process mining concerns identifying and eliminating bottlenecks [11].

A major challenge in the field of process mining is discovering process models that are detailed enough without overwhelming users with unnecessary details [2,27]. Users of process models often have diverse backgrounds and interests, including domain experts, technicians proficient in process mining tools and algorithms, business owners, and even those whose data is recorded in log files. Extracted process models can be

© The Author(s), under exclusive license to Springer Nature Switzerland AG 2023
J. Filipe et al. (Eds.): ICEIS 2022, LNBIP 487, pp. 3–21, 2023.
https://doi.org/10.1007/978-3-031-39386-0_1

difficult to interpret, particularly when they are Spaghetti process models [3]. However, unstructured process models may still be useful for specific users who are interested in exceptional and complex behaviors. In general, if no abstraction is applied to these models, they may be too complex for most users to understand [1].

Abstraction involves removing low-level details from the visualization of the discovered process [2]. It determines the level of granularity of the process model. More specifically, abstraction simplifies process models by removing edges, clustering nodes, and removing nodes to make the process model more suitable for the person looking at it [2]. Key to abstraction is finding a process model that is neither too highly abstracted (underfitting) nor too detailed (overfitting). Furthermore, the desired abstraction level also greatly depends on the skills and desired outcome of the person using the process model [28]. Higher-level management staff may need simplified process models for strategic decision-making, while those working 'in the trenches' of a process typically focus more on detailed process executions.

Several studies have addressed abstraction in process mining [7, 13, 24, 26, 38]. There also exists a range of process discovery algorithms that utilize domain knowledge in addition to event data [34]. However, little research has been done on supporting the user in specifying the domain knowledge in the context of discovered process models [34]. Current process mining practices still strongly rely on the input (e.g., assumptions and knowledge) from domain experts [15], which is of great importance for assessing the discovered process models. In this paper, we aim to find a balance between simplicity and complexity in mined process models, with a focus on the user perspective. More specifically, we are primarily concerned with finding an appropriate level of abstraction for process models that improves usability and user acceptance.

In this paper, we propose and validate a methodology for aligning the abstraction levels of discovered process models with user needs. Our goal is to discover process models that consider predictors of user acceptance. To achieve this, we analyze user perceptions, such as ease of use and usefulness [14], to determine a user's willingness to employ a process model. Building on previous work [28], we further assess and demonstrate an initial methodology. Previous research suggests that both subjective measures (e.g., usefulness and ease of use) and objective measures of complexity (e.g., number of nodes) are important for users who will ultimately employ the discovered process models. Consideration of stakeholders' needs is crucial when evaluating the quality of process models [28]. Specifically, different users may require unique process model representations based on factors such as organizational needs, background knowledge, and expertise. To validate the methodology, we apply it to a new case study and compare the results to those of the original case study presented in [28]. This allows us to evaluate the methodology's applicability and provide a broader understanding of its execution. While the original case study uses synthetic data within a logistics domain, the new case study is based on real-life event logs and concerns the reimbursement of travel costs. In summary, this study presents a novel methodology and demonstrates its effectiveness through two case studies, making it a design science research. Therefore, we follow Peffers's Design Science Research Methodology (DSRM) [31] and include a second iteration of the artifact (i.e., methodology), in addition to the results published in [28].

The remainder of this paper is organized as follows. Section 2 discusses the proposed methodology. Section 3 demonstrates the methodology by discussing the two case studies and their cross-validation. Section 4 discusses the related work. Finally, Sect. 5 concludes and gives some directions for future work.

2 A Methodology for Aligning Process Model Abstraction and Users of Process Models

The CRISP-DM methodology [35], an established methodology for data science projects, served as the foundation for the development of our proposed methodology (see Fig. 1). We expanded upon the initial version of the methodology presented in [28] by incorporating suggested tasks for each phase. Below, we provide a detailed description of each phase of our proposed methodology.

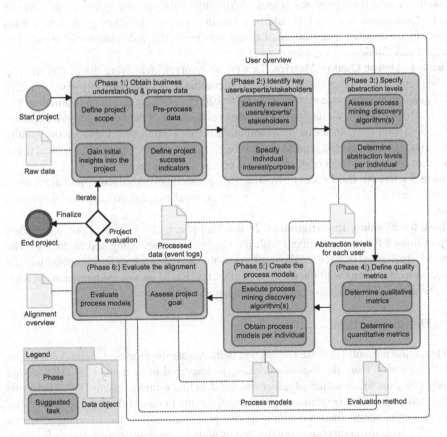

Fig. 1. A methodology for aligning stakeholders' needs to process model abstraction levels (adapted from [28]).

Phase 1 - Obtain Business Understanding and Prepare Data. The first phase involves gaining familiarity with the organizational/business context of the data. The objective is to acquire as much insights as possible into the goals of the project. Understanding the current organizational situation, including its objectives and key decision-makings, enables the user to develop a preliminary understanding of the project's aims. This phase assumes the availability of raw data that may require pre-processing before use.

Phase 2 - Identify Key Users/Experts/Stakeholders. In this phase, the business understanding is refined by specifying the key users, experts, and/or stakeholders who will evaluate the discovered process models. The organizational and business goals of each involved user are clarified to determine the appropriate levels of abstraction. Active management of relevant stakeholders and addressing their needs can benefit the organization [23,32].

Phase 3 - Specify Abstraction Levels. This phase involves defining the various abstraction levels at which the process models should be produced. Factors to consider include the properties of the process mining algorithms utilized, as well as the contextual information gleaned from previous phases. Additionally, the specific nature of the process mining algorithm that is used could be taken into account. Ultimately, this phase aims to produce a detailed specification of abstraction levels tailored to the needs of each key project stakeholder.

Phase 4 - Define Quality Metrics. This phase involves identifying the quality metrics to be employed in the evaluation process. A combination of both quantitative/objective and qualitative/subjective metrics may be utilized to ensure a thorough evaluation. For qualitative measures, we recommend using an adaptation of the Technology Acceptance Model (TAM) [14]. In terms of quantitative metrics, one could consider the quality dimensions of discovered process models (e.g., fitness), as well as metrics related to the 'size' of a process model (e.g., number of nodes, activities, etc.).

Phase 5 - Create the Process Models. This phase involves selecting and applying a process mining discovery algorithm to generate an abstracted process model for each user.

Phase 6 - Evaluate the Alignment. In the final phase, the generated process models are evaluated for their alignment with the needs of corresponding stakeholder(s). The previously defined metrics will be used assess the degree to which the complexity of the process model aligns with the user's requirements. Based on this evaluation, a decision can be made on whether to continue or finalize the data project.

3   Demonstration

This section presents two case studies that demonstrate the proposed methodology. The first case study, from the logistics domain, is described in [10] and [28]. The second study concerns the handling of travel request data and reimbursement [17]. These two case studies provide complementary perspectives due to their differing domains (logistics processes versus financial administrative processes) and datasets (synthetic versus real-life). have different characteristics and provide a complementary perspective. First, the domain differs (i.e., logistics processes and financial administrative processes). Second, the nature of both datasets differs (i.e., synthetic dataset versus real-life dataset). Conducting multiple case studies aligns with the DSRM by allowing, as it allows for further demonstration and validation of the proposed methodology. This section discusses each case study and concludes with a cross-validation of the two cases.

3.1 Case Study One: Logistics in Production Facility

The first case study examines the transportation of perishable goods via autonomous vehicles within a production facility [10]. The corresponding dataset, which is publicly available [9], encompasses information on the state of the shipments, transport units, and the environment, among other values. This logistics case study is merits exploration for several reasons: (1) it is part of research conducted within a similar project consortium, (2) logistics environments involve numerous stakeholders, rendering it an ideal candidate for demonstrating the various perspectives of our methodology, and (3) the dataset is publicly available [9]. Despite the data being simulated, the case study is still a suitable candidate for demonstrating the application of our methodology. The results of the methodological phases are discussed below.

Phase 1: Obtain Business Understanding and Data Preparation

This phase focuses on understanding the case study and characteristics of the logistics domain. Logistics processes are often complex and dynamic due to factors such as unexpected disruptions and behaviors. For a detailed description of the project scope and initial insights, we refer the reader to the work described in [10]. The goal of this project is to obtain a suitable process model abstraction for each relevant stakeholder.

The event logs contain activities related to movable transport units (autonomous vehicles) that transport smart returnable transport items (pallets) [10]. The smart pallets and vehicles are equipped with sensor technologies that record the status of products and transporters. Product quality decays over time and the rate of decay depends on factors such as vehicle and food type.

As part of data preparation, we enriched the event logs by selecting those with the warm-up period removed and with the lowest average end product decay level. We also added two attributes: decay level (DL) and vehicle identifier. We categorized the DL into four classes (see Table 1) for use in subsequent phases of this methodology. Finally, we removed incomplete traces by applying a heuristics filter plugin from the open-source tool ProM to the event logs.

Table 1. Product quality decay categories for data enrichment [28].

Quality category	Partition of
Good	$DL \geq \mu + \sigma$
Sufficient	$\mu \leq DL < \mu + \sigma$
Insufficient	$\mu - \sigma < DL < \mu$
Poor	$DL \leq \mu - \sigma$

Phase 2: Identify Key Users/Experts/Stakeholders

The utilization of a mined process model by a user may hinge on its comprehensibility in relation to the user's objectives. In this case study, we delineated a roster of representative users within a logistics organization, encompassing their respective aims and organizational interests (see Table 2).

Table 2. Identified possible process model users [28].

Stakeholder	Purpose	Organizational interest
Operational board	Identify the overall workflow of the company	Top
CFO	Get to know the overall cost picture and identifying the specific causes of high costs	Top & Middle
Planner	Identify all the steps an order undergoes and where delays occur	Middle
Driver	Find out what activities constitute to their specific task	Bottom
Exception manager	Spot exceptions and find out how they occurred	Bottom
IT expert	Find out what parts of the process require more extensive logging	Middle
Regulations expert	Make sure all steps necessary for regulation measures are taken	Bottom
Customer relations	Ensure traceability and timeliness of the orders	Bottom

We mainly focused on primary stakeholders due to their formal relationship with the organization and their expected concern about the production facility's operations [21]. We assumed that primary stakeholders also represent the needs of secondary stakeholders.

To align the purpose of our process model with each stakeholder's needs, we identified their organizational goals and interests using the work of [5,23,25,32], as shown in Table 2. In the next phase of our methodology, we will use the list of defined users.

Phase 3: Specify Abstraction Levels

We applied the fuzzy process mining discovery algorithms to extract process models. This miner can produce various abstractions of process models by adjusting the significance and correlation thresholds [2,24]. Significance refers to the relative importance of behavior, while correlation refers to the precedence relationship between two events. The fuzzy miner is suitable for obtaining various abstraction levels of process models, as the complexity of the model can be modified by adjusting the parameters.

Several thresholds can be used to simplify a process model obtained through the fuzzy miner. We consider three, as shown in Table 3. These thresholds influence the simplification process differently. Other threshold, such as the preserve threshold and the ratio threshold, will not be considered, because they affect the nodes with conflicting relations [24] and our dataset does not contain such relations.

Table 3. Considered thresholds for the fuzzy miner [28].

Threshold	Value = 0	Value = 1	Application
Utility ratio (UR)	High correlation/ low significance	High significance/ low correlation	Edge filtering
Edge cutoff (EC)	Diminishes utility ratio	Amplifies utility ratio	Edge filtering
Node cutoff (NC)	Less abstract	More abstract	Node filtering

In general, a lower threshold value results in a less abstract process model. The utility ratio (UR), however, does not directly influence the abstraction of the process model as it addresses a combination of significance and correlation. Therefore, we kept this value constant. Additionally, there were few activities with a high significance. To prevent the model from containing only a single cluster cluster, we used a relatively low node cutoff (NC) value.

We specified four abstraction levels (A, B, C, D) (see Table 4), where level A corresponds to the most simplified process model and level D is the most complex. Also, we show an indicative organizational level (corresponding to the levels specified in Table 2).

Table 4. Abstraction levels [28].

Abstraction level	UR	EC	NC	Expected organizational level
A	0.5	1.0	0.4	Top
B	0.5	0.8	0.25	Middle
C	0.5	0.6	0.1	Middle
D	0.5	0.4	0.0	Bottom

Phase 4: Define Quality Metrics

We opted to use both quantitative and qualitative metrics to assess the process models. Quality is defined in terms of fitness, perceived usefulness, and perceived ease of use. Our quantitative metrics include fitness, level of detail (measured as the percentage of nodes preserved in the process model), and the number of nodes, edges, and clusters displayed. Fitness measures how well the behavior described in an event log is represented in the process model [4]. This score provides insights into the quality of a process model, as a process model is that does not accurately represent reality may raise concerns among its users.

For a subjective assessment of process models, we combined TAM with an expert analysis. We consulted a panel of experts familiar with the case study to evaluate the process models from the perspective of different stakeholders. TAM is a widely used approach for assessing the acceptance of new technology and has been applied in process mining research (e.g., [22, 36]). The standard TAM provides a series of questions about the perceived usefulness and ease of use from an end-user's perspective. In phase 6 of our methodology, we discuss how we adapted the questionnaire to account for the quality assessment of multiple users.

Phase 5: Create the Process Model(s)

Following the abstraction levels specified in Sect. 3.1, we applied the fuzzy miner algorithm to discover the process models. Figure 2 illustrates the resulting process models, which exhibit varying levels of complexity. Notably, the complexity of the obtained process model as reflected in the number of nodes and edges.

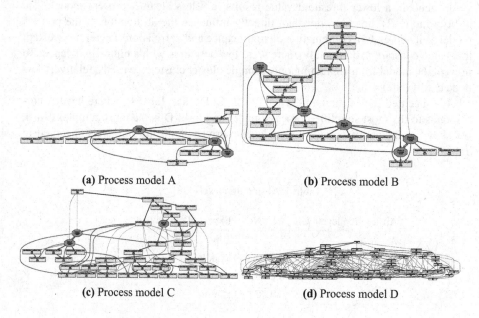

(a) Process model A   (b) Process model B

(c) Process model C   (d) Process model D

Fig. 2. Process model abstractions for case study one [28].

Phase 6: Evaluate the Alignment

In this phase, we evaluate the discovered process models using defined quality metrics to determine their effectiveness. This provides insights into the preferences of different stakeholders for specific process models. The results of the quantitative and qualitative assessments are discussed separately below.

Quantitative Results. Table 5 presents the results of the quantitative analysis. The data indicate that process model A, with a high level of abstraction, has a relatively low number of nodes and edges, while a process model D, with a lower level of abstraction, has more nodes and edges. The discovered process model D contains the most activities and edges, resulting in a high fitness. Nevertheless, this model might be perceived as

Table 5. Evaluation of process models [28].

Process model	Fitness	Detail	Nodes	Edges	Clusters (nodes)
A	99.84%	52.44%	13	21	3 (28, 24, 8)
B	96.26%	69.17%	20	40	5 (24, 12, 8, 5, 3)
C	91.99%	88.71%	38	93	3 (19, 6, 2)
D	99.00%	100%	73	150+	0

cluttered. Note that model C has clustered nodes and a lower fitness level, possibly due to removed edges or missing nodes within the cluster. Process model B contains even more clustered nodes and a reduced number of edges, with relatively small clusters that may be easier to comprehend. Despite the lower fitness of models B and C compared to models A and D, their fitness is still considered sufficient for demonstration purposes. However, important information may be omitted. Lastly, process model A has the fewest nodes and edges, leading to insufficient detail despite its relatively high fitness, potentially causing incomprehensibility.

Qualitative Results. The results of our expert analysis using the TAM approach are presented in Table 6. The table shows the perceived usefulness and ease process model, as evaluated by eight experts in the logistics domain. Each expert assumed the role of two stakeholders, as specified in Table 2, and scored the process models (on a scale from 1 to 5) based on their designated stakeholder's perspective. The average scores are shown in Table 6.

The results indicate that process model A received the lowest score for usefulness, possibly due to the limited level of detail. The other process models received scores above 3 for usefulness, suggesting that the level of abstraction is suitable for the stakeholders.

In terms of ease of use, experts found process model D difficult to understand, as anticipated by stakeholders at top or middle organizational levels. However, some stakeholders in lower-level roles found process model D relatively easy to use, as evidenced by their individual scores. The other models received an average score above 3, indicating that stakeholders generally perceived them as easy to use.

To align stakeholders with the complexity of a process model, several factors can be considered, including perceived usefulness, ease of use, stakeholder type, organizational influence, or other prioritization indicators. For instance, model C can be excluded due to its low fitness, and the scores on perceived usefulness and ease of use can be discussed among stakeholders to achieve final alignment (as shown in Table 7).

Table 6. Results of the Technology Acceptance Model (scale 1-5) [28].

Construct	Average			
General Information				
1. I have much experience with business process modelling in general	3.19			
2. I have much experience with process mining in general	2.63			
Model	**A**	**B**	**C**	**D**
Usefulness				
3. The information presented in this model is useful for my daily job	1.94	3.13	3.56	3.19
4. The model is suitable for gaining new insights about the business process	1.56	3.31	3.81	3.50
5. The model contains detailed information about the business process	1.38	2.75	4.13	4.69
6. The model helps forming an understanding of the business process in general	2.31	3.88	3.56	2.25
Average	**1.80**	**3.27**	**3.77**	**3.41**
Ease of Use				
7. The model is understandable when taking a first look at it	4.00	3.56	2.88	1.56
8. It is easy to learn understanding this model	4.06	3.50	3.06	1.75
9. It is easy to explain this model to other persons inside the organization	3.75	3.69	3.06	1.50
10. Someone without experience in process mining is able to understand this model	3.50	3.44	2.56	1.50
11. I will use the information obtained from this model in my daily job	1.88	3.31	3.63	2.63
12. This model helps me achieve my purpose inside the organization	1.69	3.31	3.81	2.94
Average	**3.15**	**3.47**	**3.17**	**1.98**

It is important to note that, in our case study, experts familiar with the logistics domain and the semantics of process models were used instead of actual users due to time and project constraints. Additionally, multiple experts were involved and provided with a description of the case study and stakeholder information. This case study serves as an example of the application of the proposed artifact rather than an exhaustive evaluation of a specific case.

Table 7. An example of preferred process model abstractions [28].

Stakeholder	Abstraction level	Stakeholder	Abstraction level
Operational board	B	Exception manager	D
CFO	B	IT expert	B
Planner	B	Regulations expert	D
Driver	B	Customer relations	B

3.2   Case Study Two: Reimbursement of Travel Cost

The second case study examines a dataset of travel requests and reimbursements for employees at the Eindhoven University of Technology (TU/e) in the Netherlands [16]. The dataset includes anonymous information on travel requests (for international and domestic travel), prepaid travel reimbursement requests, and non-travel requests [17]. The focus of the second case study is to demonstrate the application of the methodology in a different domain using a different dataset.

Phase 1: Obtain Business Understanding and Data Preparation

The initial phase necessitates an understanding of the domain's primary processes and data preparation. We begin by briefly describing the travel cost reimbursement process. Employees can request reimbursement for travel costs and related expenses [17]. There are two types of travel categories: domestic and international.

Travel is categorized into two types: domestic and international. Prior permission is not required for domestic trips, allowing employees to undertake such travel and request reimbursement afterward. In contrast, international travel requires prior approval through a travel permit [17]. The employee submits a request to the travel administration for approval. If approved, the request is forwarded to the budget owner and supervisor [17]. In some cases, the budget owner and supervisor may be the same person, and director approval may also be necessary. If the request is denied, the employee may resubmit or withdraw it. If approved, payment is made.

As the dataset is relatively clean, no additional data filtering performed. Similarly, no data enrichment was due to the event log's real-life nature, which provided rich data complexity, including concurrent and sequential relations. These insights are key for subsequent phases, underscoring the importance of the first phase.

Phase 2: Identify Key Users/Experts/Stakeholders

In this case study, we will not explore the identification of stakeholders. Instead, we will illustrate how stakeholders can be engaged in thix context.

The reimbursement process occurs within a large organization, primarily involving internal stakeholders. These stakeholders may include individuals seeking to gain insights into the process through a process mining project, such as comparing management's intended procedure with the actual execution of the process. Relevant stakeholders may comprise employees applying for reimbursement, staff members responsible

Table 8. Abstraction levels for the second case.

Abstraction level	UR	EC	NC	PT	RT
A	0.5	1.0	0.5	0.8	0.5
B	0.5	0.8	0.3	0.6	0.5
C	0.5	0.6	0.15	0.4	0.5
D	0.5	0.4	0.0	0.2	0.5

for verifying or validating applications, and the university's financial department. As in the first case study, it is advisable to assign organizational levels to each stakeholder, although this may be omitted if clear distinctions in organizational interests are not discernible.

Phase 3: Specify Abstraction Levels

In this case study, we again use the fuzzy miner to discover process models, with abstraction levels shown in Table 8. Unlike Table 3, we include additional thresholds: the preserve threshold (PT) and the ratio threshold (RT), both used for abstracting conflicting/concurrent relations within an event log [24]. Our dataset contains such relations, so we account for them in our abstraction. This highlights the importance of properly understanding the data in the first phase. We keep RT at 0.5 for uniform treatment of concurrent relations. We adjust some threshold values, such as NC, based on the complexity of the event log and process models (see Table 8), while keeping others, such as EC and UR, unchanged.

Phase 4: Define Quality Metrics

In this case study, we employ a strictly quantitative analysis. While we acknowledge the importance of qualitative and subjective analyses, they fall outside the scope of this particular investigation. Our quantitative approach utilizes the following metrics: fitness, level of detail, number of nodes, number of edges, and number of visible clusters within the model.

Phase 5: Create the Process Model(s)

During the modeling phase, we generated process models at various levels of abstraction. In contrast to our first case study, we utilized 'best' edges rather than 'fuzzy' edges. The fuzzy miner algorithm generates 'fuzzy' edges to accurately represent real-life behavior within the process model. The resulting process models are depicted in Fig. 3.

Phase 6: Evaluate the alignment

In the final stage, we evaluated the generated models using the results obtained from the previous phases. The outcomes of the quantitative analysis are presented in Table 9.

Table 9. Evaluation of process models.

Model	Fitness	Detail	Nodes	Edges	Clusters (nodes)
A	87.15%	52,55%	8	28	3 (14, 2, 28)
B	84.85%	65.02%	11	46	7 (2, 2, 8, 15, 7, 2, 3)
C	83.02%	81.65%	19	71	9 (2, 2, 2, 6, 4, 2, 3, 5, 5)
D	70.2%	100%	51	100+	0

The relatively low fitness observed in all four models is likely attributable to the nature of the dataset and the restrictions imposed by the fuzzy miner. For instance, the exclusion of fuzzy edges in the models may have contributed to a decrease in fitness. However, a trend of increased fitness was observed as models became less detailed, potentially due to the fact that abstraction tends to increase fitness by grouping behavior. Since behavior within a cluster is not modeled, it is less likely to contain constructs that negatively impact fitness. These quantitative findings provide initial insights into aligning process model abstraction levels with user needs.

3.3   Cross-Analysis of the Case Studies

We demonstrated the proposed methodology through two case studies. The first case study, originating from the logistics domain, utilized a dataset from a simulation study. The second case study examined the application of reimbursements within an organization using real-life data, which was comparatively richer in terms of complexity. Upon applying the proposed methodology, we identified similarities, differences, and challenges between the two case studies. Below, we will discuss the initial insights obtained from these case studies.

Phase 1: Obtain Business Understanding and Prepare Data

The first dataset is relatively straightforward due to the controlled environment in which it was simulated. However, we enhanced the dataset by applying a heuristics filter and enriching the event logs to capture more meaningful behaviors for stakeholders. In contrast, the second case study did not undergo filtering and enrichment. We found that the first phase is essential for the subsequent phases, as it enables a deeper understanding of the datasets and their originating domains.

Phase 2: Identify Key Users/Experts/Stakeholders

This research emphasizes the ease of understanding the mined process models, both in terms of their complexity and simplicity. The first case study identified users and their

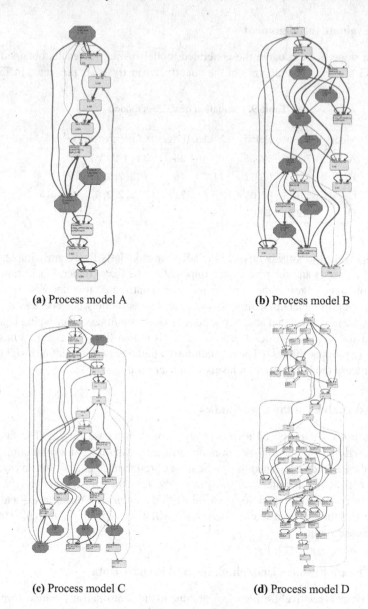

(a) Process model A   (b) Process model B

(c) Process model C   (d) Process model D

Fig. 3. Process model abstractions for case study two.

abstraction levels. In the second case study, we provide examples of internal stakeholders that can be considered. Defining abstraction levels for individual users beforehand can be pose challenges, such as (1) stakeholders having diverse intentions and goals for a process model, (2) the need to consider a range of internal and external stakeholders with varying perspectives and backgrounds, (3) the limitations of process mining algorithms not being specifically designed for a particular stakeholder, and (4) the possibility of needing multiple process model abstractions.

Phase 3: Specify Abstraction Levels

As demonstrated in both case studies, the fuzzy miner enables the specification and adjustment of the process discovery algorithm's parameters. This allowed us to obtain an initial set of process models with different abstraction levels tailored to our requirements. In the first case study we used three parameters that affected the obtained process model. In the second case study, additional thresholds were included in the specification of the abstraction levels due to the complexity and diverse type of relations in the dataset. This highlights that the definition of abstraction levels is case study-specific and requires interpretation by the practitioner(s). Additionally, the nature of the data must be considered.

Phase 4: Define Quality Metrics

In the first case study, the simplicity and complexity of the mined process models were assessed using both quantitative and qualitative metrics. In contrast, the second case study relied solely on quantitative analysis. As discussed in Sect. 2, employing various types of analyses is crucial for aligning user needs with different levels of abstraction.

Phase 5: Create the Process Models

The comprehensibility of the discovered models can be enhanced by tuning them based on the complexity of the event logs. In the first case study, we incorporated additional behavior into the process model using 'fuzzy edges', although this may not accurately reflect real-life behavior. No additional behaviors were introduced in the second case study. Practitioners must determine the most suitable process mining algorithm and its tuning parameters. Other techniques, such as filtering event logs, can also be employed to obtain process model abstractions [34].

Phase 6: Evaluate the Alignment

The final phase of our proposed methodology involves evaluating the alignment between the process model and its intended users. The evaluation method, as discussed in phase 4, influences the outcome of this phase. In the first case study, we employed both quantitative and qualitative analyses, while the second case study relied solely on quantitative indicators. Our findings suggest that quantitative metrics are suitable for use in various domains and provide a preliminary evaluation of each process model's 'quality'. The selection of appropriate qualitative metrics may depend on the group of stakeholders involved in the case study. Future work may further explore the relationship between qualitative and quantitative indicators.

4   Related Work

Event data can be recorded at varying levels of granularity, which can result in process models that are either too complex or too simplistic. The literature does not provide a definitive answer on the appropriate level of detail, but the resulting process models should be understandable [13]. In reality, granular event logs often produce spaghetti or lasagna process models [1]. To address these complex and (semi-)unstructured models, various approaches have been proposed. The discovery of process models from complex event logs has been extensively researched, but event logs may suffer from quality and accessibility deficiencies. The quality dimensions proposed by [6] can serve as a guide when creating process models from complex event logs. Pre-processing techniques can ensure an appropriate level of granularity [38], and other approaches can determine the level of detail based on the target audience for the process model, resulting in tailored views for specific audiences such as customers [12] or healthcare providers [29].

Numerous algorithms have been developed for event log abstraction in process mining [30, 38]. A systematic taxonomy has been proposed to compare and contrast existing discovery approaches [34]. Clustering algorithms have been used to reduce over-generalization in process models [30] and to cluster heterogeneous datasets for specific purposes [8]. External domain knowledge has been used to make process models more understandable for business users [2, 7, 24]. Methods have been developed for transforming event logs into preferred abstraction levels for analysts [18], and frameworks have been proposed to describe process models at an activity level [19]. However, there is a need for future research on methods to determine the main processes for specific purposes and stakeholders [33]. Our study takes into account the perspective of various stakeholders when assessing process models and is in line with the client-server-based application proposed by [37], which aims to gradually abstract fine-grained event logs without losing essential information.

5   Conclusions and Future Work

The increasing use of process mining in industry has raised concerns about the comprehension of process models by various stakeholders. Differences in understanding, background, and expertise among process model users must be taken into account. In this paper, we propose and validate a methodology for aligning process model users with the abstraction levels of discovered process models, building on the initial work presented in [28]. This methodology aims to identify an appropriate abstraction level for a given user. We applied the Design Science Research Methodology (DSRM) to address the generalizability and usability of our approach by conducting two case studies: one in logistics and one in reimbursement processes. Cross-validation of both case studies revealed that following the phases and tasks outlined in our methodology provides insights into how the complexity of a process model aligns with the needs of the user. Our methodology offers an initial set of actions to achieve alignment of user needs and process model abstraction levels, but its execution may vary across different applications and the tasks presented are not exhaustive.

The literature presents numerous opportunities for process model abstraction. Our methodology for the logistics case study was evaluated using expert reasoning and a synthetic dataset that mimics reality. However, future evaluations using real users and datasets may provide valuable insights into the creation and abstraction techniques of process models from the perspective of stakeholders. While we provided examples of tasks to be included in the various phases of our methodology, these tasks are not exhaustive or complete. Future work will focus on developing more comprehensive methods for achieving desired process models for for specific stakeholders, potentially through the use of ranking mechanisms such as multi-criteria decision-making methods. Future validation of the methodology should be conducted through large-scale, real-life case studies. Other avenues for future research may include extending the methodology to provide abstraction possibilities during multiple phases.

Acknowledgements. Rob Bemthuis has received funding from the project DataRel (grant 628.009.015), which is (partly) financed by the Dutch Research Council (NWO).

References

1. van der Aalst, W.M.P.: Process Mining: Data Science in Action. Springer, Heidelberg (2016). https://doi.org/10.1007/978-3-662-49851-4
2. van der Aalst, W.M.P., Gunther, C.W.: Finding structure in unstructured processes: the case for process mining. In: Seventh International Conference on Application of Concurrency to System Design (ACSD), pp. 3–12. IEEE (2007). https://doi.org/10.1109/ACSD.2007.50
3. van der Aalst, W.M.P.: Process mining: discovering and improving spaghetti and lasagna processes. In: 2011 IEEE Symposium on Computational Intelligence and Data Mining (CIDM), pp. 1–7. IEEE (2011). https://doi.org/10.1109/CIDM.2011.6129461
4. van der Aalst, W.M.P., de Medeiros, A.K.A., Weijters, A.J.M.M.: Process equivalence: comparing two process models based on observed behavior. In: Dustdar, S., Fiadeiro, J.L., Sheth, A.P. (eds.) BPM 2006. LNCS, vol. 4102, pp. 129–144. Springer, Heidelberg (2006). https://doi.org/10.1007/11841760_10
5. Anthony, R.N.: Planning and control systems: A framework for analysis. Division of Research, Graduate School of Business Administration, Harvard (1965)
6. Augusto, A., Conforti, R., Dumas, M., La Rosa, M., Polyvyanyy, A.: Split miner: automated discovery of accurate and simple business process models from event logs. Knowl. Inf. Syst. **59**(2), 251–284 (2019). https://doi.org/10.1007/s10115-018-1214-x
7. Baier, T., Mendling, J., Weske, M.: Bridging abstraction layers in process mining. Inf. Syst. **46**, 123–139 (2014). https://doi.org/10.1016/j.is.2014.04.004
8. Becker, T., Intoyoad, W.: Context aware process mining in logistics. Procedia CIRP **63**, 557–562 (2017). https://doi.org/10.1016/j.procir.2017.03.149
9. Bemthuis, R., Mes, M.R.K., Iacob, M.E., Havinga, P.J.M.: Data underlying the paper: using agent-based simulation for emergent behavior detection in cyber-physical systems (2021). https://doi.org/10.4121/14743263
10. Bemthuis, R., Mes, M., Iacob, M.E., Havinga, P.: Using agent-based simulation for emergent behavior detection in cyber-physical systems. In: 2020 Winter Simulation Conference (WSC), pp. 230–241. IEEE (2020). https://doi.org/10.1109/WSC48552.2020.9383956
11. Bemthuis, R.H., van Slooten, N., Arachchige, J.J., Piest, J.P.S., Bukhsh, F.A.: A classification of process mining bottleneck analysis techniques for operational support. In: Proceedings of the 18th International Conference on e-Business (ICE-B), pp. 127–135. SciTePress (2021). https://doi.org/10.5220/0010578601270135

12. Bernard, G., Andritsos, P.: CJM-ab: abstracting customer journey maps using process min- ing. In: Mendling, J., Mouratidis, H. (eds.) CAiSE 2018. LNBIP, vol. 317, pp. 49–56. Springer, Cham (2018). https://doi.org/10.1007/978-3-319-92901-9_5

13. van Cruchten, R.M.E.R., Weigand, H.: Process mining in logistics: the need for rule-based data abstraction. In: 2018 12th International Conference on Research Challenges in Informa- tion Science (RCIS), pp. 1–9. IEEE (2018). https://doi.org/10.1109/RCIS.2018.8406653

14. Davis, F.D.: Perceived usefulness, perceived ease of use, and user acceptance of information technology. MIS Q. **13**, 319–340 (1989). https://doi.org/10.2307/249008

15. Diba, K., Batoulis, K., Weidlich, M., Weske, M.: Extraction, correlation, and abstraction of event data for process mining. Wiley Interdisc. Rev. Data Min. Knowl. Discov. **10**(3), e1346 (2020). https://doi.org/10.1002/widm.1346

16. van Dongen, B.: BPI Challenge 2020 (2020). https://doi.org/10.4121/uuid:52fb97d4-4588- 43c9-9d04-3604d4613b51

17. van Dongen, B.: BPI challenge 2020: travel permit data (2020). https://doi.org/10.4121/uuid: ea03d361-a7cd-4f5e-83d8-5fbdf0362550

18. Fazzinga, B., Flesca, S., Furfaro, F., Masciari, E., Pontieri, L.: Efficiently interpreting traces of low level events in business process logs. Inf. Syst. **73**, 1–24 (2018). https://doi.org/10. 1016/j.is.2017.11.001

19. Fazzinga, B., Flesca, S., Furfaro, F., Pontieri, L.: Process discovery from low-level event logs. In: Krogstie, J., Reijers, H.A. (eds.) CAiSE 2018. LNCS, vol. 10816, pp. 257–273. Springer, Cham (2018). https://doi.org/10.1007/978-3-319-91563-0_16

20. Gandomi, A., Haider, M.: Beyond the hype: big data concepts, methods, and analytics. Int. J. Inf. Manage. **35**(2), 137–144 (2015). https://doi.org/10.1016/j.ijinfomgt.2014.10.007

21. Gibson, K.: The moral basis of stakeholder theory. J. Bus. Ethics **26**, 245–257 (2000)

22. Graafmans, T., Turetken, O., Poppelaars, H., Fahland, D.: Process mining for Six Sigma. Bus. Inf. Syst. Eng. **63**(3), 277–300 (2021). https://doi.org/10.1007/s12599-020-00649-w

23. Greenley, G.E., Foxall, G.R.: Multiple stakeholder orientation in UK companies and the implications for company performance. J. Manage. Stud. **34**(2), 259–284 (1997). https://doi. org/10.1111/1467-6486.00051

24. Günther, C.W., van der Aalst, W.M.P.: Fuzzy mining – adaptive process simplification based on multi-perspective metrics. In: Alonso, G., Dadam, P., Rosemann, M. (eds.) BPM 2007. LNCS, vol. 4714, pp. 328–343. Springer, Heidelberg (2007). https://doi.org/10.1007/978-3- 540-75183-0_24

25. Iacob, M.E., Charismadiptya, G., van Sinderen, M., Piest, J.P.S.: An architecture for situation-aware smart logistics. In: 2019 IEEE 23rd International Enterprise Distributed Object Computing Workshop (EDOCW), pp. 108–117. IEEE (2019). https://doi.org/10. 1109/EDOCW.2019.00030

26. Kumar, M.V.M., Thomas, L., Annappa, B.: Distilling lasagna from spaghetti processes. In: Proceedings of the 2017 International Conference on Intelligent Systems, Metaheuristics & Swarm Intelligence, pp. 157–161. ACM (2017). https://doi.org/10.1145/3059336.3059362

27. Leemans, S.J.J., Goel, K., van Zelst, S.J.: Using multi-level information in hierarchical pro- cess mining: Balancing behavioural quality and model complexity. In: 2020 2nd International Conference on Process Mining (ICPM), pp. 137–144. IEEE (2020). https://doi.org/10.1109/ ICPM49681.2020.00029

28. Maneschijn, D., Bemthuis, R., Bukhsh, F., Iacob, M.: A methodology for aligning process model abstraction levels and stakeholder needs. In: Proceedings of the 24th International Conference on Enterprise Information Systems (ICEIS), pp. 137–147. SciTePress (2022). https://doi.org/10.5220/0011029600003179

29. Mans, R.S., van der Aalst, W.M.P., Vanwersch, R.J.B., Moleman, A.J.: Process mining in healthcare: data challenges when answering frequently posed questions. In: Lenz, R.,

Miksch, S., Peleg, M., Reichert, M., Riaño, D., ten Teije, A. (eds.) KR4HC/ProHealth - 2012. LNCS (LNAI), vol. 7738, pp. 140–153. Springer, Heidelberg (2013). https://doi.org/10.1007/978-3-642-36438-9_10

30. De Medeiros, A.K.A., et al.: Process mining based on clustering: a quest for precision. In: ter Hofstede, A., Benatallah, B., Paik, H.-Y. (eds.) BPM 2007. LNCS, vol. 4928, pp. 17–29. Springer, Heidelberg (2008). https://doi.org/10.1007/978-3-540-78238-4_4

31. Peffers, K., Tuunanen, T., Rothenberger, M.A., Chatterjee, S.: A design science research methodology for information systems research. J. Manage. Inf. Syst. **24**(3), 45–77 (2007). https://doi.org/10.2753/MIS0742-1222240302

32. Post, J.E., Preston, L.E., Sachs, S.: Managing the extended enterprise: the new stakeholder view. Calif. Manage. Rev. **45**(1), 6–28 (2002). https://doi.org/10.2307/41166151

33. dos Santos Garcia, C., et al.: Process mining techniques and applications-a systematic mapping study. Expert Syst. Appl. **133**, 260–295 (2019). https://doi.org/10.1016/j.eswa.2019.05.003

34. Schuster, D., van Zelst, S.J., van der Aalst, W.M.P.: Utilizing domain knowledge in data-driven process discovery: a literature review. Comput. Ind. **137**, 103612 (2022). https://doi.org/10.1016/j.compind.2022.103612

35. Wirth, R., Hipp, J.: CRISP-DM: towards a standard process model for data mining. In: Proceedings of the 4th International Conference on the Practical Applications of Knowledge Discovery and Data Mining (2000)

36. Wynn, M.T., et al.: ProcessProfiler3D: a visualisation framework for log-based process performance comparison. Decis. Support Syst. **100**, 93–108 (2017). https://doi.org/10.1016/j.dss.2017.04.004

37. Yazdi, M.A., Ghalatia, P.F., Heinrichs, B.: Event log abstraction in client-server applications. In: Proceedings of the 13th International Joint Conference on Knowledge Discovery, Knowledge Engineering and Knowledge Management (KDIR), pp. 27–36. SciTePress (2021). https://doi.org/10.5220/0010652000003064

38. van Zelst, S.J., Mannhardt, F., de Leoni, M., Koschmider, A.: Event abstraction in process mining: literature review and taxonomy. Granular Comput. **6**(3), 719–736 (2021). https://doi.org/10.1007/s41066-020-00226-2

A Storage Location Assignment Problem with Incompatibility and Isolation Constraints: An Iterated Local Search Approach

Nilson F. M. Mendes, Beatrice Bolsi(✉), and Manuel Iori

Department of Sciences and Methods for Engineering, University of Modena
and Reggio Emilia, Reggio Emilia, Italy
{nilson.mendes,beatrice.bolsi,manuel.iori}@unimore.it

Abstract. Centralised warehouses are a widespread practice in the healthcare supply chain, as they allow for the storage of large quantities of products a short distance from hospitals and pharmacies, allowing both a reduction in warehouse costs and prompt replenishment in case of shortages. However, for this practice to lead to effective optimization, warehouses need to be equipped with efficient order fulfillment and picking strategies, as well as a storage policy that takes into account the specific procedures of the various types of items, necessary to preserve their quality. In this framework, we present a storage location assignment problem with product-cell incompatibility and isolation constraints, modeling goals and restrictions of a storage policy in a pharmaceutical warehouse. In this problem, the total distance that order pickers travel to retrieve all required items in a set of orders must be minimised. An Iterated Local Search algorithm is proposed to solve the problem, numerical experiments based on simulated data are presented, and a detailed procedure is provided on how to retrieve and structure the warehouse layout input data. The results show a dramatic improvement over a greedy full turnover procedure commonly adopted in real-life operations.

Keywords: Storage allocation · Healthcare supply chain · Iterated local search · Warehouse management

1 Introduction

Pharmaceutical or equipment shortages in healthcare services strongly influences healthcare services, being often the cause of interruptions and attendance delays, increasing the chance of putting patients' lives at risk. Constant item replenishment and the use of high inventory levels constitute the traditional approach to avoid this problem [2,50]. However, this solution has been often considered expensive and hard to be managed, as it requires large dedicated spaces into facilities and workload of healthcare personnel [52].

Recently, more efficient strategies have been adopted, as acquisitions through *Group Purchasing Organisations* (GPO) and wholesalers. Centralised warehouses sharing to store together products to be distributed to different customers located in the same geographical area are one of the most successful approaches. By allowing healthcare facilities to share a common structure to store a large volume of products, centralised warehouses are particularly useful. Also, they fasten the delivery process when they are needed. This enables a constant material flow, a reduction in the personnel costs, a reduced storage space in the customer facility and a lower work burden over healthcare workers.

These benefits strongly depend on the warehouse reliability and capability of delivering the ordered products in the short terms defined by the customers. In turn, this reliability directly requires an efficient warehouse internal organisation, which is a result of a good *storage location policy*.

A storage location policy is a general strategy to assign *Stock Keeping Units* (SKU) to storage positions inside a warehouse. It aims at optimising a metric (e.g. total time or distance travelled to store and retrieve SKUs, congestion, space utilisation, pickers ergonomic), while considering issues like product re-allocations efforts, demand oscillation, picking precedence and storage restrictions.

Commonly, the metric adopted to compare these strategies is the distance travelled by the pickers to retrieve all products in a list of orders. The relevance of this metric is due to the fact that picking operations accounts for around 35% of the total warehouse operational costs [53] and the energy/time spent to reach a product location is a waste of resources that must be minimised. In other frameworks, the evaluation may also consider issues like congestion, picker ergonomic, product storage conditions and total space utilisation, that can also lower the warehouse operation efficiency.

In this study, we address a problem originating from real life operations of a pharmaceutical products distributor. As a main goal, the optimisation of a dedicated storage allocation policy in a picker-to-parts warehouse, i.e., a warehouse where each product is assumed to have a dedicated/fixed position and any time an order request is received, pickers move to reach product location to retrieve order items. In more detail, we deal with a *Storage Location Assignment Problem with Product-Cell Incompatibility and Isolation Constraints* (SLAP-PCIIC). In this problem, some products cannot be assigned to certain locations due to reasons like ventilation or refrigeration (*product-cell incompatibility*), and some other products need to be isolated from the unconstrained ones due to contamination or toxicity concerns (*isolation*).

In our *Iterated Local Search* (ILS) algorithm, a set of orders and a warehouse layout are taken in input and a list of assignments of products to locations that minimises the total distance travelled to fulfill the orders is returned. Our main contribution relies on extensive instance tests of the ILS algorithm proposed in [25] with larger instances, an enriched and updated literature review and a deeper exposition of input warehouse data processing. Inferring the distance matrix from warehouse layout input data, though necessary for most of SLAP algorithms initialization, is not an easy task: in order to improve algorithm usability, we dedicated Sect. 4 to a detailed description of this process. Also, Figs. 1 and 2 are integrated for easier readability and faster comprehension.

The remainder of the paper is organised as follows: in Sect. 2, an extended literature review is presented; Sect. 3 provides a detailed problem description; Sect. 4 presents all the data processing done before the optimisation starts; Sect. 6 describes the various instance sets used to test the ILS algorithm; Sect. 5 presents the ILS algorithm; Sect. 7 describes the numeric experiments carried out, results are provided and then the conclusions are drawn in Sect. 8.

2   Literature Review

The *Storage Location Assignment Problem* (SLAP) is a generalisation of the well know Assignment Problem in which a set of elements must be assigned to a specif position inside a storage area. It is also related with the Quadratic Assignment Problem (QAP), mainly due to the methods used to evaluate the solutions [43]. Most of times, SLAP variants have complex constraints and objective functions that includes considerations about warehouse layout, picking policy, picker routing policy and order batching [13], that together can make the solution evaluation slow or imprecise.

The SLAP constraints are mostly related with strategic or tactical decisions taken by warehouse managers, as warehouse dimensions and layout, shelves capacities, storage policy, facility function, etc. [48]. By their own nature, these decisions are hard to be changed, disregarded or relaxed, due to the costs, training or setup times involved.

The most important constraint in this sense is the warehouse layout. It basically defines the number and position of available storage locations, as well the physical barriers that guide or reduce picker's mobility. It also implicitly defines if the picker will need machines to recover products allocated on higher places, tight corridors [9] or with mobile racks [16], causing a increase on picking time or even internal congestion.

Warehouse layouts are usually classified according two main characteristics: the number of blocks (single block or multi block) and the presence (or absence) of stacked/high level storage. In a warehouse organized in blocks, each block can be defined as a set of identically long, parallel and aligned shelves separated by aisles (i.e. corridors). Each aisle traverses a pair of shelves along its whole extension without being crossed by any other aisle (called in this case a cross-aisle). Stacked storage (or floor stacking storage), on its turn, is the configuration in which the allocation of products can be organized in stacks with one product directly put over another or in shelves vertically divided [31]. In low level warehouses all the shelves are only low headed and the items can be picked by a walking picker. Consequently, if some items locations are not reachable for a picker without the help of an elevator/stair the warehouse is classified as a high level storage warehouse [11,55].

The definition of warehouse layout is followed by the decision about the storage policy, that can be divided in three main groups: *random storage*, *dedicated storage* and *class based storage* [53,57]. A random storage policy allocates products in empty positions inside the warehouse using a random criteria (e.g. closest open location), without including any evaluation in the decision process. Dedicated storage goes in the opposite sense, by ranking the products according to some criteria - popularity, turnover, *Cube per Order Index* (COI) - putting the best ranked products close to the accumulation/expedition locations. Finally, a class based storage separates products in groups according to some popularity criteria and tries to optimize the position and size of each class [44].

Several studies report that random storage policies lead to a better space utilisation, due to frequent reuse of storage positions, but increase the travelled distances to pick the products [30] and require higher searching times or control over product locations [36]. Dedicated storage, on its turn, reduces the picking distance, but cause the rise with re-allocations costs (because of demand fluctuations) and space utilisation, as empty spaces can be reserved to products not currently available. Class based policies are a balance between random and dedicated storage strategies, but they require more strategic efforts to define the number of groups and their positions on the warehouse.

Random storage is rarely studied in the literature, being more an operational and practical approach used as a benchmark to evaluate other methods [33]. One noticeable exception is [36], which proposes a heuristic to dynamically allocate pallets in a storage area considering stacking constraints in order to reduce the total area used. Pallet stacking is also discussed in [31], which proposes a bi-objective mathematical model and a constructive algorithm.

A dedicated storage policy is considered in [13], in which exact distance evaluations are used to define the product assignment. [19] proposes a non-linear model and an ILS to address a storage allocation problem in a multi-level warehouse considering the compatibility between product classes. In [53], by departing from an S-shape routing policy and multi-level storage a two-phase algorithm is created to assign items to locations, and a multi-criteria approximation is used to evaluate the solutions. Other notable works regarding dedicated to storage are [3], which propose a storage assignment and travel distance estimation to design and evaluate a manual picking system and [5], which propose a model to describe a storage assignment problem, solving it to proven optimality for small instances.

Class based policies are studied in [37], [30] and [29]. In [37], class boundaries are defined based on the picking travel distance in a two-block and low-level warehouse where returning routing policy is used. The second uses a Simulated Annealing algorithm to define classes and assign locations to them inside a warehouse considering simultaneously space and picking costs. The work in [29] extends [30] by using a branch-and-bound algorithm instead of a heuristic, and by including space use reduction in the considered metrics.

In [35], the authors propose a data-mining based algorithm that uses association rules to define product dedicated positions in order to minimise the travelled distance in a picker-to-parts warehouse. Similar approaches are presented in [54] for optimizing storage with correlations related with *bill of materials*(BOM) picking, [26] for class based policy, [10] for cluster assignment, [15] for product classification and storage, and [21] for frequent item set grouping and slot allocation. In [23], an ABC classification and product affinity method is used to define the best positions for the products. A set partitioning approach is presented in [22] to allocate entire areas, instead of specific locations, to group of products.

Some warehouses can also adopt some kind of zoning storage, in which each zone contains only a subset of the items stored (defined by any criteria) and each picker only retrieve and transport products stored in a single zone. This approach leads to reduced travel distances, but requires an additional work on order consolidation to put together products picked in the different zones. A more recent approach, more suited to large e-commerce retail warehouses, is the mixed-shelve storage, in which one product can be stored in several locations to be easily retrieved from any region of the warehouse, even if it requires large use of automation to find each item [7,38,55].

Another high level decision that affects the storage allocation strategy is the the picking system. The two main systems are pickers-to-parts and parts-to-pickers. In the former, pickers depart from a consolidation/expedition point and perform a tour to pick each product in their location, manually or by using some vehicle. In the latter, an automated storage and retrieve system, composed by one or more automated guided vehicles, picks ordered items and delivers them in the place where they will be prepared to be dispatched [3]. Pickers-to-part systems usually are easier and cheaper to implement and are more robust to changes, as most of times it is operated by humans, that adapt better to new situations than machines. Parts-to-picker however make the picking activity faster and more regular along long working shifts.

Picker-to-parts systems are addressed in [34], [32] and [33]. The first models the warehouse operation as a Markov chain in order to calculate the expected distance travelled by the picker in three different zoning cases. The second propose a heuristic for a case where congestion effects are considered. The last use a genetic algorithm to define the best workload balance among the pickers.

Parts-to-pickers are the subject of a large set of works. In one of the most recent studies [27] describe an integrated cluster allocation (ICA) policy, that considers both affinity and correlation in order to minimize retrieval time. This approach is extended in [28], who include the concept of inventory dispersion and propose mathematical models to define product positions in a warehouse with robot based order picking. In [20], new routing algorithms are proposed based on autonomous vehicles communication capabilities provided by new *Internet of Things (IoF)* technologies;

We can also highlight the existence of the pick-and-pass system (also called progressive zoning system [33]), in which each picker is responsible to retrieve a specific subset of products in an order and then deliver the incomplete order to the next picker, until all the products to be put together and dispatched. This system is commonly associated with a zone storage policy.

Among the metrics commonly used to evaluate assignment quality we can cite: shipping time, equipment downtime, on time delivery, delivery accuracy, product damage, storage cost, labour costs, throughput, turnover and picking productivity [40,45,56]. By a large margin, however, the most commons are picking travel time and travel distance [40]. Their popularity is directly connected to an easier evaluation, representation and visualization. As disadvantage, we can mention the need of solving one or more instances of the *Travelling Salesman Problem* (TSP) or *Vehicle Routing Problem (VRP)* - depending on the presence of picker capacity constraints or multiple simultaneous picking tours. This additional optimization problem can let the overall process slow, depending of the number of assignments tested and the methods used to solve the TSP/VRP. In this sense, it is worth to mention that the travel distance metric has an additional advantage of allowing disregarding issues related with congestion, handling time, picker speed or searching delays.

In this context, several methods to optimise order picking routes have been proposed, both inside SLAP variants studies or in independent researches. As pointed out in [13], routing problems in warehouses can be seen as special case of the Steiner

Travelling Salesman Problem, that in some layouts can be solved to optimality ([24, 39] and [8, 42]) but in general layouts is mostly solved using heuristics ([9, 12, 41, 47]). For difficult layout warehouse configurations, exact algorithms for optimize picking route do not exist because the dynamic programming approaches used on single-block warehouses are not easy to be generalized for two or more cross-aisles [6]. The few exact methods available are used to solve problems with already defined warehouse allocations (as can be seen also in [18] and [40]) and they are mostly algorithms based on a graph theoretic algorithm for single-block warehouses [24].

Nonetheless, it is known that in real life operations pickers tend to deviate from optimal or non-intuitive routes [14]. To simulate this behaviour, many studies consider simple heuristics for picker routing as return point method, largest gap method, returning point or S-shape routing [11, 14, 21]. These heuristics also simplify order picking evaluation as they quickly allow the computation of travelling distances in deterministic problems or the evaluation expected travelling distances in stochastic problems [13]. Some operational constraints can also make pickers deviate from optimal routes. It is the case in [49], which describes a warehouse by imposing a weight precedence rule during the picking, so that the heaviest products are retrieved before lighter ones. Congestion avoidance is either a factor that impacts on route definition. For the interested reader, an extensive analysis of these routing algorithms is presented in [11].

Another commonly used technique to deal with the optimisation of order picking routes is the order batching. It consists in performing the picking of one or more small orders in a single route, with a posterior products separation on the consolidation/expedition area.

More complex problems are surveyed in [17]. In [51], batching, routing and zoning are combined to optimise the warehouse operation. A discussion on storage allocation with product picking precedence can be found in [57], whereas [46] defines a problem where allocation and routing must be decided jointly in order to avoid congestion during the picking.

The SLAP-PCIIC, discussed in these paper is a storage allocation problem in which products have their assignment locations limited by their characteristics and mutual compatibility. Studies that deal with joint optimization of storage location assignment and picking routes, as we propose here, are less common in literature than those ones considering each problem separately. We can cite [43] and [5], who propose *Mixed Integer Programming* (MIP) model to solve their problems.

The SLAP-PCIIC also considers high level storage, non regular blocks, and warehouses divided in one or more interconnected pavilions. Furthermore, instead of performing an indirect evaluation of the order picking travel distance, we directly optimize order picking routes during the location assignment optimization. Constraints similar to the ones discussed in this paper can be found on [1] and [4]. The first one, the storage of temperature-sensitive products is discussed, mainly food and pharmaceutical ones, and a dynamic policy to allocate products is proposed, by observing the different temperatures inside the warehouse, as well seasonal variations, aiming to minimise the picking traveled distance and maximise the product safety. The second propose a mathematical model to define a warehouse layout to store hazardous chemicals products.

Compatibility constraints are addressed in [19], who proposes an *Iterated Local Search* (ILS) algorithm to optimize product allocation in a multi-layer warehouse.

3   Problem Description

The SLAP can be shortly described as follows: given a set P of products to be stored, a tuple $\omega = (O_1, \ldots, O_n)$ of (non necessarily distinct) orders, in which an order $O_i \subseteq P$ is the subset of products to be picked up by a picker in a single route, and a set L of locations in a warehouse, define an assignment (i.e., an injective function) $g : P \to L$ in which an evaluation function $z = v(g, \omega)$, to be described next, is minimised. The warehouse is received in input in the form of a graph, and the travel distance d_{ij} between any pair of locations $i, j \in L$ is computed by invoking the Dijkstra algorithm. In other words, given the evaluation function v that maps a value to each ordered pair consisting of a set of orders and an assignment of products to locations, define the assignment that minimises v.

In the SLAP-PCIIC, the SLAP variant described here, due to incompatibility and isolation constraints, g is, on the one hand, relaxed to a possible partial assignment, but on the other hand it is subjected to the constraints of the location-product eligibility, i.e., each product is assigned to at most a single location and each location receives at most a single product while respecting the incompatibility and (strong) isolation constraints. In this framework, $v(g, \omega)$ is defined as the sum of minimum travelled distance to pick all the products in each order (designated by $D(g, \omega)$), plus the non negative penalties for non desirable or missing allocations, (designated by $\Phi(g)$). Notice that if all products are allocated, then we have no contribution in the penalty Φ caused by missing allocations, meaning that the lack of product assignment to locations is highly deprecated. Following the company operational rules, it is assumed that the warehouse uses a picker-to-parts picking policy (the picker visits the products locations) and orders splitting/batching are not allowed, making each order an individual and independent route. These assumptions make it possible to decompose the distance $D(g, \omega)$ as the sum of the minimal travelling distances to pick the products in each order O in the ω tuple, designated by $d(g, O)$. With these considerations, and indicating with $\mathcal{G}$ the set of relaxed admissible assignments $g : P \to L$, the SLAP-PCIIC objective function can be described as:

$$z = \min_{g \in \mathcal{G}} \sum_{i=1}^{n} d(g, O_i) + \Phi(g) \tag{1}$$

It can be noticed that to evaluate each one of the $d(g, O_i)$ it is necessary to solve another optimisation problem, more specifically a variant of the TSP that calls for the minimization of the travelled distance. Namely, if there is a single accumulation/expedition point, each product is assigned to (at most) a unique location and $O' = \{p_1, \ldots, p_{|O'|}\} \subseteq O \subseteq P$ is the requested order deprived of those products lacking of location, then $d(g, O)$ is the minimum distance to depart from the accumulation/expedition point, visit all the locations $(g(p_1), \ldots, g(p_{|O'|}))$ in the best possible sequence and then come back. Conversely, in the SLAP-PCIIC we allow the presence of more than one accumulation/expedition point, so the picker can depart from any of these points and return to another if this operation reduces the total distance travelled $D(g, \omega)$. The algorithms can be easily adapted to deal with the case in which the expedition points are, instead, fixed.

Defining an optimal picker routing in the scenario above is relatively simple if the warehouse is organised in blocks of identical and parallel shelves. However, in the SLAP-PCIIC, the shelves can have different sizes, cell quantities, orientations and positioning and also be located in different pavilions. To deal with this setting, a regular distance matrix containing the distances between each pair of locations is considered as the input of the distance minimisation method, disregarding any further information about the warehouse organisation.

The second part of the objective function, the penalty value $\Phi(g)$, is the sum of two terms: the number of unassigned products $\Phi_1(g)$ and the number of undesired allocations $\Phi_2(g)$. In this sense, the possible configurations of the function g are limited by a set F of assignment incompatibilities and a set I of isolation constraints. Each assignment incompatibility $f \in F$ is a hard constraint (i.e., it must be strictly respected) composed by a tuple of three values (p, n, c) representing a product p, the nature n of the incompatible location (cell, shelf or pavilion) and the code c of the incompatible location, respectively. For instance, the incompatibility $(p_1, \text{"shelf"}, k_1)$ defines that product p_1 cannot be allocated on shelf k_1.

An isolation constraint is based on the product classification. Given a set T of types, representing the most relevant product characteristic to the storage (toxic, radioactive, humid, etc.), an isolation constraint $\iota \in I$ is a tuple of three values (t, n, s) specifying that products of type $t \in T$ should be allocated in an isolated $n \in \{\text{"cell"}, \text{"shelf"}, \text{"pavilion"}\}$ with an enforcement $s \in \{\text{"weak"}, \text{"strong"}\}$. The enforcement s defines if the isolation is a hard constraint or can be relaxed with a penalty.

4 Input Data Processing

A common assumption in studies about SLAP is the regularity of the warehouse layout. Most of the times, the warehouse is represented by sets of parallel and identical shelves that can be accessed through aisles between them and cross-aisles that allow moving from one aisle to another. These sets are called blocks, and most of the literature about SLAP or picker routing considers the presence of one or more blocks in the warehouse.

However, for several reasons, this assumption creates problems when a proposed algorithm needs to be released to production environment. In many facilities, for example, physical or operational barriers are present (like build columns and machines), so layouts cannot be represented as simple blocks.

In this section, we describe the format we created to represent general warehouse layouts, aiming at allowing the use of our algorithm to a large number of warehouses. Furthermore, we present the data processing performed to get the distance matrix of all the relevant positions in the warehouse, and thus to make it possible to use TSP algorithms to evaluate the total distance travelled by pickers to retrieve all the products in a set of orders.

4.1 Warehouse Input Format

The main objective in proposing a new format to describe the warehouse layout is to allow a quick description of different types of layouts. More specifically, a good format should provide: *(a)* easiness of transcription in a spreadsheet or text file; *(b)* readability; *(c)* direct representation on relational databases; *(d)* a simple and robust representation of internal valid paths; *(e)* possibility of defining pavilions and connections between them; *(f)* possibility of defining multiple accumulation/expedition points; *(g)* possibility of defining heterogeneous shelves and cells.

We defined points *(a)*, *(b)* and *(c)* aiming at a future implementation in a decision support system and also at an easy utilisation of the system, as most users are used to text files or spreadsheets and most developers are comfortable in using relational databases.

The robustness to represent paths (point (d)) was considered due to the existence of several operational rules that limit the traffic inside warehouses, mainly when it involves large vehicles or robots. In the proposed format, it is possible to define rectilinear corridors (with a start position, length, direction and sense) and rectilinear segments (with start and end position) to connect two corridors (see Fig. 1). Non rectilinear corridors or connections were left out due to the variable number of points needed to define them and also because of the difficult evaluation of their length. We also defined that corridors must be parallel to one of the Cartesian axes, not allowing in this way oblique corridors.

Points *(e)* and *(f)* were adopted to take into consideration larger warehouses, in which internal divisions are common and operations can be less centralised. Notwithstanding, the pavilions must be always represented as rectangles, as allowing other formats would significantly increase the representation complexity.

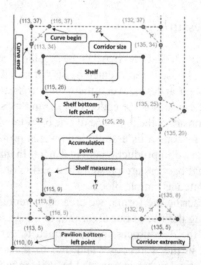

Fig. 1. Partial warehouse representation with main elements information. The dashed lines are corridors, and the dashed arrows are curves

Point *(g)* is modelled to allow describing warehouses with a very diversified set of stored items, from large machines that are positioned on pallets to small tools that can be stored in cabinet drawers.

The resulting format is a union of seven tables (Pavilions, Shelves, Cells, Corridors, Curves and Pavilion Exits, Accumulation/Expedition Points), as presented in Fig. 2. Each pavilion is composed of a code, a point (with coordinates x and y over a Cartesian plane) representing the bottom-left extremity, width and length (the height is not important to our problem, but can be easily included if needed). By its turn, each shelf has either a code, a bottom-left point, the number of rows and columns, the size of the cells (width, length and height) and the block code indicating where it is located. Each cell has a code, width, length, row and column at the shelf, a reference to the shelf and a number of vertical levels (1 or more). Corridors have a code, an initial point, a direction (horizontal or vertical), a sense (up-down, bottom-up, left-to-right, right-to-left, both), a length and a reference to the block where it is located. Each curve has a code, a reference to both corridors it connects, an initial point and a final point (in this case the length is calculated by the application). Each block exit contains a point, a width, a code and a reference to the blocks it connects. Finally, accumulation/expedition points are composed of a code and a bi-dimensional coordinate.

Products and orders use similar structures, but with less data. Each product is represented by a tuple containing code, description and type (size and weight are not considered in this problem) and each order with a tuple containing a code and a deadline. A list of items is assigned to each order, where each item contains a product code and a quantity.

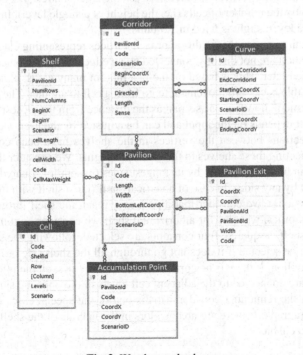

Fig. 2. Warehouse database.

4.2 Distance Matrix Extraction

Once all input information has been loaded, it is necessary to process it to get useful information for the algorithm. Basically, from a warehouse layout description we must extract a distance matrix containing the distances from each delivery point or storage location to all the others, and a compact representation of cells, shelves and blocks to check if the allocation prohibition and isolation constraints are being respected. The representation of warehouse structures (i.e. cells, shelves and blocks) was done simply using standard object-oriented classes, with no relevant improvements, so it is not detailed in the text.

To build the distance matrix we follow three steps: *(1)* transform the warehouse layout in a directed graph; *(2)* separate the vertices that represent storage locations and accumulation/expedition points from those that represent the paths in the warehouse; and *(3)* run iteratively a shortest path algorithm to determine distances between vertices.

The first step, the conversion of the warehouse in a graph, starts by creating vertices to represent accumulation/expedition points and inferior levels of the cells (i.e. those located closer to the ground, that are the connection points between shelves and corridors). The vertices are positioned in the central point of the cell level. After that, vertices to represent high levels in cells are created, always in the level central point. In detail, when there is more than one level in a cell, each vertex representing a cell level is connected by two arcs, one in each sense, to the level immediately above and below. In other words, when a cell is divided in five vertical levels, the third level is connected with the fourth and the second levels, but not with the first or fifth levels. The edge length correspond to the vertical distance between the level centres, that is the height of the cell divided by the number of cells (i.e. the height of a single level). In the instances of our study, the level height is fixed in 1.5 units.

It is important to notice that in this approach vertices representing external adjacent cells in a same shelf are not directly connected by an edge, except in some special cases that are described later. This is based on the fact that, in many cases, a picker needs to do a non-negligible backward movement to go from a cell to another. The length of this backward movement, however, is as small as the distance from the closest corridor path to the shelf, so it depends on the input and can be adjusted by the user.

Once connections between the vertices in the shelves have been completed, the process of connecting these shelves to the corridors begins. We consider that each rectangular shelf can be accessed only by its longest sides (except for squared shelves, that can be accessed by any side). In case of a vertical shelf (i.e. a shelf with the longer side parallel to the y-axis), we consider that a path may reach the shelf through its lateral (left/right) extremities, whereas for a horizontal shelf we consider bottom/up extremities. When the shelf is squared, four corridors are selected, following the same logic. A corridor can not be selected if it does not go through all the shelf side or if it is located in a different pavilion. If there is no corridor in one of the sides, all the inferior cell levels in that side are connected to the adjacent cell levels in the opposite side, to become accessible from the remaining corridor. This is one of the special cases mentioned in the previous paragraph. If there are no corridors in the laterals of the shelf, this shelf is considered unreachable.

After the adjacent corridors be defined, each inferior cell level vertex is connected by a pair of edges (one in each sense) to a vertex created over the corridor exactly in front of the cell vertex. The new vertices created on corridors in the previous step are then merged with the vertices that represent the extremities of the curves (that are always over corridors) to define all the valid paths in the warehouse. Two consecutive vertices on a corridor are connected by two edges in a two-way corridor (designated by the word "both" in the field "sense" on corridor input data) and one edge otherwise. To avoid confusions, we defined that there is not a curve on corridor interception points, thus it is not right to suppose that it is possible to move from a corridor to another through these points unless a curve passing on it is defined.

Following the procedure, each vertex representing an accumulation/expedition point is connected to the closest non-storage vertex in the pavilion in which it is located by two edges, one in each sense. This connection can be a traversal one if there is no possibility of establishing a horizontal or vertical one.

To connect two different pavilions, we first create a vertex to represent each pavilion exit and then connect it to the closest vertex in each pavilion. The connection with pavilions vertices are done in both senses, with the same rules adopted to connect the accumulation/expedition points described in the previous paragraph.

After all these steps, we obtain a graph that represents the connections between all the relevant points in the warehouse. During the graph building process, all vertices are associated with the location they represent, so it is possible to associate each edge with the distance between its incident vertices in the warehouse. Furthermore, each vertex receives a label that is used to define if it is a storage vertex or a vertex representing an accumulation/expedition point, thus being relevant in the distance matrix. The graph built is then passed to an algorithm that calculates the distance matrix.

The algorithm that calculates the distance matrix is a loop, where at each iteration a Dijkstra shortest path algorithm is executed departing from one storage or accumulation/expedition vertex. In this way, all the graph vertices are considered in the shortest path evaluation.

5   Iterated Local Search

In this section, we present the ILS algorithm that we developed to solve the SLAP-PCIIC. First, a constructive algorithm is invoked by ILS to generate an initial solution, and then three different neighborhood structures are explored to attempt improving the solution.

Two main phases constitute the constructive algorithm (see Algorithm 1). In the first phase (described in lines from 6 to 23), locations are assigned only to the products without isolation constraints (i.e., products that do not belong to any tuple $\iota \in I$) or to those with isolation constraints involving an enforcement $s =$ "$weak$". In the second phase (described in lines from 25 to 40), it assigns locations to isolated products, according to their types. This procedure, by helping controlling $\Phi(g)$ value, promotes the allocation of a greater number of products, since the constrained products are usually fewer and could lead to a large number of unusable locations.

Request frequency is the first product sorting criterion used by the algorithm. Initially, all the products are sorted by descending order of popularity, i.e., the products more frequently required are put first, and those less frequently after, where popularity is inferred from the tuple of orders ω. In a similar way, the storage location sorting criterion is the distance from the closest accumulation/expedition point (from lowest to highest). Then, the most popular product is assigned to the closest empty location, whenever this assignment is allowed. Instead, if an assignment is forbidden, the algorithm explores iteratively the next location, until an allowed position is found or the loop reaches the last position. If a product belongs to a strongly isolated type, it is not assigned during this phase.

In the second phase of the constructive algorithm, all the products presenting strong isolation constraints are divided by type (as shown in line 25 in Algorithm 1) and, similarly, the isolated warehouse locations are grouped by level (block, shelf or cell), in line 26. At this point, the allocation is done according to the structure size (from largest to smaller), starting from the isolated blocks and finishing with the isolated cells. At each step, the product types that have the corresponding isolation level are selected while preserving the frequency order. Then, each type is assigned to the isolated area where it is possible to maximise the total frequency, without worrying with the internal assignment optimisation. The list of available positions and products is updated at each step. The algorithm ends either when all the available isolated spaces are occupied or all the products are assigned.

It is significant to notice that, due to incompatibility and isolation constraints, it may be impossible to assign all the products to the warehouse locations. This may happen even when the number of available locations is higher than the number of products. Moreover, as a heuristic method, the constructive algorithm may be not able to find an initial feasible assignment even when it exists. In both cases, the solution evaluation procedure penalizes the objective function according to the number of unassigned products (i.e., the value of $\Phi_1(g)$).

However, a valid solution can still report one of the side effects of allocating groups of isolated products together in the warehouse. The first is assigning relatively good positions to several products with a low number of requests due to the existence of some very popular products in the set, that are responsible of a skewed popularity of that group. The second effect is approximately the opposite, i.e., due to a low group average popularity, very popular products can be allocated in bad positions. Both these problems are analogous to those that are reported in class or zone based storage location problems [30,37].

According to the procedure described in Sect. 5.1, the objective function of an initial solution is calculated, immediately after being created by the greedy algorithm. Then, the ILS heuristic enters in a loop devoted to the exploration of the solution space (see lines 4 to 25 of Algorithm 2). This loop is composed by three neighbourhood structures that are combined as a single local search, and a perturbation method.

Within a loop iteration, a small set of neighbours of the current solution (denoted by g) is evaluated by each neighbourhood structure and the one with lowest objective function value is stored, thus using a best improvement criteria. Then, denoting by g^w the best solution in the loop, and by g^* the best global solution, they are compared, and every time g^w is better, it is assigned to g^* and the number of iterations without improvement (line 22) is reset. Finally, a perturbation of g^* is performed at the end of

---

Algorithm 1. Greedy algorithm.

1: $g \leftarrow \emptyset$ ▷ Start with an empty assignment
2: $L* \leftarrow distanceSort(L)$ ▷ Locations are ordered by distance
3: $A \leftarrow L*$ ▷ Available locations
4: $P \leftarrow sortByFrequence(P)$
5:   ▷ First part of greedy algorithm
6: **for each** $p \in P$ **do**   ▷ For each product in P
7:     **if** $isStronglyIsolated(p)$ **then**
8:         continue
9:     **end if**
10:     $l \leftarrow firstAvailable(A)$
11:     **while** true **do**
12:         **if** $l == null$ **then**
13:             break
14:         **end if**
15:         **if** $isForbidden(l, p)$ **then**
16:             $l \leftarrow nextAvailable(A)$
17:             continue
18:         **end if**
19:         $g \leftarrow g \cup (p, l)$
20:         $A \leftarrow A\backslash\{l\}$
21:         break
22:     **end while**
23: **end for**
24:   ▷ Second part of greedy algorithm
25: $\Lambda \leftarrow stronglyIsolatedProductsByType(P)$
26: $\Psi \leftarrow availableIsolatedStructures(A)$
27: $\mu \leftarrow allocateOnIsolatedBlock(Y, \Psi)$
28:   ▷ First allocation. Assigns products isolated by block
29: $g \leftarrow g \cup \mu$
30: $\Psi \leftarrow updateAvailableIsolateStructures(\Psi, \mu)$
31: $\Lambda \leftarrow updatestronglyIsolatedProductsByType(\mu, P)$
32:   ▷ Second allocation. Assigns products isolated by shelf
33: $\mu \leftarrow allocateOnIsolatedShelf(Y, \Psi)$
34: $g \leftarrow g \cup \mu$
35: $\Psi \leftarrow updateAvailableIsolateStructures(\Psi, \mu)$
36: $\Lambda \leftarrow updatestronglyIsolatedProductsByType(\mu, P)$
37:   ▷ Third allocation. Assigns products isolated by cells
38: $\mu \leftarrow allocateOnIsolatedCell(Y, \Psi)$
39: $g \leftarrow g \cup \mu$
40: $\Psi \leftarrow updateAvailableIsolateStructures(\Psi, \mu)$

---

the iteration. The loop stops if the number of iterations without improvement reaches the value of the input parameter *iterations without improvements* (IWI).

In order to avoid non significant improvements, we enforce the comparison criterion, by assuming that an assignment g_1 is better than an assignment g_2 if and only if the objective function value of g_1 is at least 0.1% lower than that of g_2 (i.e., the ratio of between difference of solution values, $v(g_1, O) - v(g_2, O)$, and the old solution value $v(g_2, O)$ must be smaller than $\delta = -0.001$).

In the local search we use as neighbourhood structures simple location swaps between two products. The first neighbourhood (*mostFrequentLocalNeighbourhood*) is obtained by swapping the assignments of two products belonging to the subset of 20% most required products. The second (*insideShelfNeighbourhood*) is obtained by swapping the assignments of two products assigned to locations in the same shelf, working as an intensification of the search. The third neighbourhood (*insidePavilionNeighbourhood*) is a wider search: the locations of pairs of products assigned to the same pavilion are swapped.

Algorithm 2. ILS algorithm.

1: $g^* \leftarrow initialSolution(L, P)$ ▷ Get greedy assignment
2: $g \leftarrow g^*$
3: $nonImprovingIter \leftarrow 0$
4: **while** $nonImprovingIterations < IWI$ **do**
5:      $g^w \leftarrow g$ ▷ Initialize the best solution on loop
6:      $g' \leftarrow mostFrequentLocalNeighbourhood(g)$
7:      **if** $\frac{v(g^w, O) - v(g', O)}{v(g^w, O)} \geq \delta$ **then**
8:          $g^w \leftarrow g'$
9:      **end if**
10:     $g' \leftarrow insideShelfLocalNeighbourhood(g)$.
11:     **if** $\frac{v(g^w, O) - v(g', O)}{v(g^w, O)} \geq \delta$ **then**
12:        $g^w \leftarrow g'$
13:     **end if**
14:     $g' \leftarrow insidePavilionNeighbourhood(g)$
15:     **if** $\frac{v(g^w, O) - v(g', O)}{v(g^w, O)} \geq \delta$ **then**
16:        $g^w \leftarrow g'$
17:     **end if**
18:   ▷ Update best global solution
19:     $nonImprovingIter \leftarrow nonImprovingIter + 1$
20:     **if** $\frac{v(g^*, O) - v(g^w, O)}{v(g^*, O)} \geq \delta$ **then**
21:        $g^* \leftarrow g_w$
22:        $nonImprovingIter \leftarrow 0$
23:     **end if**
24:     $g \leftarrow perturbation(g^*)$
25: **end while**

The neighbourhood structures work following the same steps: (i) randomly choose two products, (ii) check if the swap between these products is valid, (iii) evaluate the change in objective function and store the best solution found.

Three rules are used to check the swap validity: (1) all swaps that assign a product to a forbidden position are not allowed; (2) no swaps are allowed between a product belonging to a strongly isolated type and a product not belonging to a strongly isolated type; (3) no swaps between products of different types are allowed. We remark that validity does not imply feasibility. In fact, while the two first rules were introduced to avoid infeasible solutions in the search, the third was created to filter the moves in order to avoid the recalculation of the penalties related to weak isolated types. It can be noticed that the third rule makes the second redundant: second and third rules are complementary, indeed. We tested two algorithm versions in the numeric tests, one with rules (1) and (2) (*type-free* swap policy), and the other with rules (1) and (3) (*same-type* swap policy).

In order to evaluate the solution after a swap, we reevaluate the distances of the routes affected by that swap and the total of products with weak isolation constraints allocated in altered areas. As the number of swaps performed during the algorithm execution is huge and the number of orders can easily reach some thousands, the procedures to evaluate them must have a strong performance.

When all neighbourhoods have been explored and the global best solution has possibly been updated, the best global solution undergoes a perturbation. This perturbation consists in $|P|/20$ unconstrained and valid swaps, chosen randomly and applied in a loop. The resulting solution is then used as the initial solution in the next ILS iteration.

5.1   Solution Evaluation

Concerning the evaluation of the routing distance, we combine two ideas: first, we make use of different TSP algorithms depending on the size of the instance and, second, we keep track of routes that passes from each position, and we take advantage from this by recalculating only those picking distances of routes containing products involved in the swap. In this case, initially the algorithm retrieves all the routes affected and checks if a route passes from both locations where the products are currently allocated. If the route has both locations on it, the reevaluation is not necessary (because the set of locations to be visited does not change), otherwise the route is reevaluated with the new location in place of the old one.

The evaluation is controlled by the parameter $TSP(\alpha, \beta)$. In this parameter, the constants α and β are two integers representing the order size thresholds used to choose each algorithm method for estimating the minimum distance to visit the product locations in a route. Let $|O|$ be the number of items in an order O in the tuple ω, if $|O| \leq \alpha$ an exhaustive search is run (i.e., all the possibilities are tested and then the distance found is optimal). Otherwise, if $\alpha < |O| \leq \beta$, a closest neighbour algorithm is used to initialise the route and a subsequent quick local search is performed. In this local search, $(|O| - 1)$ swaps among two consecutive locations are tested in $(|O| - 1)$ iterations (always departing from the first to the last product) and the current solution is updated when the swap reduces the smaller distance. Finally, if $|O| > \beta$, only the closest neighbour heuristic is performed.

The complexity of the route evaluation method described above is easily seen to be quadratic, as the exponential time approach is limited depending on the number of visited points. Though not being able to guarantee an optimal route, it is better suited to our purpose of using non block-based warehouse layouts than algorithms based on the method proposed in [39].

We use Algorithm 3 below in order to explain how the evaluation of penalties is done by breaking weak isolation constraints. This pseudo code shows the process for shelves, but, in practice, it can be easily replicated for cells and pavilions.

First, the algorithm gets all the allocated products and groups them by shelves (line 4). Products are grouped by type for each shelf (line 9) and then the algorithm counts the number of different product types assigned to the shelf. If there is only one type assigned to the shelf or if no assigned product has an isolation constraint $\iota \in I | n = $ "$shelf$" (see isolation constraint definition on Sect. 3), no penalty is applied (lines 11 to 13). Otherwise, the actual penalty evaluation is performed (lines 14 to 22).

The penalty strategy is based on the division of the products assigned to a structure (which can be a shelf, a cell or a pavilion) into two groups, one with isolation constraints and another without. The group with less assignment in this structure is defined as the minority, while the other one is said to be the majority.

The method tries to push the search through the predominant configuration by penalizing minority groups. For example, if the shelf is mostly occupied by products belonging to types without isolation constraints, it penalizes the products with isolation constraints (lines 17 and 18) that are the minority. In the same way, it penalizes products belonging to types without isolation constraints if they are the minority in the shelf (lines 19 and 20). Moreover, instead of simply counting the number of products and

Algorithm 3. Isolation penalty evaluation after a swap.

```
1: totalPenalty ← 0
2: S ← allShelves(L)
3: T^w ← WeakIsolationTypes()
4: P_s ← productsAllocated(g, s)   ∀s ∈ S
5: for each s ∈ S do
6:     P_i ← {p | p ∈ P_s, type(p) ∈ T^w}
7:     P_f ← {p | p ∈ P_s, type(p) ∉ T^w}
8:     T_s ← {type(p) | p ∈ P_s}
9:                                              ▷ Group products with isolation constraints by type
10:     H_t ← {p ∈ P_i|type(p) = t}  ∀t ∈ T^w
11:     if |P_i| = 0 or |distinct(T_s)| = 1 then
12:         continue
13:     end if
14:     penalty ← 0
15:     x ← max_{t∈T^w}(|H_t|)                   ▷ Type with max cardinality
16:     r ← x
17:     if |P_f| ≥ |P_i| then
18:         penalty ← W_{pen} * |P_i|^2/|P_s|
19:     else
20:         penalty ← W_{pen} * (|P_f|^2 + r)/|P_s|
21:     end if
22:     totalPenalty ← totalPenalty + penalty
23: end for
```

to distinguish among similar assignments, the proportion of products belonging to the minority/majority over the total number of products is considered. This decision was taken due to the fact that only counting the products was causing no changes after a swap in the total penalty.

6   Instance Sets

In order to test the algorithm performances, we built up several instance sets, which are to be described in this section. We created three warehouse layouts W_1, W_2 and W_3 with several differences between them, not only regarding the number of positions, but also regarding how these positions are distributed in the area. A short description of the warehouses is provided in Table 1. Visual representations of the warehouses are provided in Figs. 3, 4 and 5. The shelves are defined by the rectangles with circles on the corners. The dashed lines are the corridors and the dashed arrows are the connections between the corridors. The accumulation/expedition points are represented by isolated circles, as well as the connection between two pavilions is represented by a triangle (see the detail in the centre of Fig. 5). In these figures, it is possible to notice some of the aforementioned characteristics of these generalised warehouses, as the presence of heterogeneous shelves or blocks, the mandatory moving sense and the multiple accumulation/expedition points.

The experiments were performed in three steps. In the first step, the goal was the analysis of the algorithm performance by considering only the travelled distance and thus its suitability at dealing with directly calculated distances. In the second step we tested the performance by considering the incompatibility and isolation constraints, in order to check if these constraints were well handled in the algorithm. The third step

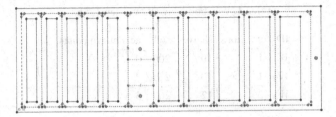

Fig. 3. Layout of the warehouse W_1.

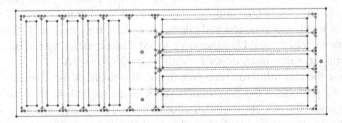

Fig. 4. Layout of the warehouse W_2.

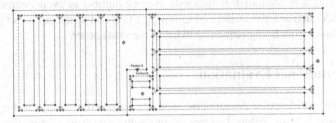

Fig. 5. Layout of the warehouse W_3.

was aimed at analysing the algorithm performance over a set of large instances, the word large referring to a higher number of products and orders.

By using the instance set I_1, we tested both the insertion of products and the algorithm parameters. We experimented the insertion of two sets of products in the warehouses, one with 100 and the other with 200 products. For each product set, we tested scenarios with $|\omega| = 500$, 1000 and 5000 orders. For each combination of warehouse, number of products and number of orders, five realistic instances were created, for a total of $|I_1| = 3 \cdot 2 \cdot 3 \cdot 5 = 90$ instances. In all these instances, we used cells with only one position/level, as showed in Table 1.

Regarding the algorithm parameters, we investigated two values: the maximum number of IWI, used as stopping criteria, and the TSP thresholds described in Sect. 5. Three values of maximum number of IWI were tested, and three different combination of the two TSP thresholds. Additionally, we tested the local search with and without incompatibility of swaps between products of different types. In this way, $3 \cdot 3 \cdot 2 = 18$ algorithm parameter combinations were experimented.

Table 1. Warehouse layout overview [25].

ID	pavilions	shelves	cells	accum/exp points
W_1	1	10	200	3
W_2	1	10	240	3
W_3	2	12	260	3

To test the algorithm capabilities in managing incompatibility and isolation constraints, we used a subset of the initial instances, using 3 variations to each combination of warehouse, number of products and number of orders, for a total of $3 \cdot 2 \cdot 3 \cdot 3 = 54$ instances. For each of these 54 instances, we tested 5 incompatibility and isolation constraints, leading to a new instance set I_2, with $|I_2| = 5 \cdot 54 = 270$.

To analyse the algorithm run time growth and also to check its efficiency in a large warehouse, we performed the experiments summarised in Table 4. In these instances, there are no isolated types or prohibited allocations and the warehouse is an expanded version of the warehouse W_3, as described in Sect. 7.

In all the instances mentioned above, the number of products by order was determined following a Poisson distribution with average number of events equal to 6. The products inside each order were defined following a uniform distribution, but preventing the same product from being requested twice in the same order.

7   Computational Evaluation

In this section, we describe the numeric experiments performed to test the efficiency of the proposed ILS algorithm. In these experiments, we observed the algorithm performance on heterogeneous sets of instances, by monitoring features of interest such as the average run time, the local search capacity of improving the initial solution, the influence of the parameters on the final solution, and the solution quality.

The algorithm was implemented in C++17 language, with source code being compiled with -O3 flag. The tests were performed on a Dell Precision Tower 3620 computer with a 3.50GHz Intel Xeon E3-1245 processor and 32GB of RAM memory, running the Ubuntu 16.04 LTS operational system. All the executions were performed on a single thread, without any significant concurrent process.

The results obtained for the I_1 instance set are presented in Table 2, grouped by the TSP evaluation parameters. Each line in this table represents the average over five instances of the computational results with a specific algorithm parameter setting. Column z_0 gives the average initial solution value produced by the greedy algorithm, column z_{best} gives the average solution value produced by the ILS, column time(s) reports the average run time of the ILS in seconds, and column $\%G$ shows the average percentage gain of z_{best} with respect to z_0.

We start our analysis from the effect of the TSP evaluation parameters on the solution value. As the constructive algorithm is not affected by these parameters, it provides always the same initial solution to a given instance, but with a different objective function value, due to the difference in the solution evaluation.

As expected, using exact evaluations (and better heuristics) in more routes leads to a decrease in the objective function values, but considering the initial solution value, this variation is inferior to 5% in total from the most precise to the less precise evaluation, which suggests that simple heuristics are sufficiently efficient to check the quality of a storage assignment. This evidence is in accordance with what is stated in the literature and practised in real scenarios, mainly by the pickers, that tend to follow the most intuitive and sub-optimal routes [12, 14]. On the other hand, the improvement on evaluation precision is unequivocally counterbalanced by a significant run time increase if the whole algorithm is considered. The total run time using the $TSP(7, 11)$ configuration is over the double of the total run time using the $TSP(5, 9)$ configuration.

An interesting effect of the TSP evaluation parameter in the solution is the negative correlation between the evaluation precision and the improvement obtained by the local search. In other words, if we increase the number of routes that are evaluated by an exact method (or better heuristics), we get less improvement over the greedy solution. A possible cause of this result is the higher probability of a travel distance over-estimation when evaluating good-quality solutions if the evaluation precision is low. This could guide the search to a region with low-quality solutions, increasing the convergence time without improving the solution quality.

In the same sense, it is noticeable that an increase in the number of products and orders in the instances also reduces the gains over the initial solution. This is an expected result, as it is harder to balance all picker route distances if there are more routes to balance and more points to visit among the different routes.

When comparing the results of scenarios with the same evaluation parameters but different IWI values, it is possible to notice that a higher IWI only slightly improves the average gains over the initial solution (on average less than 0.7 percentage points increase for each two more IWI), even with significantly higher run times. Among the hypotheses to explain this behaviour, we can cite a bad performance of the perturbation method to escape from local optima, a too quick convergence of the method to values close to an average local optimum, the existence of too many local optima, or the existence of several regions of the solution space with very similar characteristics causing repetitive searches.

It is interesting to notice that when comparing "type-free swap" with "same-type swap" results, the latter on average gets better solutions with a higher run time. We expect that a more restricted swap could lead to a quick conversion and thus a worse solution. As the differences between solutions are relatively small, while between run times are relevant, we can suppose that the method, when using a more restricted local search, performs more but smaller improvements. This explanation is compatible with concerns about meta-heuristic parameter calibration, in which a developer tries to balance the size of intensification and diversification steps in order to find a better algorithm performance.

In Table 3, we show the algorithm results for instances with isolation and allocation incompatibility constraints. In the table, Φ_0 gives the average value of the initial penalties, Φ_{best} the average value of the penalties in the best solutions produced by the ILS, $\%G_z$ the average percentage gain over the initial objective function values, and $\%G_\Phi$ the average percentage gain over the initial penalties. In Table 3, the block *Isolation 1* refs to instances containing one type of weak isolation constraints, the block *Isolation 2* refers to instances containing one type of weak isolation and one type of strong isolation constraints. Besides that, both the blocks *Isolation 1* and *Isolation 2* in the table are subjected to allocation incompatibilities.

Table 2. Computational results on instance set I_1 [25]. Each line is an average on 90 instances.

Swap policy	IWI	$TSP(5,9)$				$TSP(6,10)$				$TSP(7,11)$			
		z_0	z_{best}	time(s)	%G	z_0	z_{best}	time(s)	%G	z_0	z_{best}	time(s)	%G
Same-type	6	1675434.0	1025435.0	683.0	41.88	1644702.6	1059339.3	788.0	38.27	1604009.9	1149117.5	1519.7	30.85
	8	1675434.0	1016371.8	830.8	42.70	1644702.6	1047753.8	991.4	39.19	1604009.9	1137799.8	1709.8	31.47
	10	1675434.0	1007464.1	1038.8	43.34	1644702.6	1042574.2	1126.9	39.70	1604009.9	1130933.5	2078.8	32.10
Type-free	6	1675434.0	1035816.2	589.0	41.39	1644702.6	1065611.6	722.7	37.59	1604009.9	1153347.8	1224.9	30.18
	8	1675434.0	1030057.3	709.2	41.85	1644702.6	1059577.0	893.1	38.21	1604009.9	1144094.8	1525.9	30.75
	10	1675434.0	1024599.0	866.0	42.26	1644702.6	1055123.4	1090.6	38.56	1604009.9	1143325.2	1858.5	31.20

Table 3. Computational results for instances with isolation and incompatibility constraints on instance set I_2 [25]. Each line is an average on 3 instances. Parameters used: 8 IWI, $TSP(7,11)$. Swap policy: *type-free*.

Id	$|P|$	W	$|\omega|$	Isolation 1							Isolation 2					
				z_0	Φ_0	z_{best}	Φ_{best}	$%G_z$	$%G_\Phi$	time(s)	z_0	Φ_0	z_{best}	Φ_{best}	$%G_z$	$%G_\Phi$ time(s)
1			500	405287.9	51048.9	225120.3	37610.6	44.32	25.38	276.5	632041.1	344081.8	503372.6	317109.3	20.36	7.84 288.2
2		W_1	1000	749806.0	49873.4	430050.9	31858.6	42.63	35.43	713.7	1187816.0	621282.7	977614.8	590412.1	17.64	4.96 499.6
3			5000	3558644.9	49217.2	2193491.7	52373.3	38.36	-6.53	2839.9	5659723.9	2795037.9	4846198.5	2786653.8	14.37	0.30 2119.7
4	100		500	383964.6	54282.0	240575.6	42760.2	37.33	20.93	261.0	613230.2	353249.2	534527.6	341856.0	12.82	3.15 203.0
5		W_2	1000	704103.1	51699.1	469837.4	42271.7	33.28	17.90	680.6	1162098.5	644019.8	1025637.5	619162.5	11.77	3.90 489.7
6			5000	3332955.4	54780.4	2276342.0	52617.7	31.70	2.89	3005.5	5555117.6	2933844.9	4960398.8	2911717.1	10.71	0.76 2388.9
7			500	371185.3	58468.7	251748.4	36607.4	32.19	36.99	334.2	575261.8	346403.8	518039.3	334616.3	9.94	3.39 402.6
8		W_3	1000	673638.7	49932.3	504716.8	38462.8	25.07	21.97	611.3	1096946.6	632364.3	1001514.8	621692.1	8.70	1.73 519.2
9			5000	3221278.3	58714.3	2444756.2	46692.6	24.10	19.00	3519.7	5296059.8	2928326.8	4907968.1	2915739.4	7.32	0.43 2897.5
10			500	438229.7	27666.7	326534.0	26333.3	25.48	4.87	281.8	728354.9	363274.5	576015.8	297922.2	20.98	18.17 278.5
11		W_1	1000	854100.0	32666.7	633261.7	24000.0	25.83	26.33	766.9	1223601.2	503051.5	1022193.0	412985.7	16.44	17.91 622.5
12			5000	4208520.8	51535.1	3145456.1	59535.1	25.26	-25.94	2824.4	5413919.5	1718497.8	4499844.1	1673331.1	16.88	2.62 3092.1
13	200		500	425402.5	30307.5	306401.6	25061.0	27.97	14.48	254.7	711924.2	362643.5	546403.6	277711.3	23.24	23.43 282.3
14		W_2	1000	814109.8	28141.2	622948.2	23876.2	23.47	15.01	760.5	1210574.2	515456.9	993881.1	447135.4	17.88	13.18 678.9
15			5000	4046685.1	32227.3	2971407.6	30121.8	26.57	6.28	3593.7	5265289.3	1724671.7	4412597.4	1637695.8	16.19	5.02 3621.2
16			500	431234.1	32694.1	356368.3	23407.3	17.36	28.40	441.0	711317.5	360032.5	611663.2	297630.6	13.98	17.25 229.6
17		W_3	1000	833826.9	31042.3	689633.4	24334.8	17.29	21.15	775.5	1208464.6	504536.9	1055921.7	426959.7	12.61	15.21 532.4
18			5000	4165773.8	30477.8	3403163.1	31035.4	18.31	-2.27	3409.6	5349024.8	1733807.5	4761622.8	1645191.5	10.98	5.11 3864.5
Totals				1731243.0	42656.8	1257468.8	42760.2	37.33	20.93	1486.4	2523801.4	1100055.3	2180204.3	341856.0	12.82	3.15 1362.0

We can first observe a satisfactory improvement over the initial solution obtained by the local search, although smaller than that observed in the previous results. Nevertheless, in instances without hard isolation constraints, this metric raises to 19.4% and 15.1% in blocks *Isolation 1* and *Isolation 2*, respectively. These results may suggest that if the constructive algorithm does not handle well isolation constraints, then local search has troubles in improving the solution.

We notice that the local search reduces proportionally less the penalty value than the overall objective function value, except in instances with 200 products in warehouse W_3

Table 4. Computational results on large instances. Parameters used: 8 IWI, $TSP(7, 11)$. Each line is an average on 5 instances.

Id	$|P|$	$|\omega|$	z_0	z_{best}	$\%G_z$	time(s)
1	400	5000	4150015.0	3438747.4	3.43	6727.4
2	800	2000	1737677.0	1467340.2	3.11	7035.9
3	800	10000	8945296.0	7596591.4	3.02	44700.2
4	1600	2000	1722730.4	1518573.6	2.37	23162.5
5	1600	10000	9025347.8	7749001.4	2.83	166274.7

(lines 16 to 18 of Table 3), suggesting that the method could be giving more relevance to travel distance.

The average run time is smaller than one hour for almost all instance settings, except on instances on lines 15 and 18 and *Isolation 2*. It suggests that the method converges in an acceptable time even in more realistic and complex mid-size instances.

In Table 4, z_0, z_{best}, $\%G_z$ and time represent the average values for five different instances. In this case, the gain over the initial solution obtained by the search is between 2% and 4%. This interval is far inferior to those observed in smaller instances, as expected, but still relevant in a real life operation. The run time rises approximately in a linear way with respect to the number of orders (6.37 times for instances with 800 products and 7.18 times for instances with 1600 products), but very fast with respect to the number of products to be allocated, getting to an average of 48 h for instances with 1600 products. It suggests that the amount of order data is not a critical factor in algorithm scalability.

An interesting result can be noticed when observing the initial solution value. While this seems to be directly proportional to the number of orders (as more orders mean more routes), it does not change with the number of products. This can be explained by the proportional growth of warehouse capacity per area, which makes the average picking density to remain similar, causing a similar route distances.

8   Conclusion

In this paper, we study the Storage Location Assignment Problem with Product-Cell Incompatibility and Isolation Constraints. This generalises the classic Storage Location Problem, by introducing the possibility of preventing the placement of certain products in certain positions and keeping certain product families isolated. In a pharmaceutical logistic operators context, this problem is often encountered when aiming at improving warehouse performance while maintaining flexibility in defining warehouse layout. Additionally, the work also described how to process the warehouse input data, which can be useful in several other studies that, as this one, need to deal with non-conventional warehouse layouts or for situations where design changes also need to be evaluated.

To solve the problem, an Iterated Local Search method is proposed and tested on a large heterogeneous set of instances, demonstrating its suitability in providing good quality solutions within a satisfactory run time. Analysis enlighten that variations on the stopping criteria involve significant changes in the run time, but just slight changes in the solution quality. On the other hand, the variations in the parameters related to the TSP solutions (to determine the pickers' routes) prove to be very influential in both run time and solution quality. Large instances are well handled by the ILS algorithm, though with higher run times, with this raise strongly related to the number of products to be allocated and less to the number of orders considered. Instances with isolation and incompatibility constraints present lower improvements in the local search phase, but still relevant for commercial purposes.

In the future, we aim to investigate more deeply the influence of initial allocation of strongly isolated types on the overall performance of the optimization method. General weighting of the different terms in the objective function 1 should be addressed, after an extensive analysis of several factors, e.g. incidence of the distribution of the various types of penalties among products, or the ratio between warehouse dimensions and the number of orders. This could also lead to the development of interesting multi-objective optimization algorithms. Also, the suitability of the method to more dynamic situations should be investigated, mainly when different sources of information are available for the orders. It could be interesting to create stronger strategies to avoid route re-evaluation or discard non-improving swaps, as this could reduce significantly the run time and allow a broader search.

References

1. Accorsi, R., Baruffaldi, G., Manzini, R.: Picking efficiency and stock safety: a bi-objective storage assignment policy for temperature-sensitive products. Comput. Ind. Eng. **115**, 240–252 (2018)
2. Aldrighetti, R., Zennaro, I., Finco, S., Battini, D.: Healthcare supply chain simulation with disruption considerations: a case study from northern Italy. Glob. J. Flex. Syst. Manag. **20**(1), 81–102 (2019)
3. Battini, D., Calzavara, M., Persona, A., Sgarbossa, F.: Order picking system design: the storage assignment and travel distance estimation (SA&TDE) joint method. Int. J. Prod. Res. **53**(4), 1077–1093 (2015)
4. Bo, D., Yanfei, L., Haisheng, R., Xuejun, L., Cuiqing, L.: Research on optimization for safe layout of hazardous chemicals warehouse based on genetic algorithm. IFAC-PapersOnLine **51**(18), 245–250 (2018). https://doi.org/10.1016/j.ifacol.2018.09.307, 10th IFAC Symposium on Advanced Control of Chemical Processes ADCHEM 2018
5. Bolaños Zuñiga, J., Saucedo Martínez, J.A., Salais Fierro, T.E., Marmolejo Saucedo, J.A.: Optimization of the storage location assignment and the picker-routing problem by using mathematical programming. Appl. Sci. **10**(2), 534 (2020)
6. Bottani, E., Casella, G., Murino, T.: A hybrid metaheuristic routing algorithm for low-level picker-to-part systems. Comput. Ind. Eng. **160**, 107540 (2021)
7. Boysen, N., De Koster, R., Weidinger, F.: Warehousing in the e-commerce era: a survey. Eur. J. Oper. Res. **277**(2), 396–411 (2019)
8. Cambazard, H., Catusse, N.: Fixed-parameter algorithms for rectilinear Steiner tree and rectilinear traveling salesman problem in the plane. Eur. J. Oper. Res. **270**(2), 419–429 (2018)
9. Chen, F., Xu, G., Wei, Y.: Heuristic routing methods in multiple-block warehouses with ultra-narrow aisles and access restriction. Int. J. Prod. Res. **57**(1), 228–249 (2019)

10. Chuang, Y.F., Lee, H.T., Lai, Y.C.: Item-associated cluster assignment model on storage allocation problems. Comput. Ind. Eng. **63**(4), 1171–1177 (2012)
11. De Koster, R., Le-Duc, T., Roodbergen, K.J.: Design and control of warehouse order picking: a literature review. Eur. J. Oper. Res. **182**(2), 481–501 (2007)
12. De Santis, R., Montanari, R., Vignali, G., Bottani, E.: An adapted ant colony optimization algorithm for the minimization of the travel distance of pickers in manual warehouses. Eur. J. Oper. Res. **267**(1), 120–137 (2018)
13. Dijkstra, A.S., Roodbergen, K.J.: Exact route-length formulas and a storage location assignment heuristic for picker-to-parts warehouses. Transp. Res. Part E Logist. Transp. Rev. **102**, 38–59 (2017)
14. Elbert, R.M., Franzke, T., Glock, C.H., Grosse, E.H.: The effects of human behavior on the efficiency of routing policies in order picking: The case of route deviations. Comput. Ind. Eng. **111**, 537–551 (2017)
15. Fontana, M.E., Nepomuceno, V.S.: Multi-criteria approach for products classification and their storage location assignment. Int. J. Adv. Manuf. Technol. **88**(9–12), 3205–3216 (2017)
16. Foroughi, A., Boysen, N., Emde, S., Schneider, M.: High-density storage with mobile racks: picker routing and product location. J. Oper. Res. Soc. **72**(3), 535–553 (2021)
17. van Gils, T., Ramaekers, K., Caris, A., de Koster, R.B.: Designing efficient order picking systems by combining planning problems: state-of-the-art classification and review. Eur. J. Oper. Res. **267**(1), 1–15 (2018)
18. Gu, J., Goetschalckx, M., McGinnis, L.F.: Research on warehouse operation: a comprehensive review. Eur. J. Oper. Res. **177**(1), 1–21 (2007)
19. Guerriero, F., Musmanno, R., Pisacane, O., Rende, F.: A mathematical model for the multi-levels product allocation problem in a warehouse with compatibility constraints. Appl. Math. Model. **37**(6), 4385–4398 (2013)
20. Keung, K., Lee, C., Ji, P.: Industrial internet of things-driven storage location assignment and order picking in a resource synchronization and sharing-based robotic mobile fulfillment system. Adv. Eng. Inform. **52**, 101540 (2022). https://doi.org/10.1016/j.aei.2022.101540
21. Kim, J., Méndez, F., Jimenez, J.: Storage location assignment heuristics based on slot selection and frequent itemset grouping for large distribution centers. IEEE Access **8**, 189025–189035 (2020). https://doi.org/10.1109/ACCESS.2020.3031585
22. Kress, D., Boysen, N., Pesch, E.: Which items should be stored together? A basic partition problem to assign storage space in group-based storage systems. IISE Trans. **49**(1), 13–30 (2017). https://doi.org/10.1080/0740817X.2016.1213469
23. Li, J., Moghaddam, M., Nof, S.Y.: Dynamic storage assignment with product affinity and ABC classification-a case study. Int. J. Adv. Manuf. Technol. **84**(9–12), 2179–2194 (2016)
24. Lu, W., McFarlane, D., Giannikas, V., Zhang, Q.: An algorithm for dynamic order-picking in warehouse operations. Eur. J. Oper. Res. **248**(1), 107–122 (2016)
25. Mendes, N., Bolsi, B., Iori, M.: An iterated local search for a pharmaceutical storage location assignment problem with product-cell incompatibility and isolation constraints. In: Proceedings of the 24th International Conference on Enterprise Information Systems - Volume 1: ICEIS, pp. 17–26. INSTICC, SciTePress (2022). https://doi.org/10.5220/0011039000003179
26. Ming-Huang Chiang, D., Lin, C.P., Chen, M.C.: Data mining based storage assignment heuristics for travel distance reduction. Expert Syst. **31**(1), 81–90 (2014)
27. Mirzaei, M., Zaerpour, N., De Koster, R.: The impact of integrated cluster-based storage allocation on parts-to-picker warehouse performance. Transp. Res. Part E Logist. Transp. Rev. **146**, 102207 (2021)
28. Mirzaei, M., Zaerpour, N., de Koster, R.B.: How to benefit from order data: correlated dispersed storage assignment in robotic warehouses. Int. J. Prod. Res. **60**(2), 549–568 (2022). https://doi.org/10.1080/00207543.2021.1971787

29. Muppani, V.R., Adil, G.K.: A branch and bound algorithm for class based storage location assignment. Eur. J. Oper. Res. **189**(2), 492–507 (2008)
30. Muppani, V.R., Adil, G.K.: Efficient formation of storage classes for warehouse storage location assignment: a simulated annealing approach. Omega **36**(4), 609–618 (2008)
31. Öztürkoğlu, Ö.: A bi-objective mathematical model for product allocation in block stacking warehouses. Int. Trans. Oper. Res. **27**(4), 2184–2210 (2020)
32. Pan, J.C.H., Shih, P.H., Wu, M.H.: Storage assignment problem with travel distance and blocking considerations for a picker-to-part order picking system. Comput. Ind. Eng. **62**(2), 527–535 (2012)
33. Pan, J.C.H., Shih, P.H., Wu, M.H., Lin, J.H.: A storage assignment heuristic method based on genetic algorithm for a pick-and-pass warehousing system. Comput. Ind. Eng. **81**, 1–13 (2015)
34. Pan, J.C.H., Wu, M.H.: A study of storage assignment problem for an order picking line in a pick-and-pass warehousing system. Comput. Ind. Eng. **57**(1), 261–268 (2009)
35. Pang, K.W., Chan, H.L.: Data mining-based algorithm for storage location assignment in a randomised warehouse. Int. J. Prod. Res. **55**(14), 4035–4052 (2017)
36. Quintanilla, S., Pérez, Á., Ballestín, F., Lino, P.: Heuristic algorithms for a storage location assignment problem in a chaotic warehouse. Eng. Optim. **47**(10), 1405–1422 (2015)
37. Rao, S.S., Adil, G.K.: Optimal class boundaries, number of aisles, and pick list size for low-level order picking systems. IIE Trans. **45**(12), 1309–1321 (2013)
38. Rasmi, S.A.B., Wang, Y., Charkhgard, H.: Wave order picking under the mixed-shelves storage strategy: a solution method and advantages. Comput. Oper. Res. **137** (2022). https://doi.org/10.1016/j.cor.2021.105556
39. Ratliff, H.D., Rosenthal, A.S.: Order-picking in a rectangular warehouse: a solvable case of the traveling salesman problem. Oper. Res. **31**(3), 507–521 (1983)
40. Reyes, J., Solano-Charris, E., Montoya-Torres, J.: The storage location assignment problem: a literature review. Int. J. Ind. Eng. Comput. **10**(2), 199–224 (2019)
41. Roodbergen, K.J., Koster, R.: Routing methods for warehouses with multiple cross aisles. Int. J. Prod. Res. **39**(9), 1865–1883 (2001)
42. Scholz, A., Henn, S., Stuhlmann, M., Wäscher, G.: A new mathematical programming formulation for the single-picker routing problem. Eur. J. Oper. Res. **253**(1), 68–84 (2016)
43. Silva, A., Coelho, L.C., Darvish, M., Renaud, J.: Integrating storage location and order picking problems in warehouse planning. Transp. Res. Part E Logist. Transp. Rev. **140**, 102003 (2020)
44. Silva, A., Roodbergen, K.J., Coelho, L.C., Darvish, M.: Estimating optimal ABC zone sizes in manual warehouses. Int. J. Prod. Econ. **252**, 108579 (2022)
45. Staudt, F.H., Alpan, G., Di Mascolo, M., Rodriguez, C.M.T.: Warehouse performance measurement: a literature review. Int. J. Prod. Res. **53**(18), 5524–5544 (2015)
46. Thanos, E., Wauters, T., Vanden Berghe, G.: Dispatch and conflict-free routing of capacitated vehicles with storage stack allocation. J. Oper. Res. Soc. 1–14 (2019)
47. Theys, C., Bräysy, O., Dullaert, W., Raa, B.: Using a tsp heuristic for routing order pickers in warehouses. Eur. J. Oper. Res. **200**(3), 755–763 (2010)
48. Trab, S., Bajic, E., Zouinkhi, A., Abdelkrim, M.N., Chekir, H., Ltaief, R.H.: Product allocation planning with safety compatibility constraints in IoT-based warehouse. Procedia Comput. Sci. **73**, 290–297 (2015)
49. Trindade, M.A., Sousa, P.S., Moreira, M.R.: Ramping up a heuristic procedure for storage location assignment problem with precedence constraints. Flex. Serv. Manuf. J. **34**(3), 646–669 (2022)
50. Uthayakumar, R., Priyan, S.: Pharmaceutical supply chain and inventory management strategies: optimization for a pharmaceutical company and a hospital. Oper. Res. Health Care **2**(3), 52–64 (2013)

51. Van Gils, T., Ramaekers, K., Braekers, K., Depaire, B., Caris, A.: Increasing order picking efficiency by integrating storage, batching, zone picking, and routing policy decisions. Int. J. Prod. Econ. **197**, 243–261 (2018)
52. Volland, J., Fügener, A., Schoenfelder, J., Brunner, J.O.: Material logistics in hospitals: a literature review. Omega **69**, 82–101 (2017). https://doi.org/10.1016/j.omega.2016.08.004, http://www.sciencedirect.com/science/article/pii/S0305048316304881
53. Wang, M., Zhang, R.Q., Fan, K.: Improving order-picking operation through efficient storage location assignment: a new approach. Comput. Ind. Eng. **139**, 106186 (2020)
54. Xiao, J., Zheng, L.: Correlated storage assignment to minimize zone visits for BOM picking. Int. J. Adv. Manuf. Technol. **61**(5), 797–807 (2012)
55. Xu, X., Ren, C.: A novel storage location assignment in multi-pickers picker-to-parts systems integrating scattered storage, demand correlation, and routing adjustment. Comput. Ind. Eng. **172** (2022). https://doi.org/10.1016/j.cie.2022.108618
56. Zhang, R.Q., Wang, M., Pan, X.: New model of the storage location assignment problem considering demand correlation pattern. Comput. Ind. Eng. **129**, 210–219 (2019)
57. Žulj, I., Glock, C.H., Grosse, E.H., Schneider, M.: Picker routing and storage-assignment strategies for precedence-constrained order picking. Comput. Ind. Eng. **123**, 338–347 (2018)

Using IoT Technology for the Skilled Crafts

André Pomp$^{(\boxtimes)}$, Andreas Burgdorf, Alexander Paulus, and Tobias Meisen

Chair of Technologies and Management of Digital Transformation, University of Wuppertal, Wuppertal, Germany
{pomp,burgdorf,paulus,meisen}@uni-wuppertal.de
http://www.tmdt.uni-wuppertal.de

Abstract. The Internet of Things (IoT) is a rapidly growing technology that connects everyday objects to the internet, allowing them to send and receive data. This technology has the potential to revolutionize the way that craftsmen work, by providing them with access to real-time information and automation capabilities. For instance, devices that are equipped with sensors can transmit data about their usage and performance, providing craftsmen with valuable insights into their work. Moreover, IoT can also enable craftsmen to remotely monitor and control their tools and machines, increasing efficiency and productivity. In addition, IoT can help craftspeople to better manage their inventory and supply chain. By attaching sensors to raw materials and finished products, craftspeople can track the movement and status of their goods in real-time, improving the accuracy of their forecasts and reducing the risk of stock-outs. Thus, adopting the IoT technology has the potential to greatly enhance the capabilities of the skilled crafts, allowing them to work more efficiently and effectively, and to better serve the needs of their customers.

The aim of the IoT4H project is to make IoT technology usable for craftsmen. For that, we develop a platform that helps craft businesses in exploring and implementing IoT solutions. The platform will help craftsmen to identify suitable IoT use cases and business models for their trades, provide an overview of required components, and allow them to set up use cases without technical expertise. It will also collect and manage data and allow craft businesses to share new use cases with other platform users. So far, the project has already identified many different IoT use cases for the skilled crafts sector, which are presented in this paper. In addition, initial platform architecture considerations have also been made based on these use cases and are presented in this paper.

Overall, the results are intended to assist craft businesses from a wide range of trades to identify and implement new business models in such a way that they can be integrated into the existing processes of the businesses.

Keywords: Internet of things · Semantic Systems Engineering · Usability · Skilled crafts

1 Introduction

Digitalization is both an opportunity and a challenge for a large number of industries. Examples of how the potential of advancing digitalization can be exploited can be seen

J. Filipe et al. (Eds.): ICEIS 2022, LNBIP 487, pp. 48–63, 2023.
https://doi.org/10.1007/978-3-031-39386-0_3

in modern agriculture, industrial production, and the transport and logistics sector. In these industries, numerous enterprises have already been able to gain wide experience with digital and digitized products as well as services, from which a wealth of novel business models has emerged. At the same time, experience has shown that digital transformation cannot be taken for granted and involves a wide range of challenges. The Internet of Things (IoT) [1] plays a key role here and it is rising in different domains, such as manufacturing [15], smart cities [7] or smart home [16]. The focus of the underlying technologies is on mapping and describing the physical, real world in a digital image by generating digital shadows or even digital twins [2]. The fundamental step in this process is the incorporation of sensors into products, machines and systems with the purpose of creating a digital image from the collected data through a range of analysis and learning processes. In turn, the digital image obtained in this way creates the basis for optimizing existing products, services and business processes. In addition, in many cases the digital image also forms the basis for generating new digital or digitalized business models. The *skilled crafts* sector, in which value creation is determined by the manufacture of material products or the provision of services on demand for the material product, is hence predestined for the introduction of IoT technologies into the daily work routine. Mapping the physical world in a digital representation opens up numerous application areas [14]. That makes it all the more remarkable that although IoT technologies are increasingly finding their way into the consumer market, the focus of developments is completely bypassing the skilled crafts. As a result, there has been little or no awareness of the potential of IoT among traditional small and medium-sized craft businesses - even though the technologies are available and craft businesses are expressing interest in using IoT, for example, in public workshops and hackathons.

By contrast, large industrial companies, as representatives of industrial mass production that are often in direct competition with the skilled crafts, are increasingly equipping their products with IoT-capable sensor and communications technology and offering corresponding applications for use by end customers. These companies focus on their area of application, which prevents a holistic view and correspondingly advanced overarching digitalization. For example, customers can query the operating status of their heating systems and directly contact the manufacturer in the event of possible malfunctions, but not the local craftsman. This fragmented IoT platform landscape of individual industrial providers does not provide a breeding ground from which regional craft businesses can benefit or participate due to the lack of access to data. Other central challenges for these craft enterprises arise, for example, from the difficult access to the topic and the related major uncertainties. Accordingly, for regional craft businesses, i) assessing the potential of IoT solutions and their impact on their operations, and ii) acquiring the knowledge and skills to develop and operate adequate IoT solutions are key obstacle factors.

To address these challenges, the goal of the IoT4H[1] project is to design and implement a vendor-independent end-to-end IoT Platform, which will be developed from the skilled crafts for the skilled crafts. This portal, called *IoT4H platform*, should enable German craft businesses to i) identify suitable IoT use cases and resulting business models for their trade as well as their products or services and share them with other

---

[1] http://www.iot4h.de.

businesses, ii) get an overview of the required components (e.g. sensors) needed to implement the use cases, iii) set up the identified use cases in a simple and user-friendly way via the portal, and iv) collect, manage and use the resulting data for their own business purposes.

The two most important aspects for the development of the IoT4H Platform deal with the identification of suitable IoT use cases and the creation of a corresponding crafts-friendly IoT portal as a prototype, which can be used by the craft businesses for the implementation of their IoT use cases. A large number of different IoT sensors will be used to implement the use cases. The resulting data will be stored in a central data lake and managed via it. The main research focus is on developing ways how the accessibility of IoT technology can be enabled for the craftsmen. In particular, the development of recommendation and search systems based on machine learning methods as well as the design of corresponding visual user interfaces are in focus. Another important aspect is the investigation of approaches on how the IoT technology and its application potential can be didactically prepared for the craft businesses and communicated to them in a target group-oriented or -adaptive way. Based on the IoT4H Platform, different participating craft businesses from a wide variety of trades aim to identify and develop IoT-based use cases and prototypes, which either optimize an internal process or enable the implementation of an additional (digital) business model specific to their trade and business.

In this way, craft businesses improve their work organization and design with digital transformation technologies. Ensuring that the data generated on the developed IoT4H platform belongs to the craft enterprises themselves guarantees that they can use it profitably. Meanwhile, craft businesses can already benefit from implemented IoT use cases from other (craft) businesses. As soon as a new use case is identified and implemented via the developed IoT4H platform, it is also available to other companies. This facilitates knowledge transfer both within the own trade as well as across trades, so that the full potential of IoT use cases for the skilled crafts is exploited. Summarizing, the developed IoT4H platform enables the strengthening of craft services in Germany by enabling the development of modern IoT-based business models, which, for example, results in new services for the end customer of the respective craft businesses.

Please note that this paper extends our previous work [11]. Compared to our previous work, this paper covers the following extensions. First, we extended the related work section with new solutions that we identified. Second, we present our first results regarding possible use cases for the skilled crafts that we identified in four hackathons, the architecture of the IoT4H platform and our IoT living lab (cf. Sect. 4).

The rest of the paper is organized as follows: Sect. 2 will show related work and discuss the issues of current IoT platforms for technical inexperienced users. Afterwards, we present the methodology and the approach that we follow in order to develop the IoT4H Platform in Sect. 3 and present our first results in Sect. 4. Finally, we give a short conclusion in Sect. 5.

2   Related Work

The related work can essentially be divided into i) the area of technical implementation (IoT platforms and existing IoT solutions) and ii) the area of existing platforms in the Skilled Craft Sector that push the issue of digitalization.

2.1   Technical Implementation and Existing IoT Solutions

Today, IoT platforms are ubiquitous. Prominent tech companies such as Amazon[2], Google[3], Microsoft[4] or Deutsche Telekom[5] already offer scalable IoT platforms that enable the implementation of any IoT use case - presuming users have the technical programming skills and already know exactly which use case might fit their business. In addition, there are a number of other platforms (Bosch[6], Upswift[7], Influxdata[8] or Device Insight[9]) which also offer IoT platforms for customers. These platforms focus on the collection and analysis of raw data generated in the IoT. However, these platforms target either users with advanced technical knowledge or large companies with their own IT departments.To produce the required raw IoT data, users must identify and implement the IoT use cases themselves and knowledge transfer between users does not take place as these platforms do not focus on this. By implication, this means that every craft business must acquire the necessary skills itself, identify its use cases and then either also implement these in a time-intensive manner via an IoT platform or commission IT experts to do so - without knowing a priori the economic benefits. Moreover, there exist also concepts for semantic IoT and data platforms, such as [4,6,10] or [12]. However, they only deal with semantically describing sensor data and do not focus on IoT use cases.

In addition to the existing IoT platforms, which are mainly aimed at companies from the Industry 4.0 sector, there are already various IoT-based solutions that have been developed for the skilled crafts. Hilti [8] and Bosch [3], for example, offer IoT-based solutions that enable craftsmen to track the tools in use. The company Doka [5] offers sensor technology and the associated software to measure and predict the drying time of cement. In the context of monitoring moisture damage in roofs, Saint Gobain has developed Isover GuardSystem [13], a solution that allows flat roofs to be monitored in real time and alerts the owner in the event of moisture damage. While on the one hand these examples show the potential already inherent in IoT solutions for the skilled crafts, on the other hand, it becomes clear that all these solutions are isolated applications which are offered by one manufacturer and can only be used in this context. So if a craft business now wants to use all these different solutions, it

---

[2] https://aws.amazon.com/de/iot/.
[3] https://cloud.google.com/iot-core.
[4] https://azure.microsoft.com/en-us/overview/iot/.
[5] https://iot.telekom.com/en.
[6] https://developer.bosch-iot-suite.com/.
[7] https://www.upswift.io/.
[8] https://www.influxdata.com/influxdb-cloud-iot.
[9] https://www.device-insight.com/.

must first learn of their existence (for example, via trade fairs, representatives of the companies or via other craft businesses) and then buy these solutions from the various large companies in each case. At the same time, the craft business loses control over its data, because it usually becomes the property of the large company. On the one hand, this prevents reuse for other IoT use cases. On the other hand, it shows that there is not yet a solution tailored to the needs of the skilled crafts that enables them to have an overview and all-encompassing use of the Internet of Things. There are craft businesses that try out IoT solutions in hackathons or even implement them on their own initiative (e.g., Holzgespür[10]) and offer them to their customers - but this remains more the exception than the rule - even though the IoT offers numerous possibilities.

2.2   Digital Platforms for the Skilled Crafts

Over the last few years, numerous platform-based business models have emerged in the digital craft sector that have significantly changed communication between customers, crafts, trade and industry. The current study by the Ludwig Fröhler Institute for Skilled Crafts Sciences (LFI) [9] shows that more than 100 transaction-oriented platforms are active in the craft value chain in Germany. Here, previous platforms can be divided into the categories of partner brokers, franchisers, infrastructure providers, advertising platforms and online stores. Partner mediators, like MyHammer[11] or Blauarbeit[12], focus thereby on the switching of craft businesses for services desired by the final customer. Businesses, which do not co-operate with the respective platforms, remain thereby outside. Franchiser, like Myster[13] or Banovo[14], even go one step further. Here, the local craftsman's business is only used as an executing instance - all offers and contract arrangements run via the respective platform. The consequence of this, of

Fig. 1. Adding IoT Platforms as missing Platform for the Skilled Crafts [11].

---

[10] https://www.holzgespuer.de/.

[11] https://www.my-hammer.de/.

[12] https://www.blauarbeit.de/.

[13] http://myster.de/.

[14] https://www.banovo.de/.

course, is that the local craft company is no longer in control of the business itself and must therefore submit to the rules of the platform operators. Infrastructure providers, such as Helpling[15] or Caroobi[16], on the other hand, increase the freedom of craft businesses. Advertising platforms and online stores are then already trying to attract customers via new channels. Io-Key[17] and Trotec TTSL[18] are first commercial approaches to provide professional IoT devices in combination with data access. The LFI study shows that platforms have become an important tool in the skilled crafts for establishing new business models and facilitating interaction with customers. However, the study only looks at transaction-oriented platforms. Data-centric platforms, which essentially include all Infrastructure-as-a-Service (IAAS), Software-as-a-Service (SAAS), Platform-as-a-Service (PAAS) and other specialized platforms, such as IoT platforms, were not considered. The preceding section has already shown that technical solutions do exist, but they are not tailored to the skilled trades and their needs.

In summary, we can state that there is no existing IoT platform that enables craft businesses to i) find out about the IoT, ii) identify relevant use cases and associated sensor technology, and iii) subsequently implement and operate the use cases relevant to their own business themselves.

3   Methodology and Approach

As described in Sect. 2, there is, to the best of our knowledge, no platform that focuses on enabling craft businesses to implement IoT use cases (cf. Fig. 1), as these i) only add value for technically skilled individuals, ii) do not provide an overview of which IoT use cases are implementable, relevant or already exist, iii) do not enable knowledge transfer between users, and iv) are usually only in English. These disadvantages are accompanied by the fact that small and medium-sized craft businesses - unlike large enterprises - typically do not have the financial resources to establish the appropriately required technical staff to identify, set up and manage IoT use cases. This means that relevant IoT use cases, such as, *Monitoring the degree of drying of screed*, cannot be efficiently implemented with the existing solutions on the market, as these solutions do not address the explicit needs of craft businesses. At the same time, however, our experience shows, among other things, that these are precisely the kind of use cases that support craft businesses in their day-to-day work, or in which they see potential for the digitalization of craft businesses. Accordingly, the goal for the developed IoT4H Platform is to show businesses similar scenarios from IoT projects that have already been carried out by other craft businesses and to provide information on sensor technology and information processing which is accessible to craftsmen. For the above example, the portal could display, for example, *Moisture sensors that can be installed in the screed*. After the required sensors have been purchased by the craftsman's company, they can be integrated directly into the portal and the corresponding analyses and monitoring solutions

---

[15] https://www.helpling.de/.

[16] https://caroobi.com/.

[17] https://autosen.com/en/Industry-4-0-solutions/io-key.

[18] https://uk.trotec.com/produkte-services/software/ttsl-remote-control-monitoring/.

can be carried out. In the aforementioned example, a craftsman's company would be notified when the desired degree of drying has been reached, so that the follow-up work can be carried out earlier than expected, if necessary.

Compared to the current state of the art, the IoT4H Platform presented here fills an important gap. It represents an end-to-end solution that is explicitly adapted and optimized for the skilled crafts and can be used in a wide variety of trades. Through an explicit focus in development of the IoT4H Platform on intuitive usability (queries using natural language, simple plug-and-play connection of sensors, automatic recommendation of useful use cases), a complexity of the solution is achieved that is suitable for craftsmen without appropriate technical training and is missing in the current market. Continuous expansion of the use cases is also intended to ensure that all companies always have the opportunity to keep up with the latest technology. If new use cases arise in a trade, they can be implemented via the IoT4H Platform and also made directly available to other craft businesses (knowledge transfer). If a business identifies and implements a new use case, such as the *detection of leaking roofs* via the IoT4H Platform, it is possible to make this use case available to other businesses. This means that other companies can see what sensor technology is needed and gain access to the implemented use case. This mechanism allows other businesses to see use cases that have already been successfully implemented by others. On the one hand, this mechanism thus strengthens confidence in the use of IoT technology. On the other hand, it enables individual craft businesses to independently develop new business models for their own operations. At the same time, the business that provides a use case can also benefit.

3.1   Approach

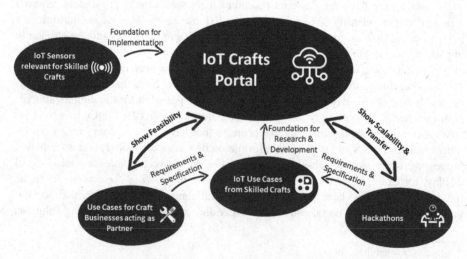

Fig. 2. Main Aspects for the development of the IoT4H Platform [11].

Figure 2 shows the main aspects for the development of the IoT4H Platform and how they are related to each other. All in all, we follow an agile approach for developing the IoT4H Platform. This first requires the identification of potential use cases in the skilled crafts that can be implemented with the help of current IoT technology. In order to create a broad coverage of use cases for different trades, the IoT4H Platform will be developed in close cooperation with seven application partners from seven different trades. Together we define IoT-based use cases at the beginning of the development phase that are to be implemented with the help of the IoT4H Platform. On the one hand, these use cases serve to derive requirements for the technical development of the platform and, on the other hand, they serve as initial use cases that can be used by other crafts businesses. By conducting six different hackathons with a large number of craft businesses that are not involved in the development, it will be evaluated, on the one hand, that already defined use cases are transferable to other businesses and/or trades, and on the other hand, the agility of the further development will be demonstrated, as further use cases will be identified and added to the IoT4H Platform. In this context, we also explore in particular which didactic concepts are suitable to bring craft businesses closer to IoT technology. Both the execution of hackathons and the implemented use cases (lighthouse projects) can then be used specifically to promote the dissemination of the IoT4H Platform to transfer it to the broad public. In the implementation of the IoT4H Platform, technical research is needed in particular on how data processing pipelines for heterogeneous data streams can be implemented for a wide variety of IoT use cases by people who do not deal with IoT solutions in detail in everyday life. For this purpose, setting up IoT use cases by means of intuitive user interfaces is seen as a central building block. In the following, we will elaborate these important aspects in more detail.

3.2   Identification and Evaluation of IoT Use Cases

At the beginning, two hackathons will be conducted with the aim of identifying a broad mass of potentially relevant IoT use cases and establishing an understanding of the necessity as well as the potential of using IoT in the skilled crafts (cf. Sect. 4). In doing so, the participants of the hackathons will be supported and accompanied by experienced scientists and developers. In order to ensure target group-adaptive communication, a didactic concept will be developed that will enable craft businesses to gain added value from and access to IoT technology. In addition to the use cases that arise during the hackathons, each craft business of the seven application partners from seven different trades will define at least one further use case relevant to this enterprise in a workshop.

The evaluation of the IoT4H Platform is based on four further hackathons. In contrast to the first two hackathons, which serve to identify user cases, these evaluation hackathons will be conducted with less assistance. This will show whether the skilled craftsmen are able to implement their use cases independently using the IoT4H Platform. At the same time, this will iteratively identify additional requirements and use cases. In this way, the development of the portal will be continuously guided in the right direction and its usability will also be ensured by craft businesses not involved in the development of the IoT4H Platform. Based on the experience of hackathons already held (cf. Sect. 4), it can be assumed that additional previously unknown use cases will be identified in each hackathon.

3.3 IoT-Sensor Screening

Based on the identified IoT use cases, an overview will be created of which IoT sensors are relevant for the identified use cases, how they function and to what extent they already have industry approval. In addition, all further sensor technology that might be relevant to the skilled crafts will be identified in a further search and included in the sensor technology catalog so that it can later be included in the IoT4H Platform. This systematic approach will ensure that a wide range of use cases for the skilled crafts can be covered by the IoT4H Platform. This screening will be continuously updated. The results will be made available to the public in a separate database.

3.4 IoT4H Platform

Based on the identified use cases and an overview of relevant sensors, technical requirements for the IoT4H Platform are formulated and a technical architecture is designed. This includes the definition of interfaces for data acquisition, decisions on technologies to be used, and the identification of suitable data analysis and AI-based learning methods needed to implement the use cases.

Next, procedures will be developed to connect the identified sensors to the IoT4H Platform via plug-and-play solutions. For this purpose, modules are needed that record and read out the data from the sensors and forward it to the IoT4H Platform. To solve the problems of heterogeneity in terms of data format and meaning (semantics), the data added to the portal will be annotated with semantics based on a craft IoT ontology that is to be defined before. Based on the semantic annotation, the data will be converted into a syntactically uniform data format and will be stored in a data lake.

In order to be able to flexibly implement the different use cases of the craft businesses on the basis of the data to be collected, modular data processing is implemented. This is based on the semantic information defined when adding the sensors. In addition, generic data processing building blocks are implemented that allow different processing steps, such as converting units or merging data streams. In this way, flexible data processing is created that enables the fusion of different sensor data streams and thus provides a basis to adequately implement a wide range of different IoT crafts use cases.

One of the most important building blocks of the IoT4H Platform will be a recommendation and search engine. This engine is being designed and implemented to enable knowledge transfer between the craft businesses and to encourage them to implement further IoT use cases or identify them for themselves. The search engine provides an efficient way to identify use cases that have already been implemented by other businesses based on a natural query language. The recommendation system, on the other hand, suggests further useful use cases to businesses based on their profile and trade as well as the use cases implemented so far, which have been implemented by other businesses with similar profiles. This ensures that businesses are always kept up to date and are notified when new IoT use cases emerge that are of interest to them. State of the art machine learning techniques will be used to develop the recommendation and search engine.

In order to enable a simple and intuitive operation for the craft businesses, a user interface is implemented that allows to realize own use cases even without technical training. This interface should create the possibility to identify, implement and monitor desired use cases. In addition to the findings from the first two hackathons, the requirements and wishes of the craft businesses for the user interface will be recorded in a workshop. On this basis, a user interface will be developed, which will be continuously adapted with the crafts businesses in an agile approach.

4 Preliminary Results

4.1 Identification of IoT Use Cases for the Skilled Crafts

In order to identify a broad set of IoT use cases for the skilled crafts, we conducted four hackathons of which two were part of the IoT4H project. Each hackathon was conducted with our four main partners (Kreishandwerkerschaft Rhein-Erft, Wirtschaftsförderung Rhein-Erft GmbH, Umlaut SE, Mittelstand-Digital Zentrum Handwerk). Altogether, we had more than 40 participants from different craft businesses and trades. In each hackathon, we focused on exploring and developing potential use cases for either improving processes within the own business or setting up new IoT-based business models. Each hackathon was organized in a two-day workshop format, with craft businesses identifying IoT-based use cases on the first day and prototyping them with scientists from our institute and developers on the second day.

The aim of implementing the prototypes together with the craftsmen was to show them, using concrete examples that are relevant to them and that were identified by themselves, (i) which sensors are available, (ii) what they can measure and how they record data, (iii) how the data can then be sent to a backend and how they can be stored and processed accordingly. For this purpose, we have provided the craftsmen with a large set of different sensors (cf. Fig. 3). After selecting sensors which were relevant for their desired use cases, the craftsmen had to connect the sensors either to a Raspberry Pi or to an Arduino. When using the Raspberry Pi, the craftsmen were able to record and process the data themselves by using NodeRED[19]. From NodeRED, the data was then send via MQTT to a cloud instance on which we then developed tailored visualizations and business logic for the identified use cases (cf. Fig. 4). In case of using Arduino, the programming part was conduced by the scientists and developers and the data was then also transmitted to the cloud instance via MQTT. In this way, the craftsmen were able to understand exactly what was happening, how the data was transmitted and how it could actually be processed. However, without the help of the advisors, the craftsmen would not have been able to implement the prototypes.

A central result of the four hackathons conducted is the large number of different IoT use cases that participants from various trades identified and assessed as relevant for their business. In total, 75 different IoT use cases were identified in different domains.

---

[19] https://nodered.org/.

Fig. 3. Selection of sensors for the craftsmen [11].

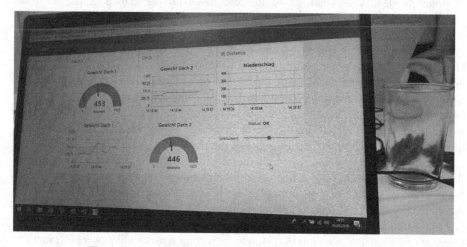

Fig. 4. Developed dashboard for one of the IoT use cases [11].

Figure 5a shows the exemplary brainstorming dashboard from the last hackathon. In addition to some obvious use cases that are already covered by the state of the art but were unknown to the craftsmen (e.g., tracking of tools or moisture penetration on the roof), many interesting new use cases were also identified. For example, two businesses in the construction sector have identified a use case for monitoring the tightness of basements during and after the construction phase. Craftsmen from the glazier or metal trades developed use cases to equip their products with additional sensor technology, for example, in order to track the use of these products - and thus to be able to optimize their products in the future. Businesses from the electrical, sanitary-heating-air-conditioning or roofer sectors had ideas to offer novel services to their customers in the form of "predictive maintenance" by using IoT technology. A roofing company, for example,

(a) Overview of use cases.

(b) Concrete developed prototype.

Fig. 5. Overview of the brainstorming for the different use cases as well as a concrete use case for roofers.

would like to equip gutters with water flow sensors. Based on the amount of fallen rain and the flow rate in the gutter, it is possible to compute whether the gutter is clogged and whether maintenance is required, which can then be offered in advance. Another roofing business has dealt in its use case with the occupational health and safety of employees, or to be more precise, with the UV exposure of roofers. For this purpose, a prototypical cap was developed (cf. Fig. 5b) in the hackathon, which measures the UV exposure by means of a UV sensor and then summarizes and checks the daily, monthly and annual dose of the roofers in the associated application. At the end of each hackathon, all developed prototypes were presented to all participants.

Altogether, the hackathons have proven themselves to be a beneficial tool for introducing craftsmen to IoT topics. In particular, the great potential of IoT for the skilled crafts could be experienced by the participants in person, but it also became obvious that the implementation of own IoT use cases cannot be achieved by the craftsmen themselves without any training or advice. The support of developers and scientists has always been needed.

4.2   IoT4H Platform

The use cases that we identified during the four hackathons show how different and heterogeneous use cases can be. For the development of the IoT4H platform, however,

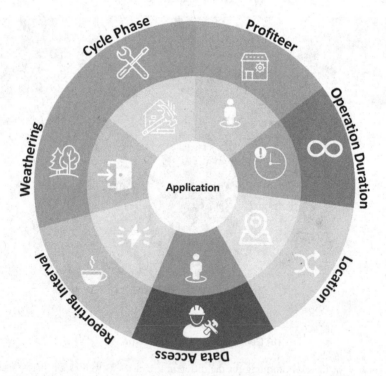

Fig. 6. Overview of the different areas that are relevant for the IoT4H platform.

all characteristics of the use cases must be covered. In total, we identified seven different main characteristics for a use case (cf. Fig. 6). First of all, the place of use must be identified for a use case. This can be indoors or outdoors. It is also important to identify in which phase a use case is applied. Is it to be applied in the construction phase of a building or is it more of a sensor for an everyday use case. Next, it is necessary to consider who the profiteer is for a use case. This can either be the business itself or the customer of a business. Further, it must be considered how long a use case will run. Is it a one-time application (e.g. during the construction phase) or should the use case run permanently for the customer? Then the location for the sensor system can either change or it is constantly the same location. In addition, a distinction must be made for each use case whether only the craftsman's business has access to the data or whether the customer who uses the sensors also has access. It must also be possible to set how often the data is to be transmitted to the IoT4H platform for each use case. Is once a day sufficient or do the data have to be transmitted and evaluated in real time?

Based on these scenarios, we initially focused the construction of the IoT4H platform architecture on use cases within homes. This can apply both during the construction phase and in daily business. The IoT4H ecosystem essentially provides for four active components (cf. Fig. 7). The simplest components are sensors that simply measure data. These sensors are either connected to IoT4H-Beacons or they are directly connected to the IoT4H-Cube. The beacons are used to record the readings from the sensors and store them temporarily if necessary. The beacons then transmit the recorded

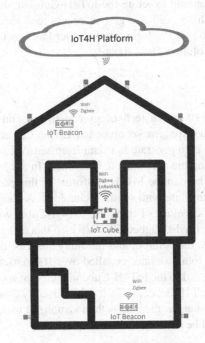

Fig. 7. Demonstration of the usage of Cube, Beacons ans sensors inside a house connected to the IoT4H platform.

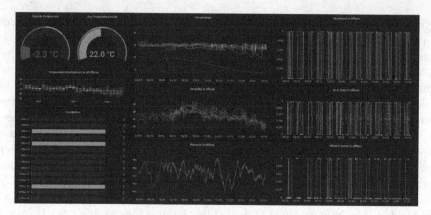

Fig. 8. Dashboard of the implemented living lab.

data to the IoT4H cube, which is the central gateway of the IoT4H ecosystem. The central IoT4H-Cube then forwards the data to the IoT4H Platform, where the data is finally analyzed, stored and visualized. The cube provides different radio technologies to communicate with its beacons. These include, for example, WiFi, Zigbee, ZWAVE, etc. The beacons, on the other hand, are equipped with different radio technology depending on the use case. The sensors are connected to the beacons either by cable or by radio.

At present, we are planning to set up the IoT4H-Cube on the basis of Homeassistant[20]. In the long term, this would also give us access to a wide range of smart home sensor technology. The IoT4H Platform, on the other hand, is to be developed on the basis of the open source solution Thingsboard[21].

4.3   IoT Living Lab

To build up a foundation of sensor technology and to test and evaluate the scenario described in the last chapter, we have set up an IoT Living Lab at our institute building. The goal of the living lab is to generate IoT data from sensors, send the data to IoT4H beacons, which in turn send the data to an IoT4H-Cube. In the long term, the data will then be transferred from there to the IoT4H platform. To this end, we have equipped a total of 14 offices at our institute with sensor technology. A Raspberry Pi with Home-assistant is installed in each office. This is the IoT4H-Cube, which is equipped with different radio technology (WiFi, Zigbee, ZWAVE). To this end, each office has motion sensors, vibration sensors, temperature and humidity sensors, door and window sensors, and CO2 sensors. In total, we have installed over 100 sensors at our institute. The sensors are directly connected to the IoT4H-Cube in their current state. We have not yet developed and deployed IoT4H beacons. Each cube currently transmits data to a central database. Based on this database, the data is then visualized using Grafana (cf. Fig. 8). In the future, this part will be done by the IoT4H platform.

---

[20] https://www.home-assistant.io/.
[21] https://thingsboard.io/.

5   Conclusion

In this paper, we presented our concept for the development of a manufacturer-independent IoT portal enabling small and medium-sized craft enterprises from the skilled crafts sector to realize IoT solutions on their own. We presented a concept for the implementation of the IoT4H platform, designed to enable craftsmen to implement IoT use cases on their own. The central building blocks here are sensors, IoT4H-Beacons, the IoT4H-Cube and the IoT4H platform. In addition, we showed the results of the first hackathons and the setup of our IoT living lab. The next steps are now to organize further hackathons, identify more IoT use cases and implement the IoT4H platform.

References

1. Ashton, K., et al.: That 'internet of things' thing. RFID J. **22**(7), 97–114 (2009)
2. Atzori, L., Iera, A., Morabito, G.: The internet of things: a survey. Comput. Netw. **54**(15), 2787–2805 (2010)
3. Bosch: Track My Tools (2020). http://bosch-trackmytools.com
4. Cambridge Semantics: Anzo Smart Data Discovery (2016). http://www.cambridgesemantics.com/
5. Doka: Smart Tools for Construction (2020). https://www.doka.com/en/news/press/in-the-digital-fast-lane
6. Dorsch, L.: How to bridge the interoperability gap in a smart city (2016). http://blog.bosch-si.com/categories/projects/2016/12/bridge-interoperability-gap-smart-city-big-iot/
7. Gyrard, A., Serrano, M.: A unified semantic engine for internet of things and smart cities: from sensor data to end-users applications. In: 2015 IEEE International Conference on Data Science and Data Intensive Systems, pp. 718–725. IEEE (2015)
8. Hilti: on!Track (2020). https://ontrack.hilti.com
9. Ludwig-Fröhler-Institut für Handwerkswissenschaften: Plattformen im Handwerk (2019). https://bit.ly/3ruEbeI
10. Palavalli, A., Karri, D., Pasupuleti, S.: Semantic internet of things. In: 2016 IEEE Tenth International Conference on Semantic Computing (ICSC), pp. 91–95 (2016). https://doi.org/10.1109/ICSC.2016.35
11. Pomp, A., Burgdorf, A., Paulus, A., Meisen, T.: Towards unlocking the potential of the internet of things for the skilled crafts. In: Filipe, J., Smialek, M., Brodsky, A., Hammoudi, S. (eds.) Proceedings of the 24th International Conference on Enterprise Information Systems, ICEIS 2022, Online Streaming, 25–27 April 2022, vol. 1, pp. 203–210. SCITEPRESS (2022). https://doi.org/10.5220/0011066100003179
12. Pomp, A., Paulus, A., Burgdorf, A., Meisen, T.: A semantic data marketplace for easy data sharing within a smart city. In: Proceedings of the 30th ACM International Conference on Information & Knowledge Management, pp. 4774–4778 (2021)
13. Saint Gobain: ISOVER GuardSystem (2020). https://www.isover.de/guardsystem
14. Sisinni, E., Saifullah, A., Han, S., Jennehag, U., Gidlund, M.: Industrial internet of things: challenges, opportunities, and directions. IEEE Trans. Ind. Info. **14**(11), 4724–4734 (2018)
15. Trakadas, P., et al.: An artificial intelligence-based collaboration approach in industrial IoT manufacturing: key concepts, architectural extensions and potential applications. Sensors **20**(19), 5480 (2020)
16. Zschörnig, T., Wehlitz, R., Franczyk, B.: A personal analytics platform for the internet of things-implementing kappa architecture with microservice-based stream processing. In: International Conference on Enterprise Information Systems, vol. 2 (2017)

Implementation Solutions for FAIR Clinical Research Data Management

Vânia Borges(⊠) , Natalia Queiroz de Oliveira , Henrique F. Rodrigues ,
Maria Luiza Machado Campos , and Giseli Rabello Lopes

Universidade Federal do Rio de Janeiro, Rio de Janeiro, Brazil
{vjborges30,hfr}@ufrj.br, {natalia.oliveira,
mluiza}@ppgi.ufrj.br, giseli@ic.ufrj.br

Abstract. The COVID-19 pandemic and its global actions have emphasized the importance of more detailed clinical data for studies of diseases and their effects. However, collecting and processing these data and publishing them according to rules defined by the responsible institutions is challenging. This paper addresses a modular, scalable, distributed, and flexible platform to promote such processing and publication in a FAIR manner. In addition, we describe approaches adopted to automate the FAIRification phase activities. A proposal for a FAIR Data Point on the cloud is also presented, aiming to improve the scalability of the metadata repository.

Keywords: VODAN Brazil · FAIRification · ETL4FAIR · Clinical Research Platform · FAIR Data Point

1 Introduction

During the COVID-19 pandemic, the volume of information available on the Web regarding the cases reached levels well above expectations, but most of this information referred to totals of people infected, hospitalized, recovered, and deaths. Although relevant, it hardly contributes to understanding the disease behavior and evolution, limiting the search for solutions to face the phenomena. In this aspect, the COVID-19 pandemic has reinforced the importance of clinical data for research conducted on diseases, especially those responsible for viral outbreaks. For these cases, more detailed patients´ data contemplating clinical characterization, as well as administered medications, test results, and outcomes, are fundamental to the scientific community. However, while valuable for more detailed studies in clinical research, they are not always available [1].

This context demonstrates the importance of an infrastructure to provide anonymous and structured clinical data from electronic health records (EHRs) or clinical research in hospitals. Implementing such an infrastructure is not a simple task. To be effective and efficient, it must address the issues associated with processing and publishing these data. Examples of such processing issues in extracting and collecting this type of data are: (i) privacy protection issues of personal data stored in the EHR, aligned to the Personal Data Protection Law; and (ii) complexity in processing the free-text fields of the EHR,

J. Filipe et al. (Eds.): ICEIS 2022, LNBIP 487, pp. 64–82, 2023.
https://doi.org/10.1007/978-3-031-39386-0_4

making data extraction difficult. Furthermore, regarding the publication of these data, we observed (i) the challenges for the development and implementation of a federated infrastructure to support this process; and (ii) the difficulty in providing linked data and metadata with different semantic artifacts to facilitate reuse by researchers.

In particular, for the COVID-19 pandemic, in order to accelerate and promote cooperation between different initiatives concerning research results, the Virus Outbreak Data Network Initiative (VODAN-IN) was conceived [2]. This IN is an activity of the Data Together organizations comprising GO FAIR[1], Research Data Alliance[2] (RDA), World Data Systems[3] (WDS), and the Committee on Data[4] (CODATA) of the International Science Council (ISC) [3]. Through this initiative, Data Together aims to establish a global ecosystem for data and identify opportunities and needs to address data-driven science [3].

VODAN-IN aims to develop a distributed infrastructure to support data interoperability on viral outbreaks, its first context being COVID-19 [4]. To this end, it proposes the creation of a "community of communities" to design and build, in an agile manner, the infrastructure for an international interoperable and distributed data network to support the search for evidence-based answers to viral outbreak cases [4].

This initiative has been designated in Brazil as the VODAN BR Project, being part of the thematic network GO FAIR Brazil Health[5] under the GO FAIR Brazil office [5]. This project is coordinated by Oswaldo Cruz Foundation (FIOCRUZ) and has as partners the Federal University of Rio de Janeiro (UFRJ), the Federal University of the State of Rio de Janeiro (UNIRIO), and the University of Twente. In addition, the project has partnerships with the Gaffrée Guinle University Hospital, in Rio de Janeiro; the São José State Hospital of Duque de Caxias, a hospital converted during the pandemic to attend only COVID-19 cases; and the Albert Einstein Israelite Hospital, in São Paulo. As it is an initiative of the GO FAIR consortium, the (meta)data and services generated must comply with FAIR principles, i.e., they must be Findable, Accessible, Interoperable, and Reusable [6].

This article is an extension of [7], a position paper presented at the 24th International Conference on Enterprise Information Systems (ICEIS 2022). In the position paper, we introduced an overview of a scalable, distributed, flexible and modular platform based on a generic architecture, with its processes and computational assets developed to support the VODAN BR project. Moreover, we discussed the desiderata for this development and the mechanisms established to optimize it. In this extension, we detail the platform implemented advances, addressing aspects that contribute to the management and organization of its activities, including the roles responsible for them. Then, we discuss the ETL4FAIR framework, providing some examples of integrations and expansions to automatize the FAIRification phase activities in the VODAN BR. Finally, we present an experiment to convert a FAIR DP to the cloud, aiming to provide greater scalability for the repository that consolidates the project metadata.

---

[1] https://www.go-fair.org/.

[2] https://rd-alliance.org/.

[3] https://world-datasystem.org/.

[4] https://codata.org/.

[5] https://www.go-fair-brasil.org/saude.

In this paper, Sect. 2 presents a background and some proposals concerned with the publication and dissemination of clinical research data. Section 3 describes the proposed platform for providing FAIR clinical data, explaining approaches adopted to organize its activities. Section 4 discusses the ETL4FAIR framework, exemplifying solutions to address the FAIRification phase activities. Section 5 introduces the FAIR DP for the cloud aiming at scalability for research metadata. Finally, Sect. 6 concludes with some final considerations and work in progress.

2 Background

The Semantic Web is an evolution of the World Wide Web (WWW) and constitutes a set of best practices and technologies to promote the understanding of data by humans and especially by machines. For this, it adds a layer of metadata on the existing data on the Web and adopts the representation of knowledge with Linked Data (LD), proposing a global data network [8].

In order to promote the publication of LD, the Resource Description Framework (RDF) has been adopted. This representation model, including in this context the RDF Schema (RDFS), provides the required formalism and flexibility for the creation and availability of linked data and metadata. It establishes a <subject>, <predicate> and <object> structure for the data representation that can be serialized in different formats such as, for example, Turtle, JSON-LD, NTriple and XML [9].

In this context, the FAIR principles add to the established standards for LD the importance of using metadata to facilitate the discovery and understanding, especially by machines (software agents) [10]. Thus, they guide the creation of data and metadata, aiming for data findability, accessibility, interoperability, and reuse [6].

FAIRification is the name for the process to make data FAIR. A generic workflow, applicable to any type of data and metadata, was proposed to facilitate this process [11]. This workflow is organized in three phases [11]: (i) pre-FAIRification, (ii) FAIR-ification, and (iii) post-FAIRification. These phases are divided into seven steps. The pre-FAIRification phase consists of 1) identifying the FAIRification objective, 2) analyzing data, and 3) analyzing metadata. The proper FAIRification phase involves the steps to 4) define a semantic model for data (4a) and metadata (4b), 5) make data (5a) and metadata (5b) linkable, and 6) host FAIR data. Finally, the post-FAIRification phase is responsible for 7) assessing FAIR data.

A suitable infrastructure for FAIRification should make (meta)data FAIR, respecting the established workflow. In addition, it must support the creation of rich metadata and provide the treatment and visibility to generated data and metadata, including provenance information. According to [12], it must contain repositories to host, manage, organize, and provide access to data, metadata, and other digital objects. It is also important to have descriptors that aggregate different types of metadata, such as descriptive, access rights, and transactions. Platforms with this infrastructure provide support to researchers without Information Technology (IT) expertise.

The dissemination of semantically enriched and machine-actionable metadata regarding research data promotes its discovery, access, and reuse. In this context, FAIR Data Point (FAIR DP) is an approach specially developed to expose digital objects metadata from any domain in a FAIR manner [13]. It defines standard methods for provisioning and accessing metadata compliant with the FAIR principles [13]. This conformity allows client applications to know how to access and interact with the provided metadata. Furthermore, as a FAIR-compliant metadata provider, it (i) enables owners, creators, and publishers to expose the metadata of FAIR-compliant catalogued resources; (ii) provides consumers and users with the discovery of information about resources of interest; and (iii) provides metadata in a machine-actionable form [13].

In the literature, we found related works to improve FAIR (meta)data management. For example, the Collaborative Open Omics (COPO) platform was developed to assist researchers in publishing their assets by providing metadata annotation support and mediating data submission to appropriate repositories [14]. FAIR4HEALTH, a European Horizon 2020 project, developed a workflow to implement FAIR principles in health datasets [15]. In addition, two tools were also designed to facilitate the transformation of raw datasets according to FAIR and preserve data privacy. The implemented platform adopts different solutions to handle the data like our work. This data is mapped to an HL7 FHIR repository according to rules defined by the data manager. However, our platform transforms the data into Case Record Form (CRF) associated with a semantic model annotated with semantic artifacts from the health domain. In addition, metadata regarding the data is generated and made available in the repository and the FAIR DP.

The Learning Healthcare System Data Commons (LHSDC) is a cloud-based orchestration environment and governance framework presented in [16]. It aims to meet standards for security, cost efficiency, and platform extensibility by enabling scalable access to a FAIR computing environment. This platform is open source, pay-as-you-go, and cost-effective, making it interoperable in an ecosystem of National Institutes of Health (NIH)-supported commons-based data enclaves and supporting big data initiatives in academic or public-private partnerships. This initiative adopts the Gen3[6] data platform, allowing researchers to submit their research data and query those in storage. In addition to data submission, our platform seeks to meet the entire workflow for generating FAIR (meta)data. In this context, FAIR DP enables human and software agents to search these data. Here it is worth highlighting our proposal for a FAIR DP in the cloud. This component, discussed in Sect. 5, can constitute federated platforms for FAIRification like the one presented in this paper, allowing greater scalability for access to metadata.

In general, the presented initiatives seek to mitigate the difficulties in (meta)data management. In addition, they highlight the importance of implementing policies for generating FAIR data to promote trustworthy health research.

Another important project is VODAN AFRICA[7]. It is funded by the Philips Foundation and aims to promote distributed access to COVID-19 data collected by the World Health Organization CRF (WHO-CRF) in African countries, serving African universities, hospitals, and research institutions. VODAN AFRICA proposes an integrated architecture with clinical and research data [17]. This architecture makes data available in closed dashboards according to two levels. The first level handles data from each clinic, and the second aggregates data from the VODAN community. In contrast to this initiative, the VODAN BR project proposes a platform for generating and publishing

---

[6] https://gen3.org/.

[7] https://www.vodan-totafrica.info/index.php.

FAIR (meta)data processable by software agents. These agents consolidate information and retrieve aggregated data according to research needs.

3 A Platform for FAIR Clinical Data

The VODAN BR platform is based on the generic architecture established for the project presented in Fig. 1. This architecture considers different mechanisms for clinical data extraction, respecting how the data is made available. Thus, data can be extracted from TXT and CSV files or the EHR database of the Hospital Units (HUs). An Extract-Transform-Load (ETL) mechanism (1) treats, anonymizes, and loads these clinical data into a staging database (2). In addition to these formats, the architecture establishes an application called eCRF4FAIR (3). This application allows recording data associated with CRFs directly into the staging database (2).

The staging database (2) is one of the fundamental elements of the architecture. It links the anonymized patient data and the original clinical data. This aspect is essential to identify treatments and research for the same patient. Furthermore, the staging database standardizes the collected data, respecting ontologies and semantic artifacts registered in it. To enable this standardization, the collected and treated data are recorded in the staging database, respecting a structure of questions and answers previously defined by CRFs stored in the database. Mechanisms for this standardization require the intensive participation of Hospital Unit (HU) researchers, considering that each HU may adopt different terminology and acronyms for recording clinical cases.

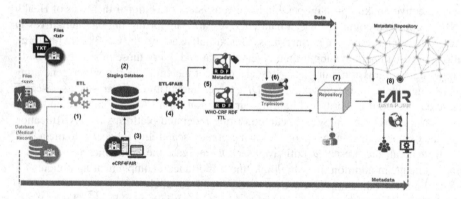

Fig. 1. Schematic View of VODAN BR Architecture [7].

From the staging database (2) onwards, the architecture proposes the adoption of an ETL4FAIR framework (4), i.e., a framework capable of: (i) extracting the treated and anonymized data and transforming them into linked data in the RDF model, associated with semantically enriched and linked metadata; (ii) generating provenance metadata and describing the datasets created in the RDF model; (iii) publishing data and metadata in the RDF model in triplestores (6); (iv) publishing the datafiles (5) associated with the data and metadata in research data repositories (7); (v) publishing the metadata of the generated datasets, as well as of their different distributions, in a FAIR DP (8).

As aforementioned, in order to automate and thus optimize the FAIRification phase activities, we adopted the ETL4FAIR framework. This framework must use solutions that work in an integrated way, either by services or by APIs. In this regard, the greater the capacity for integration and interoperability between solutions, the more efficient the automation. These automations facilitate the use of the framework by the researcher. It is worth mentioning that this architecture was designed to make data and metadata available on different platforms, contributing to the access, reuse, and interoperability of clinical data research.

An essential component of the architecture is the FAIR DP. It is a solution that composes the FAIR ecosystem, aiming to expose semantically enriched metadata so that research data and metadata gain visibility through the FAIR DP, meeting the Findability and Accessibility principles. Furthermore, the availability of data and metadata as linked data associated with ontologies and standardized semantic artifacts (vocabularies, thesaurus, taxonomies) contributes to interoperability and reuse.

Based on this architecture, the VODAN BR platform was proposed. Table 1 presents the solutions and semantic artifacts adopted for the project. In addition to these elements, an adapted FAIRification workflow organizes and guides the actions to generate FAIR (meta)data on this platform. The adapted workflow was proposed in [18] and follows the generic workflow presented in [11]. It renamed steps 6 and 7 to (6) host FAIR data and metadata and (7) to assess FAIR data and metadata to emphasize the importance of storing, publishing, and evaluating both FAIR data and metadata. This adapted FAIRification workflow [18] is adequate for our platform, providing a useful and practical set of actions and sub-actions to be pursued for clinical research data FAIRification, particularly for COVID-19.

In addition, the profiles required for the workflow execution were defined. It is important to highlight that FAIRification involves several knowledge domains and requires different types of expertise and should, therefore, be carried out by a multidisciplinary team. The mapping of each professional profile in this team has been essential to identify the various areas of expertise, such as understanding: (i) the data to be FAIRified and how they are managed; (ii) the FAIRification process (guiding and monitoring it); (iii) semantic models for data and metadata; (iv) the FAIR principles; (v) provenance; and (vi) architectural features of the software that will be used to manage the data. According to our experiments, this multidisciplinary team is composed of the following profiles and their responsibilities: Researcher - to understand the meaning of the data (e.g., medical); FAIRification Steward - to guide the team through the steps of the workflow to achieve the goals of FAIRification; Data Modelling Steward - to support the definition of semantic models for data and metadata; FAIR Data Steward - to conduct the team in compliance with the FAIR principles; Provenance Steward - to guide the team through provenance actions; Data Engineer - to develop solutions to automate the process; and IT Analyst - to install and configure solutions such as FAIR DP and triplestores.

In order to facilitate the FAIRification process understanding and the workflow organization, we used the Business Process Model and Notation (BPMN) to offer a neutral notation to all roles involved. BPMN has been available in version 2.0 since 2010 [19], and its application in industry as well as in academia is alive and well [20].

The BPMN notation has three principal types of elements: (i) actors (or actor roles) that perform the activities of the process; (ii) connections to describe the logical sequence of the process; and (iii) artifacts elaboration, being the expression of the data within or between activities of the process or just a simple annotation providing further information [21]. Due to the use of BPMN associated with the FAIRification workflow, it was possible to detail a flow of activities (actions) for orchestrating the roles, responsibilities, and technology used to execute the process.

Associated with BPMN, we adopted a factsheet to describe each action and subaction of the FAIRification workflow. As a result, it provides a better organization of the activities associated with the actions and the artifacts generated and used by them. This organization also contributes to obtaining the provenance metadata associated with the workflow. The definition of the profiles required for the FAIRification workflow, the use of BPMN, and the factsheet to better organize the workflow activities are out of the scope of this work. All these notions should be further elaborated on in future work.

Regarding the solutions presented in Table 1, it is worth emphasizing the Pentaho Data Integration[8] (PDI) as a solution for (meta)data processing and solutions integrator. PDI is a suite used for data extraction, transformation, and loading (ETL tool) that allows the development of plugins adding new functionalities. It also provides native components that enable integration with other solutions, such as Python. Because of these features, it has been chosen as the foundation of our ETL4FAIR framework. This framework is detailed in Sect. 4.

The semantic artifacts PROV Ontology[9] (PROV-O) and Data Catalog Vocabulary[10] (DCAT) are adopted for annotations, providing standardization and alignment with FAIR data environments. Furthermore, adopting the COVIDCRFRAPID[11] ontology enables the alignment of clinical data to the WHO-CRF for COVID-19. As for triple-stores, the platform considers using, for example, GraphDB[12], Virtuoso[13], Allegro[14], or Blaze Graph[15]. These solutions can be employed with ETL4FAIR to publish the triplified (meta)data. Similarly, Dataverse[16], Dspace[17], and CKAN[18] were proposed as data repositories. Finally, FAIR DP[19] is the selected solution as a metadata repository.

---

[8] https://help.hitachivantara.com/Documentation/Pentaho/9.3.

[9] https://www.w3.org/TR/prov-o/.

[10] https://www.w3.org/TR/vocab-dcat-2/.

[11] https://bioportal.bioontology.org/ontologies/covidcrfrapid.

[12] https://graphdb.ontotext.com/.

[13] https://virtuoso.openlinksw.com/.

[14] https://allegrograph.com/products/allegrograph/.

[15] https://blazegraph.com/.

[16] https://dataverse.org/.

[17] https://dspace.lyrasis.org/.

[18] https://ckan.org/.

[19] https://www.fairdatapoint.org.

Table 1. Solutions and Semantic Artifacts of VODAN BR Platform.

Architecture elements	Solution/Semantic Artefacts
ETL Tool	Pentaho Data Integration (PDI)
Staging Database	MySQL or Postgres
eCRF Application	Php/Javascript Application
ETL4FAIR	Framework ETL4FAIR
Metadata Semantic Model	PROV-O and DCAT Ontologies
Data Semantic Model	COVIDCRFRAPID Ontology
Triplestore	GraphDB, Virtuoso or Blaze Graph
Research Data Repository	Dataverse, Dspace or CKAN
Metadata Repository	FAIR DP

Figure 2 shows possible platform configurations according to the infrastructure and data access security aspects of each HU. In the figure, HUs (a), (c) and (d) process clinical data from the EHR database and publish the results using a triplestore and data repository. The first (a) uses Blazegraph and DSpace; the second (c) adopts GraphDB and CKAN; and the last (d) GraphDB and Dataverse. HUs (b) and (c) use simpler platforms. In (b), the eCRF application registers the survey and the FAIR (meta)data are published on the Virtuoso triplestore. In (c), similar to (a) and (d), clinical data are processed from the EHR database; however, the staging database is not employed. All HUs publish their metadata in the VODAN BR FAIR DP (e) hosted at FIOCRUZ. Thus, the dissemination and access to the (meta)data are performed through this FAIR DP (e), respecting the authorization rules established by each HU. For example, in the figure, HU (d) requires access authorization.

Developing solutions to support the processing and anonymization of clinical data and their conversion into FAIR (meta)data contributes to making valuable information available to the scientific community. These solutions aligned with a robust platform aim to relieve researchers of the required clinical data treatment to focus their efforts on the research itself. In addition, adopting federated networks of FAIR DP helps disseminate research, promoting agility in data discovery and reuse.

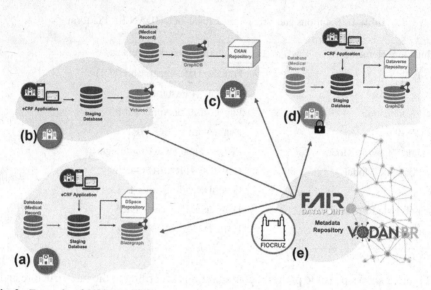

Fig. 2. Example of VODAN BR Platform configurations in a FAIR Ecosystem (Adapted from [7]).

4 The ETL4FAIR Framework

Once the process is organized and the data and metadata semantic models are defined, the activities related to FAIRification can be carried out manually or by different solutions. Adopting solutions can provide better results by reducing human error and allowing easy reproducibility. Moreover, it can provide manners to register the changes in the data throughout the process, contributing to the generation of the provenance metadata.

In this context, the ETL4FAIR framework proposes a systematic manner for the data and metadata treatment. Seen as an agnostic approach on a higher level, it harmonizes with the FAIRification workflow actions. It also establishes the tasks to be implemented from (meta)data triplification to publication in Semantic Web solutions. For the VODAN BR project, a set of solutions and services has been adopted, but there are many other solutions that could be added to this approach. Instead, adding new solutions is encouraged to support the researcher and minimize human errors.

This section presents an overview of the ETL4FAIR framework adopted in the VODAN BR project. In addition, we highlight the integrations performed with the adopted ETL solution that has broadened the scope of platform activities automation, especially in the core FAIRification process.

4.1 ETL4FAIR Overview

To support the FAIRification process, the ETL4FAIR needs to cover several processes. Firstly, the transformation and adequacy of data, then the conversion of (meta)data into linked data aligned with the FAIR process. Next, the organization and treatment of data provenance, and finally, FAIR (meta)data hosting and publishing.

The ETL4FAIR framework brings together different solutions already available and some created. Those solutions interact together, providing a unified environment. Figure 3 shows the solutions adopted by the ETL4FAIR framework to automate the mechanisms associated with the VODAN BR platform.

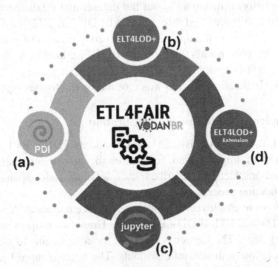

Fig. 3. Adopted solutions in the ETL4FAIR Framework of the VODAN BR Platform.

In Fig. 3, the first solution concerns the mentioned data treatment, for which the PDI ETL tool (**a**) was chosen. This is a well-known application, already used in many cases. It supports a variety of ETL activities, reducing the effort and time involved. Also, it permits a fluid interaction through a user interface and the possibility of storing and reproducing. Moreover, it has other important features: it can be expanded and can register all data transformations. All these points were considered in choosing this application.

The second focuses on data triplification, as the FAIR principles require data to be well-described, interoperable, and reusable. Those characteristics are facilitated by data represented in RDF triples. Therefore, it is crucial to organize and export data into this format. For this reason, the need for a tool to support this process emerged. In this context, the ETL4LOD+[20](**b**) was proposed. It was initially developed to assist the data triplification process, in the context of the Web of Data. However, over the years, it has evolved to provide automated support for RDF data generation and semantic annotation in other scenarios.

The third part of the solution focused on provenance, an essential aspect of reusing FAIR data. Provenance refers to the origin, processing, and transformation applied to the data. Different solutions were used to automate the collection and management of provenance metadata. In our project, the logging functionality of the PDI tool was used to record the transformations applied to the data in a relational database. The Prov-O

[20] https://github.com/Grupo-GRECO/ETL4LODPlus.

library was integrated with Jupyter notebook (**c**) to treat the recorded log. As a result, this solution supported the annotation of provenance metadata captured throughout the FAIRification process and workflow, providing information for the reproducibility and reuse of the research.

Finally, the FAIR principles emphasize the need to make (meta)data findable and accessible. In our platform, metadata about the dataset and its distributions are made available and can be easily accessed through a FAIR DP. The ETL4LOD+ Extension[21] (**d**) was developed to automate the upload and publishing process of (meta)data into the adopted solutions. This plugin adds functionality to PDI for loading triplified (meta)data into the triplestore, such as GraphDB, and the metadata schemas into the FAIR DP. In that way, the core FAIRification process was covered by the framework.

4.2   Solutions Implemented to Cover FAIRification

As aforementioned, the VODAN BR ETL4FAIR adopts PDI as its base software. It allows the execution of implemented code through native components and can be expanded by new add-ons. Here we describe some integrations and expansions performed to cover the FAIRification process.

The integrations were employed to handle provenance metadata. As expansions, we highlight ETL4LOD+ and ETL4LOD+ Extension. These two plugins add new components to PDI, called steps. The former is intended for components to treat data triplification and its association with semantic artifacts. The latter is more focused on FAIR goals related to the dissemination of data and metadata.

Running Implemented Code in PDI. We have implemented Python codes to automate the generation of provenance metadata. These codes use the Prov-O library[22], a library for W3C Provenance Data Model, in the Jupyter Notebook. For example, we can cite the implemented code to capture provenance metadata from the FAIRification ETL process. This code uses the PDI log generated when executing data transformation and publication. The activities, entities, and agents involved in the process are mapped from this log registered in a relational database.

The semantic model representing the FAIRification process provenance establishes the annotations made for each identified element. The adopted library allows defining the elements, their annotations, and the chaining of the activities, generating at the end a file with the RDF metadata, serialized in Turtle. These metadata are made available in the triplestore, and the generated file is published in the data repository.

Figure 4 presents a simplified view of the operations performed. The first operation (1) defines the entity that plays the data source role, describing its origin through annotations. Next, we define the agents involved in the transformation process (2), such as ETL4LOD+ and GraphDB. In the next operation (3), we set up access to the PDI log database and define the queries that will be employed. Based on these queries, we treat each executed activity (4), associating them with the input and output entities. Each of these elements is annotated according to the established semantic model. Finally, a file is generated (5) containing the triplified provenance metadata, serialized in Turtle.

---

[21] https://github.com/Grupo-GRECO/ETL4LODplus_Extension.

[22] https://prov.readthedocs.io/en/latest/readme.html#.

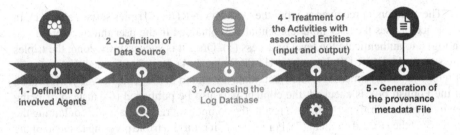

Fig. 4. Schematic view of the operations for generating the provenance metadata for the FAIRification process.

Expanding PDI. The new additions implemented via plugins are aligned with the FAIR principles. As an example, we highlight here the ETL4LOD+ Extension. This plugin has added three new components to PDI. The first is the Load Triple File. It loads triplified (meta)data in triplestores. The second is the FAIR Data Point Retriever, responsible for promoting access and information retrieval from a FAIR DP. Finally, the FAIR Data Point Loader is responsible for loading and publishing the metadata of the generated datasets in a FAIR DP.

To illustrate, we detail the process of the FAIR Data Point Loader component, represented in Fig. 5. This component aims to receive the metadata serialized in RDF, execute their upload, and enable the publication into a FAIR DP.

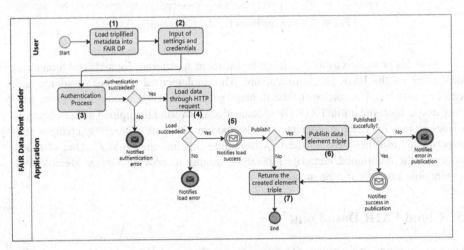

Fig. 5. FAIR Data Point Loader process.

The component receives as input the metadata in RDF, NTriples serialization (1). In sequence, it uses the user input credentials (2) obtained in the user interface presented in Fig. 6 to authenticate the level of access (3). Once it is successfully done, the triples are sent to the FAIR DP through an HTTP Request (4). Being processed with success, it returns a positive message, and (meta)data are stored in the repository (5). Moreover, if the Publish box is checked, the content will also be published (6), instead of being maintained as a draft version. As a result, the component receives a triple containing the registry of the created element (7). For instance, it returns a triple pointing to the id of the created dataset. This supports the creation of substructures as datasets and distributions in the transformation flow/pipeline sequence. The user defines the substructure to be inserted in the <Load Type> interface option.

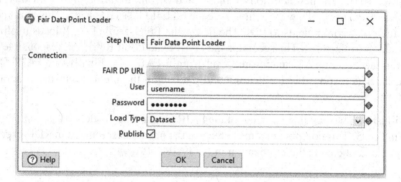

Fig. 6. FAIR Data Point Loader User Interface.

Once this process is available, it can be executed at different contexts and moments, depending on the FAIR DP infrastructure. The implemented code is free-source and can be customized according to the domain needs. The current implementation was prepared to meet the initial FAIR DP specifications, covering the upload and publication of descriptors for catalog, dataset, and distribution elements. Therefore, changes and experiments in other parts of the transformation pipeline are possible. One of those concerns the mentioned metadata repository, aiming to provide a great, flexible and expandable structure for the metadata.

5   Cloud FAIR Data Point

Here, we present an overview of an experiment that implemented and tested a FAIR DP converted into a cloud-oriented solution running on Google Kubernetes Engine[23] (GKE) from the Google Cloud Platform[24] (GCP) [22].

GCP is one of the world's most widely used public clouds, with points of presence on every continent. To serve a highly competitive market, it offers its customers a large

---

[23] https://cloud.google.com/kubernetes-engine.
[24] https://cloud.google.com/.

set of services that can be purchased on demand. These services range from well-known and expected products, such as storage and machines, to highly specific ones, such as Kubernetes Clusters as a Service and device farms, where code can be tested on specific devices. This paper covers only a tiny part of these capabilities.

GKE provides a controlled environment to deploy, manage, and scale containerized applications using Google infrastructure. As a benefit, much of the configuration work to support a cluster is done on the hosts and the network, allowing developers to focus primarily on using that cluster. In addition, it works strictly on demand, providing the tools for developers to set the specifications and resources they want, even dynamically. GKE has two modes of operation: standard and autopilot. Standard is the classic use of GKE and allows users to fully control cluster decisions. Autopilot is a practical solution that cedes control to Google algorithm. In standard mode, the payment policy is pay-as-you-go, based on infrastructure. On the other hand, autopilot runs a pay-as-you-go policy related to the number and specifications of running containers.

With the conversion of FAIR DP to a cloud-oriented solution, we aimed for an adaptable and flexible structure capable of managing the real-time demand for accesses. Thus, we aligned it with the scalability and distribution aspects emphasized in our platform. These aspects are essential when considering FAIR DP as the discovery and access point for the research metadata associated with VODAN.

5.1 FAIR DP Architecture

FAIR DP is a single product composed of four different elements working together: (i) a web server and a reverse proxy client, (ii) a Java backend application served through an application server, (iii) a NoSQL database server, and (iv) a triplestore. Each of these components is briefly explained below.

Nginx[25] represents the web server client. It is responsible for exposing the endpoints for user interaction and acts as the entire system external entry point. It translates into indirectly exposing the web page, API, and backend through its reverse proxy capabilities. The backend application that responds to Nginx can be understood as the main FAIR DP code responsible for handling the (meta)data. This application is written in Java and exposed through the Maven application server. Also, the Spring dependencies are installed through Maven to start the REST API. It is the only part directly connected to all the others, so it acts, from an external point of view, as the centerpiece of the product.

MongoDB[26] is a non-relational document database that provides JSON-like storage support and is also the NoSQL database chosen for the FAIR DP. While it is a general-purpose database notorious for its performance, MongoDB is used for the FAIR DP to store data related to user authentication, such as profiles, API Keys, and passwords. Finally, there is the component where all the metadata associated with a FAIR DP installation are stored: the RDF triplestore. The triplestore is a graph database that follows a predefined standardized model. FAIR DP supports linking different triplestore solutions to the main code, such as AllegroGraph, GraphDB, and BlazeGraph.

---

[25] https://www.nginx.com/.

[26] https://www.mongodb.com/.

5.2 FAIR DP Endpoints and Pods

In the implementation detailed in [22], all four components of a FAIR DP must run independently, be distributed, and be able to communicate with each other. Given this, each part needs its internal endpoint, used by the other components to reach it. Services are recommended to avoid the challenge of several different units that may change their IP during their lifetime. In Kubernetes, a service represents a network object that exposes a set of target Pods through a common channel and global access policy. It provides a common IP that can be resolved by DNS and reverse proxies for multiple Pods, much like a virtual IP but with more room for configuration.

Therefore, each of the parties needs its own Service that represents the access point to that part of the system. Pods are the smallest and most basic deployable objects in Kubernetes. A container must be associated with a Pod to be put into execution. A Pod can run multiple containers, although a single configuration is more common. Each Kubernetes Pod has its own unique IP address and section, and if multiple containers run on a single Pod, they share network and filesystem settings seamlessly. A Pod is allocated on a single Node and gets its computational resources (i.e., CPU and RAM) from it. It is up to the developers to determine whether a Pod can use the hosts' resources as much as it needs or must be subjected to a maximum and minimum value. Setting lower and upper limits on resource usage for a Pod is recognized as resource allocation.

In our configuration, the Nginx client is the entry point to the entire FAIR DP application, so it is the only one that requires an externally accessible endpoint, such as a LoadBalancer - which should not be used unnecessarily. Since Pods do not need to be accessed directly from the outside world, ClusterIP is always a better option. Figure 7 shows a schematic diagram of communication between Pods and from the outside world to the cluster.

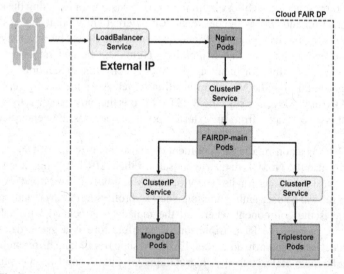

Fig. 7. Relation between the endpoints and the Pods (Adapted from [22]).

5.3   VODAN BR Cloud FAIR DP

Here, we present an architecture for a FAIR DP converted into a cloud-oriented solution. For this architecture, we used the fact that a FAIR DP is structured as a micro services-oriented application, with each part having its own Docker image. These characteristics allowed us to follow an approach to provide the most suitable configuration for each part separately. Figure 8 shows all the Kubernetes objects used in our architecture and their correlation to each other. The cloud FAIR DP evaluated in VODAN BR adopted BlazeGraph as the triplestore.

In Fig. 8, a Load Balancer service (1) establishes connections to the world outside the cluster, in front of the Nginx Pods (2), which then reverse proxies for the rest of the application. The Nginx Pods (2) have a container running the application[27] image. The number of replicas is determined by a Horizontal Pod Autoscaler (hpa) (3), using CPU usage as criteria. A Deployment Workload (deploy) (4) manages them.

Its only upstream address is the ClusterIp service (5) that routes to the FAIR DP pods (5) that run the main application code. Roughly speaking, these Pods are Nginx-like because they are also running stateless containers, a custom FAIR DP image, and are managed by a Deployment (8) and an hpa (7), but make use of two other complementary objects. They consist of the ConfigMap (cm) (9), which contains the custom application.yaml (configuration file) and is mounted as a Volume within the Pods and initContainers. These Containers are part of the boot cycle and are represented by the job logo (10) in the absence of one itself. The application.yaml used in our FAIR Data Point comes with the ClusterIP addresses of the data-related dependencies of MongoDB (11) and Blazegraph Pods (12).

As for structure, the MongoDB and Blazegraph Pods have similar design strategies and use precisely the same objects. This structure is prepared for simpler cases without handling replication, as this would increase infrastructure requirements.

Given this, to avoid and/or mitigate the damage caused by possible failures, Pods are managed by StatefulSets (sts) (13)(17) and consume Persistent Volume Claims (pvc) (14)(18). Finally, custom StorageClass (sc) (15)(19) is employed to make the lifetime of persistent volumes independent of the Pods by defining the retaining policy.

The final configuration is the result of the composition of four distinct implementations. By having each part use the configuration that best fits its role in the system and our constraints, we aim for the whole to also reflect this effort. Also, using Kubernetes as a container orchestrator allows us to manage all the disparate parts from a single centralized point while maintaining maximum independence. In other words, it allows us to handle the four distinct implementations as a single logical and connected project.

---

[27] fairdata/fairdatapoint-client:1.12.0.

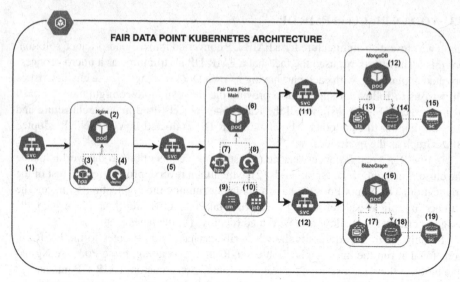

Fig. 8. FAIR Data Point Kubernetes architecture for GCP (Adapted from [22]).

6 Conclusions and Future Works

Data for more detailed clinical research studies are valuable to the scientific community but are not always available [1]. Collecting, processing, and publishing FAIR data represent a challenge that can be mitigated with an infrastructure whose mechanisms adhere to a FAIRification workflow with solutions for its automation. In this context, the implementation of FAIR platforms can assist the activities of researchers both in the dissemination of their research and in obtaining reusable FAIR data.

Automating platform-related activities speeds up FAIR data production and minimizes human error. This paper presented a modular, scalable, distributed, and flexible platform to process and publish FAIR clinical research data in Brazil. It also highlighted aspects contributing to FAIRification workflow activities management and organization experimented in the VODAN BR project. We introduced the ETL4FAIR framework to automate actions of the FAIRification phase activities by integrating different solutions. Its main goal is to optimize and facilitate the execution of the FAIRification process by the researchers.

In this paper, we discuss the role of PDI as the base solution of the ETL4FAIR Framework. In addition, we describe how the ETL4LOD+ and ETL4LOD+ Extension plugins contributed to the activities of making Linked Data into the RDF model and publishing them into data hosting solutions. The implementation of components (steps) in ETL4LOD+ Extension is still under study for publishing data and metadata files on free digital platforms used as research data repositories, such as CKAN, Dataverse, and DSpace. We also cover the automated handling of provenance metadata implemented with Prov-O Python on the Jupyter notebook.

The experience in the VODAN BR Project has reinforced the importance of FAIR DP as an essential component for disseminating research metadata. It works as a federated access point associated with search engines that promote the (re)use of FAIR (meta)data. In addition, it also supports research data requiring privacy/secrecy, such as patient data. According to its specification, FAIR DP provides an appropriate authentication and authorization infrastructure in distributed scenarios. Thus, sensitive data is only accessible when authorized, allowing "data to be as open as possible and as closed as necessary" [4].

We emphasized how a FAIR DP requires real-time access as a discovery point for FAIR clinical research metadata in the VODAN BR platform. Therefore, we proposed a FAIR DP to the cloud [22], intending to increase the scalability of the repository that hosts the project metadata.

Currently, there are ongoing studies to provide the development of new clinical trials using the eCRF application; and improve the automatic capture of provenance metadata throughout the FAIRification process, with different granularity according to the data stewards. It contributes to automating FAIRification steps, collaborating on integrating different solutions, and improving the federated FAIR data ecosystem.

References

1. Hallock, H., et al.: Federated networks for distributed analysis of health data. Front. Public Health **9**, 712569 (2021). https://doi.org/10.3389/fpubh.2021.712569
2. GO FAIR: VODAN IN Manifesto, March 2020. https://www.go-fair.org/wp-content/uploads/2020/03/VODAN-IN-Manifesto.pdf. Accessed 01 Oct 2022
3. GO FAIR: Data Together COVID-19 Appeal and Actions, March 2020. https://go-fair.org/wp-content/uploads/2020/03/Data-Together-COVID-19-Statement-FINAL.pdf. Accessed 01 Oct 2022
4. Mons, B.: The VODAN IN: support of a FAIR-based infrastructure for COVID-19. Eur. J. Hum. Genet. **28**, 1–4 (2020). https://doi.org/10.1038/s41431-020-0635-7
5. Veiga, V., Campos, M.L., da Silva, C.R.L., Henning, P., Moreira, J.: VODAN BR: a gestão de dados no enfrentamento da pandemia coronavírus. Páginas a&b: arquivos e bibliotecas 51–58 (2021). https://doi.org/10.21747/21836671/pagnespc7
6. Wilkinson, M.D., et al.: The FAIR guiding principles for scientific data management and stewardship. Sci. Data **3**(1), 1–9 (2016). https://doi.org/10.1038/sdata.2016.18
7. Borges, V., Queiroz de Oliveira, N., Rodrigues, H., Campos, M., Lopes, G.: A platform to generate FAIR data for COVID-19 clinical research in Brazil. In: Proceedings of the 24th International Conference on Enterprise Information Systems - Volume 1: ICEIS, pp. 218–225 (2022). https://doi.org/10.5220/0011066800003179. ISBN: 978-989-758-569-2. ISSN: 2184-4992
8. Heath, T., Bizer, C.: Linked Data: Evolving the Web into a Global Data Space. Synthesis Lectures on the Semantic Web: Theory and Technology, vol. 1, no. 1, pp. 1–136 (2011). https://doi.org/10.2200/S00334ED1V01Y201102WBE001
9. RDF Working Group: Resource Description Framework (RDF) (2014). https://www.w3.org/RDF/. Accessed 01 Oct 2022
10. Poveda-Villalón, M., Espinoza-Arias, P., Garijo, D., Corcho, O.: Coming to terms with FAIR ontologies. In: Keet, C.M., Dumontier, M. (eds.) EKAW 2020. LNCS (LNAI), vol. 12387, pp. 255–270. Springer, Cham (2020). https://doi.org/10.1007/978-3-030-61244-3_18
11. Jacobsen, A., et al.: A generic workflow for the data FAIRification process. Data Intell. **2**(1–2), 56–65 (2020). https://doi.org/10.1162/dint_a_00028

12. Manola, N., et al.: Implementing FAIR data infrastructures (Dagstuhl Perspectives Workshop 18472). In: Dagstuhl Manifestos, vol. 8, no. 1. Schloss Dagstuhl-Leibniz-Zentrum für Informatik (2020). https://doi.org/10.4230/DagMan.8.1.1

13. da Silva Santos, L.O.B., Burger, K., Kaliyaperumal, R., Wilkinson, M.D.: FAIR data point: a FAIR-oriented approach for metadata publication. Data Intell. **5**(1), 163–183 (2022). https://doi.org/10.1162/dint_a_00160

14. Shaw, F., et al.: COPO: a metadata platform for brokering FAIR data in the life sciences. F1000Research **9**, 495 (2020). https://doi.org/10.12688/f1000research.23889.1

15. Carmona-Pírez, J., et al.: Applying the FAIR4Health solution to identify multimorbidity patterns and their association with mortality through a frequent pattern Growth association algorithm. Int. J. Environ. Res. Public Health **19**(4), 2040 (2022). https://doi.org/10.3390/ijerph19042040

16. Saldanha, A., Trunnell, M., Grossman, R.L., Hota, B., Frankenberger, C.: Economical utilization of health information with learning healthcare system data commons. Perspect. Health Inf. Manag. **19**(Spring) (2022). https://www.ncbi.nlm.nih.gov/pmc/articles/PMC9123530/

17. van Reisen, M., et al.: Design of a FAIR digital data health infrastructure in Africa for COVID-19 reporting and research. Adv. Genet. **2**(2), e10050 (2021). https://doi.org/10.1002/ggn2.10050

18. de Oliveira, N.Q., Borges, V., Rodrigues, H.F., Campos, M.L.M., Lopes, G.R.: A practical approach of actions for FAIRification workflows. In: Garoufallou, E., Ovalle-Perandones, M.-A., Vlachidis, A. (eds.) MTSR 2021. CCIS, vol. 1537, pp. 94–105. Springer, Cham (2022). https://doi.org/10.1007/978-3-030-98876-0_8

19. OMG: Business Process Model and Notation (BPMN) - Version 2.0. Object Management Group (2010). http://www.omg.org/spec/BPMN/2.0. Accessed 01 Oct 2022

20. Recker, J.: BPMN research: what we know and what we don't know. In: Mendling, J., Weidlich, M. (eds.) BPMN 2012. LNBIP, vol. 125, pp. 1–7. Springer, Heidelberg (2012). https://doi.org/10.1007/978-3-642-33155-8_1

21. OMG, Business Process: Notation (BPMN) specification Version 2.0.2. Object Management Group (OMG) (2014). https://www.omg.org/spec/BPMN/2.0.2. Accessed 01 Oct 2022

22. Knopman, R.B.: Cloud computing as an option for higher education and research - a real case development, 151 p. Federal University of Rio de Janeiro, Rio de Janeiro (2022)

Artificial Intelligence and Decision Support Systems

Visual Interactive Exploration and Labeling of Large Volumes of Industrial Time Series Data

Tristan Langer[1]([✉])[ID], Viktor Welbers[2][ID], Yannik Hahn[1][ID], Mark Wönkhaus[1][ID], Richard Meyes[1][ID], and Tobias Meisen[1][ID]

[1] Chair of Technologies and Management of Digital Transformation, University of Wuppertal, Wuppertal, Germany
{tlanger,yhahn,m.woenkhaus,meyes,meisen}@uni-wuppertal.de
[2] Chair of Electronic Design Automation, Technical University of Munich, Munich, Germany
v.welbers@gmx.de

Abstract. In recent years, supervised machine learning models have become increasingly important for the advancing digitalization of the manufacturing industry. Reports from research and application show potentials in the use for application scenarios, such as predictive quality or predictive maintenance, that promise flexibility and savings. However, such data-based learning methods require a large training sets of accurately labeled sensor data that represents the manufacturing process in the digital world and allow model to learn corresponding behavioral patterns. Nevertheless, the creation of these data sets cannot be fully automated and requires the knowledge of process experts to interpret the sensor curves. Consequently, the creation of such a data set is time-consuming and expensive for the companies. Existing solutions do not meet the needs of the manufacturing industry as they cannot visualize large data sets, do not support all common sensor data forms and offer little support for efficient labeling of large data volumes. In this paper, we build on our previously presented visual interactive labeling tool Gideon-TS that is designed for handling large data sets of industrial sensor data in multiple modalities (univariate, multivariate, segments or whole time series, with and without timestamps). Gideon-TS also features an approach for semi-automatic labeling that reduces the time needed to label large volumes of data. Based on the requirements of a new use case, we extend the capabilities of our tool by improving the aggregation functionality for visualizing large data queries and by adding support for small time units. We also improve our labeling support system with an active learning component to further accelerate the labeling process. We evaluate the extended version of Gideon-TS on two industrial exemplary use cases by conducting performance tests and by performing a user study to show that our tool is suitable for labeling large volumes of industrial sensor data and significantly reduces labeling time compared to traditional labeling methods.

Keywords: Labeling · Time series · Unsupervised learning · Active learning · Sensor data · Visual analytics

J. Filipe et al. (Eds.): ICEIS 2022, LNBIP 487, pp. 85–108, 2023.
https://doi.org/10.1007/978-3-031-39386-0_5

1   Introduction

One of the driving factors for digitalization in the manufacturing industry is the utilization of sensor data for process automation. Capturing sensor data throughout the whole manufacturing process enables potentials for training of machine learning models to support decision making, improve product quality and optimize machine maintenance cycles [19]. In this regard, especially supervised machine learning models have proven to provide accurate prediction results in a variety of use cases [32]. Therefore, plant and machine manufacturers have begun to introduce more and more sensor technology into their production lines resulting in large volumes of data being obtained. However, despite the drastic increase in data volumes, companies nowadays still struggle to exploit the aforementioned potential and put machine learning into practical use. One of the reasons being that training supervised machine learning models suitable for use in production requires a large amount of accurately labeled data [1,4]. The process necessary to create such a data set is very time-consuming and expensive, since in most cases it cannot be performed automatically [13] and relies on expert knowledge. A key challenge in this regard is that only process experts have the necessary knowledge required to evaluate and annotate conspicuous patterns, like outliers, level shifts or regime changes within a data set [35]. Those process experts are mostly technically minded people, who are primarily experts in the evaluation and control of the production process, but do not necessarily have a sound knowledge of data science methods or even know how to program e.g. to write a labeling script. One approach to solving this problem is interactive labeling, which allow process experts to explore the data directly via a visual user interface and annotate it with labels. Unfortunately, those tools are not generally applicable to any industrial use case, as these use cases have specific requirements. Therefore, we are exploring methods to develop a tool that supports all forms of time series, is suitable for displaying large volumes of data, provides a labeling support system to speed up the labeling process, and is not only designed for one specific use case.

In order to meet these requirements, we extend our own labeling tool *Gideon-TS* that we presented in our previous work [18]. Gideon-TS is a general purpose semi-automatic labeling tool designed to fit the needs of industrial sensor data use cases. It supports integration of a wide range of time series forms (see Sect. 4.2) and individual label classes. It is able to visualize large data volumes by storing data in a dedicated time series database and by aggregating and sampling if the estimated response size of a query exceeds a maximum threshold of displayed points. Our original approach of Gideon-TS features a semi-automatic labeling support system to explore and label large data sets in a short amount of time. This is realized by splitting the data set into windows and performing unsupervised clustering to detect anomalies in them. These anomalous windows are then flagged as potential error candidates that are matched to patterns from already labeled errors via a similarity search and suggested to the user for labeling.

In summary, Gideon-TS is a semi-automatic labeling tool that is suitable for (1) integration of time series in various forms, (2) features performant visualization for large data volumes and (3) label predictions from an integrated unsupervised labeling support system.

We extend our previous work [18] with the following additional content:

- enhancement of the previous approach for semi-automatic labeling with a new active learning based approach
- adaptive aggregation for zoom functionality
- introduction of functions to label segments or time intervals
- support for timestamps in microseconds
- a more in-depth look into our architecture and implementation

Furthermore, we evaluate the extended Gideon-TS tool on the original deep drawing exemplary use case and introduce a second metal arc welding evaluation use case. Thus, we evaluate our application based on performance tests and labeling process effectiveness based on two real world industrial exemplary use cases.

The remainder of this paper is organized as follows. In Sect. 2, we introduce the two motivating exemplary use cases and give a problem statement for labeling of industrial sensor data. In the following Sect. 3 we give an overview of related work regarding visual interactive labeling of time series data in respect to requirements for industrial use cases. In Sect. 4 we describe requirements, design and implementation of Gideon-TS, followed by our extended label support system in Sect. 5. Then, we evaluate and discuss performance, usability and effectiveness of our tool in Sect. 6. Finally, we conclude our work and describe future directions of our research in Sect. 7.

2   Problem Statement

In this section, we introduce two motivating exemplary use cases that we have worked on with partners from the manufacturing industry. Furthermore, we give a problem description of the labeling of sensor data sets for machine learning tasks.

2.1   Motivating Exemplary Use Cases

First, we describe the deep drawing of metal sheets process that we use to evaluate our previous paper [18]. Deep drawing is a sheet metal forming process in which the mechanical force of a punch radially draws a metal sheet into a forming die [8]. Our data set was collected from eight strain gauge sensors that were attached to the blank holder of a smart deep drawing tool. Each of the sensors contains a strain sensitive pattern that measures the mechanical force that is exerted by the punch to the base part of the tool. Data was only recorded while the tool performed a stroke resulting in a multivariate panel data set (one sample per stroke).

Sometimes, the deformation of the sheet through great mechanical force causes cracks in the metal. In the sensor data, cracks appear as a sudden drop in the values of some of the sensors that are caused by a brief loss of pressure while the punch drives through the sheet. It is desirable to identify cracks as early as possible because they can damage the tool resulting in maintenance downtime with very high costs. Figure 1 shows two exemplary sensor curves on the left side: sensor curves of a good stroke at the top where the progression of the sensor values is smooth and no crack occurs and,

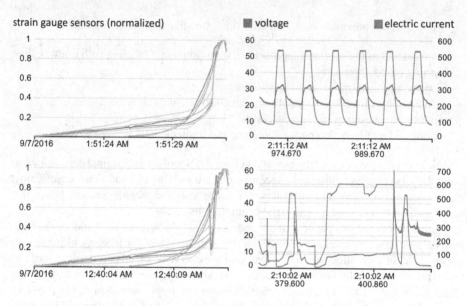

Fig. 1. Left: Two exemplary strokes of the deep drawing use case from [18] with data from eight sensors each (distinguished by color). Top shows a good stroke with a smooth progression of all strain gauge sensor curves. Bottom shows an erroneous stroke with a sudden drop of some strain gauge sensor values caused by a crack. **Right:** Two example time intervals from the new metal arc welding use case. Top shows a normal progression of a welding process with regular increase in electric current and voltage from the welding electrode. Bottom shows a long short-circuit transition what may lead to bad quality of the weld seam. (Color figure online)

at the bottom, curves of a bad stroke with a crack shown by a sudden drop in sensor values. When working on the use case, we identified three different types of curves: *clean, small crack* and *large crack*. The data set is available open source at the following URL: https://doi.org/10.5281/zenodo.7023326 and consists of about 3,400 strokes with data from the eight sensors and a timestamp, which in turn contain values from about 7,000 measurements each. This results in a data set of about 190,400,000 data points (4.4 GB).

Next, we introduce an additional use case for metal arc welding. This is a fusion welding process in which electric gas discharge is used as the energy carrier. Here, an arc is generated by an electrical discharge between voltage-carrying electrodes in an ionized gas. In detail, the arc welding process is divided into two phases. The first phase requires a low energy input and serves to melt the base material and filler, but the current density is not sufficient to detach the droplets by the pinch effect. The second phase requires a high current and thus a high current density, in which the droplet transition is initiated before it reaches a size that would trigger a short-circuit transition [14]. These two phases result in a repeating pattern which gives insights into the resulting welding quality. A long short-circuit transition may indicate a faulty welding process. Therefore, it is important to identify these short-circuit transitions to update the control parameters or identify rejected goods.

Our data was collected through a robot performing torch guidance on an immobile work piece that was locked into position on a table. Current and voltage were collected for 32 welding procedures over the same time axes in a $10\,\mu s$ interval or 100 kHz. Figure 1 shows two example time intervals of welding runs. The data set consist of 96,096,000 data points and is available open source at the following URL: https://doi.org/10.5281/zenodo.7023254.

2.2 Labeling of a Sensor Data Set for Machine Learning Tasks

We define a time series formally as a sequence of values $\{X_t, t \in T\}$ where T is the set of times at which the respective value was observed. Löning et al. [20] identify multiple types of time series data. They distinguish between *univariate* and *multivariate* time series, where univariate means that the series consists of values from only one observed dimension, i.e., $X = x_1, x_2, x_3, ...x_m$ (i.e., in our case, values from a single sensor) and multivariate means that multiple dimensions are observed over time, i.e., $X = X_1, X_2, X_3, ...X_m$ and $X_1, ..., X_m$ being $(n \times 1)$ vectors containing n values each (i.e., multiple sensors record inputs at the same points in time). In case of the presented deep drawing use case, multiple independent instances of the same kinds of measurements are observed, i.e., $X = ((x_{1,1}, ..., x_{1,m}), (x_{2,1}, ..., x_{2,m}), ..., (x_{n,1}, ..., x_{n,m}))$. These are called *panel data* and for time series usually contain several univariate or multivariate *segments* of the same observation (in our case deep drawing strokes) that do not immediately follow each other in time [9] but have an alternating large time gap between records. Alternatively, panel data can also contain segments of time series that have been extracted from a continuous measurement [15].

We define the time series labeling problem as follows: given a time series data set X of length n that consists of time intervals corresponding to discrete, well-defined behaviors, (which need not to be known in advance) and given a human oracle with knowledge to label such behavior, we want to produce an integer vector A of length n which annotates those time intervals of X [21]. With regards to panel data set, we treat each segment in the panel data as a separate time series.

3 Related Work

In this section, we give an overview of related work and tool for the problem of labeling time series data set as presented in Sect. 2. In this regard, we focus on the research gap that we aim to fill with our tool Gideon-TS. We focus on the three factors that are relevant to labeling data collected from manufacturing processes: supported forms of time series, performance for displaying large data volumes, and build-in support system to make labeling of large data sets more efficient. We extend the state of the art presented in [18] with additional related work and introduce a paragraph of related work in active learning based labeling of time series.

There are many examples of in a variety of domains where visual interactive labeling has become a common task. It has been used for text [12], images [2] or handwriting [28]. To generalize the process, Bernard et al. [4] introduce a formal concept of *Visual-Interactive Labeling (VIAL)* that describes how a system has to be designed in

Table 1. Extended overview of reviewed time series labeling tools from [18]: IRVINE [10], Label-less [36], VA tool [17], TimeClassifier [35], Label Studio [34], Curve [6] and TagAnomaly [31].

	Supported forms of time series	Use case independent	Visualization of large data sets	Label prediction for large data sets
IRVINE	multivariate segments	✗	✓	✓
Label-less	univariate segments	✗	✗	✓
VA tool	univariate segments	✓	✗	✓
TimeClassifier	univariate segments	✗	✗	✓
Label Studio	multivariate series	✓	(✓)	✗
Curve	univariate series	✓	✗	✓
TagAnomaly	univariate series	✓	✗	✓

order to efficiently label data by a human oracle. They also introduce the following label categories: categorical, numerical, relevance score or relation between two instances.

In the field of labeling of time series data there are already some preliminary works and tools. We review a representative selection of relevant related applications. An overview of the tools and their functional scope in regards to our evaluation criteria for applications in an industrial context is shown in Table 1. *IRVINE* by Eirich et al. [10] is a tool to label audio data from a manufacturing process of car bodies (multivariate data segments). It features an algorithm to support efficient labeling of this one specific use case that detects previously unknown errors in the production process. The visual representation of IRVINE is also tailored to this specific use case and shows an aggregated view of the data segments, thus, allowing visualization and labeling of larger data volumes. The *Label-Less* [36] tool by Zhao et al. is another specialized labeling approach to detect and label anomalies in a Key Performances Indicator (KPI) data set (univariate data segments). The authors present a two step algorithm to support fast labeling of large data sets. First, the algorithm uses an isolation forest to identify anomaly candidates and second, assigns them to a label category utilizing an optimized version of Dynamic Time Warping for similarity search. Label-Less features a basic user interface that shows individual segments one after another. Langer and Meisen [17] present a visual analytics (VA tool) to label industrial sensor data. They support time series as univariate segments and assist the interactive labeling process by presenting a selection of different clustering methods to the user. The user assigns labels to the resulting clusters and refines the label assignments afterwards. Finally, the VA tool also features the view of a confusion matrix trained on a selected machine learning algorithm. The VA tool is not designed to display large amounts of data as it uses plain files for data persistence and does not contain mechanisms for visualizing larger data volumes. Walker et al. [35] present the *TimeClassifier* application—a tool for interactive labeling of animal movement data. It also provides a small specialized interface for labeling segments of movement data as specific behaviour patterns. Furthermore, there are some community-driven projects that develop open source time series labeling tools. The top rated labeling tool with 7k stars on GitHub, *Label Studio* [34], supports labeling of many data types like images and text, and also multivariate time series. Label-Studio has a visualization that is tuned to be efficient when showing many data points. However, the solution only optimizes the visual component but does not include virtualization or aggregation of data, thus, we assume that there will still be problems with very large data volumes. Community projects with smaller functional scopes are *Baidu's Curve* [6] and *Microsoft's TagAnomaly* [31]. Both applications are designed to label time intervals of univariate time series. Thus for all three community-driven tools, in a panel data set each segment would have to be integrated and labeled individually. Curve and TagAnomaly also do provide basic visualization charts that do not implement mechanisms for displaying large amounts of data. Both tools use anomaly detection algorithms to suggest anomalous label candidates to the user, while Label Studio only provides an API to implement such an algorithm. As far as we can tell, all the reviewed approaches only support units down to the millisecond range, since the native date format of browsers as well as libraries building on the native date form do not include microseconds.

In addition to the development of interactive labeling tools, there is also work on the algorithmic development of efficient labeling methods. A category of algorithms the potential of which we want to investigate for our system and that has been found particularly useful for this tasks by Bernard et al. [4] are active learning based algorithms. This type of algorithm proceeds as follows: it regularly selects and presents samples from the unlabeled data set to the user in order for him or her to assign a label to the selected sample. Simultaneously, a machine learning model is continuously trained on the expanding labeled data set until it reaches a predefined accuracy. Some of the interactive applications described above (IRVINE, Label-Less and the VA tool) could also be classified as active learning based systems, as they generate labeling candidates for the user and train models based on the assigned labels. Other examples of approaches that focus on the algorithmic side of active learning in the time series domain include research from Peng et al. [25] and Souza et al. [30]. Peng et al. [25] develop a metric to measure informativeness on time series data. They aim to improve segment selection for active learning of time series classification tasks based on shapelet discovery, i.e., discovery of representative patterns in time series. An overview and comparison of nine different unsupervised active learning models is presented by Souza et al. [30]. They evaluate selected clustering based algorithms and centrality measures from graphs for labeling 24 fully unlabeled popular data sets. In more general terms, they deal with candidate selection strategies to present samples to the user and apply different models to train their system.

In summary, we reviewed existing labeling tools regarding evaluation criteria for the use in industrial use cases. In IRVINE, Label-less, the VA tool and TimeClassifier, entire segments are labeled, whereas with the community-driven solutions time intervals of a time series are annotated. All tools only support specific time series forms and they are either designed for a specific use case, do not support large data sets or do not provide a labeling support mechanism. In the following section, we present our approach for a labeling tool from our previous work [18] that meets all of the criteria discussed here. We also present extension that we made for this extended article.

4 Efficient Exploration and Labeling of Multivariate Industrial Sensor Data

In this Section, we describe our approach for exploring and labeling industrial sensor data. We elaborate on the design goals, describe the conceptualization and functional implementation of our approach proposed in an earlier conference paper [18] as well as present extended functionality.

When considering the presented exemplary use cases and the overview of related tools in Sect. 3, it is demonstrated that there is a need for a time series labeling tool that allows both the integration of large data sets and a support for data on a microsecond-scale. Therefore we present an extended and revised version of our labeling tool Gideon-TS. Our tool is suitable for the interactive and semi-automated labeling of large data volumes. It has shown to produce more accurate labels, while requiring a lower labeling time, when compared with traditional labeling methods [18]. We now extend and improve Gideon-TS by adding capabilities for active learning, allowing the integration of more granular data, and offering a more detailed visualization.

4.1 Design Goals

We designed our tool based on expert interviews and our own experience working on the two presented exemplary use cases mentioned in Sect. 2.1 [23]. In the following, we describe a list of specific design goals that we pursued in the conception, implementation and extension of our tool. Our design goals reflect the research goal of creating an efficient labeling option that meets requirements from the manufacturing industry while also accessible being accessible for domain experts.

Sensor Data Forms. In the industry, univariate series, multivariate series and pre-segmented panel data as well as whole time series occur. Our tool has to be designed to support all these forms of time series.

Data Exploration. Our tool has to allow the user to explore data freely. For this purpose, there has to be a main area for freely configurable charts. The charts must be easy to create by drag-and-drop and able to display any dimensions (sensors) and segments (samples).

Visualization of Large Data Sets. Since large amounts of data are generated quickly in the manufacturing industry, our tool has to be able to visualize even very large data sets of several gigabytes of data while remaining performant.

Support of Small Time Units. In the metal arc welding use case, data is recorded in the range of microseconds. Therefore, our tool must also be able to support small time units.

Labeling of Samples and Segments. Based on the machine learning task, our tool must be able to support different forms of labels. It must support freely configurable label classes and allow the annotation of whole segments (e.g. for subsequent classification) or time segments (e.g. for the detection of specific patterns).

Labeling Support. Our tool is designed to enable efficient labeling by domain experts even for large data sets. For this purpose, it must contain a suitable labeling support system that reduces the interaction of the expert with the data set by generating labelling suggestions in short time.

4.2 Data Formats and Forms

Gideon-TS supports two different file formats for the integration of data. The first option are JSON files where a key *time* contains timestamps of a series and all other keys contain sensor data values. The second option are files in the .ts format by sktime, which represents a file format dedicated to time series data (we refer to [20] for a detailed description of the format). We utilize *TimescaleDB* [33], a dedicated time series database based on PostgreSQL, to store the data that is uploaded to our tool. TimescaleDB can be used to speed up complex time-based queries by splitting data into chunks and distributing it over multiple so called hypertables [33]. Hypertables in TimescaleDB represent a virtual view on tables—so called chunks. Chunks are generated by partitioning data contained in hypertable by the values present in a column, where an interval determines how chunks are split, e.g., by day or hour. Queries that are send to a hypertable are then forwarded to the respective chunk with a local index resulting in a better query performance for time related characteristics of time series data retrieval [33]. For our use case, we create hypertables for dimensions, segments, and time to retrieve data faster as our tool supports explicit queries for these values when interacting with the user interface.

4.3 Visualization of Large Data Sets

Since browsers can only display a certain number of data points at a time before performance problems arise, and we deal with labeling large industrial data sets (cf. Sect. 2.1), we first describe the conceptualization of the basic functionality of our tool and the design for large data sets. This essentially corresponds to the concept of our paper [18]. Figure 2 shows a screenshot of the user interface that we designed for our labeling tool. It consists of several common components that are needed for interactive labeling which we derive from the user interfaces of the tools reviewed in Sect. 3. We extend and modify some of the components, e.g., to support the aforementioned time series forms and performance improvements. On the left side (1) our tool shows an overview of the project as a tree view with segments (here: deep drawing strokes) and dimensions (here: sensors). If the project is not a pre-segmented panel data set, we only display the dimensions/sensors. In the middle (2) there is an overview of the complete data set at the top and a custom configurable chart area (3). In this area, a user creates charts by dragging and dropping the segments and dimensions from the project area into the chart area or an existing chart. While dragging, available drop zones are highlighted above or below the existing charts and hovering over a chart will highlight it. This design aims to be intuitive to the user, as available actions are highlighted and it allows the user to freely explore the data set before the actual labeling. At the top, each chart contains its own legend to temporarily hide/show individual segments and dimensions allowing users, e.g., to compare two segments that contain similar errors. On the right side (4–6)

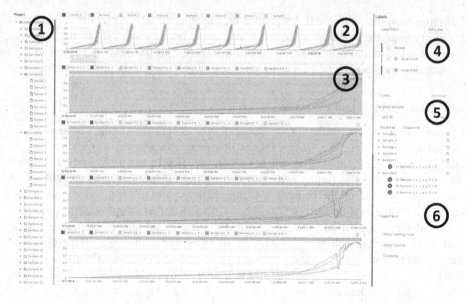

Fig. 2. General overview of our tool as an updated version presented in [18]: 1) project structure to select sensor segments and dimensions, 2) aggregated overview of all sensor data with data zoom, 3) configurable chart area to label individual segments and dimensions, 4) label class editor, 5) list of assigned labels and 6) list of algorithms for semi-automatic labeling and active learning.

are all interaction components located for actually labeling the data which we discuss in detail in Sects. 4.6 and 5.

Displaying a large number of elements within a web browser can slow it down considerably. Therefore, we use virtual scroll components from the *Angular CDK Scrolling* module [3] that renders only the elements that are in the currently visible area of the user interface. We apply those mechanisms to the tree views (project overview and labeling) and the configurable chart area to avoid performance issues. This allows us to support arbitrarily large data sets of all time series forms as trees (e.g. overview of projects with many segments and dimensions) and an arbitrary number of charts can be virtualized and displayed.

4.4 Support of Small Time Units

Our metal arc welding use case is an example of how industrial sensor data is sometimes recorded in very small time units. This is a problem for the tools we have reviewed in the related work, as the date format in browsers only supports timestamps down to the millisecond range [22]. To work around this problem, we first send x-axis times from the backend to the frontend as Python datetime timestamps in floating point number format. The following code snippet shows how we parse the number to a browser supported date and a number representing the microseconds:

```
1   function parseDate(timestamp: number): [Date, number] {
2     const date = new Date();
3     const microSeconds = Math.round(timestamp * 1000000 % 1000);
4     const jsTime = timestamp * 1000;
5     date.setTime(jsTime);
6     return [date, microSeconds];
7   }
```

From the resulting data, we create a custom timeline for displaying the timestamps of our sensor data sets. To do this, we first display a date when it first occurs and then the times in two lines. The upper line shows hours, minutes and seconds and the lower line milliseconds and microseconds. Figure 3 shows an example of a timeline created for the metal arc welding use case.

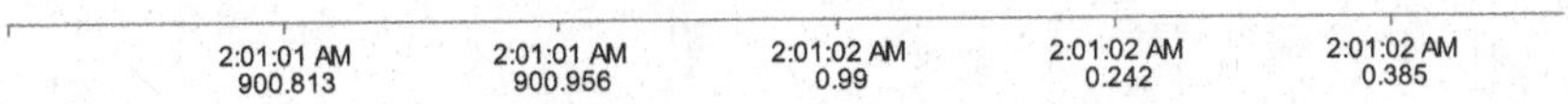

2:01:01 AM	2:01:01 AM	2:01:02 AM	2:01:02 AM	2:01:02 AM
900.813	900.956	0.99	0.242	0.385

Fig. 3. An example of an x-axis with time units in microseconds. The axis shows new Dates when they first occur and then shows the time in two lines. The first line shows hours, minutes and seconds, whereas the second line shows milliseconds and microseconds.

4.5 Aggregation and Adaptive Aggregation of Data for Large Query Results

A common problem with visualizing large data volumes in a single chart is that web applications tend to become slow and unresponsive because of the sheer number of data points that need to be rendered in a web browser. To prevent this, we implement an aggregation policy which is utilized when a configurable defined maximum number of data points is exceeded [18]. Hereby, an exploratory approach suggested that for our hardware configuration 500,000 data points was a suitable maximum. We use *Continuous Aggregates* from TimescaleDB to realize this policy. Continuous Aggregates allow us to aggregate data in predefined intervals and retrieve minimum, maximum, and average for each interval respectively [33]. For our application, we calculate the intervals by multiplying the average time delta between two timestamps in a series or segment t_{delta} with the number of total queried data points $n_{datapoints}$ divided by the defined maximum number of data points n_{max}:

$$t_{interval} = t_{delta} * \frac{n_{datapoints}}{n_{max}} \tag{1}$$

We store the total amount of data points $n_{datapoints}$ in our database during the integration process as well as segment, dimension and average time delta and can therefore request those values before a query is attempted. With this method of calculating intervals, the maximum amount of queried data points that must be visualized by the user interface is always smaller or equal to the maximum number of data points. Another advantage of utilizing Continuous Aggregates is that queries can be performed at higher speed because the calculations for aggregation only need to be performed once during the integration process and are then stored in cached views rather than recalculating the aggregates for each query [33].

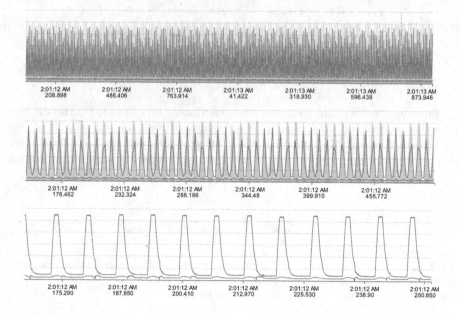

Fig. 4. Different zoom and aggregation levels: (top) low level zoom with strong aggregation, (middle) medium zoom level with some aggregation, (bottom) high level zoom with no aggregation.

Building on this aggregation methodology, which utilizes *Continuous Aggregates* from TimescaleDB, we extend it with an adaptive aggregation. In detail, this means when a user zooms into a chart of choice, we progressively lower the aggregation of the values and present the data in a higher resolution. This is feasible because the amount of data points that need to be displayed in a chart, is lowering with each consecutive zoom level. At a certain zoom level we can also dismiss aggregation entirely and present a non-aggregated view of the data, dependant on the amount of data points currently present. Therefore our labeling tool is able to offer a more detailed view when working with a large volume of data points and allows for a more robust annotation of anomalies. Figure 4 shows an example from the metal arc welding use case with three levels of zoom and aggregation from high to low level zoom with high to no aggregation.

4.6   Label Classes and Formats

To add time series labels, the user must first create label classes (typically types of errors, e.g., small crack or large crack). For this purpose we have created a label class editor in the labeling area (4). Label classes consist of a name and a severity (okay, warning, error). The severity can be used by a data scientist afterwards, e.g., to export only errors of a certain type. We also use the severity in our labeling support system to map new error candidates to already created error classes (see Sect. 5.1). When a user selects a label class by either clicking in the list entry or via number shortcut, a brush mode is activated in all displayed charts allowing him or her to create labels of the selected class. We offer two labeling modes: label segment and label time interval. Pressing the accept button above the chart applies the selected label to the whole segment, and brushing a time interval inside one of the charts creates a label for the selected interval. Each label contains the following information: the start and end time, the selected label class as well as the segment and dimensions (i.e., sensors) in the selected range. A label can never span several segments of a panel data set, since there can be any amount of time between the segments. If a brush operation would span several segments, the selection will be split creating one label per segment. The label area also contains a list of all created labels (5). Labels can be displayed either in a list sorted by time or a tree view grouped by label class or sample. Finally, labels can also be downloaded in JSON-format to use them e.g. for training a machine learning model.

5   Label Support System

In this section, we present our approach to a labeling support system. We revise our originally presented approach for semi-automatic labeling of a time series data set presented in [18] and extend the approach with an active learning component to speed up the overall labeling process.

5.1   Semi-automated Unsupervised Labeling Support

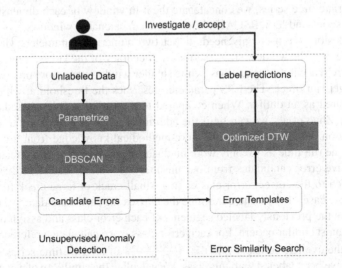

Fig. 5. Overview of our labeling support system as presented in [18]. The system consists of two steps: first, we identify candidate errors using DBSCAN. Second, we match those candidates to patterns extracted from already assigned labels using an optimized version of DTW.

An overview of our two-step label prediction approach is presented in Fig. 5. We derive the general process of our support system from Zhao et al. [36], who use feature extraction and an isolation forest to detect outliers in univariate KPI samples and then assign them to error patterns using an optimized version of dynamic time warping. Feature extraction of time series requires specific knowledge of data science methods for time series, which we do not necessarily expect process experts in the manufacturing industry to have. Therefore, instead of using isolation forest for outlier detection, we developed a similar density based approach based on DBSCAN clustering, thus, eliminating the need for feature engineering and replacing it with a threshold parameter EPS that we expect to be more intuitive for domain experts.

For unsupervised anomaly detection in our work, we employ DBSCAN clustering because it provides an efficient way to detect outliers at a fast runtime. DBSCAN is a density based clustering method, that identifies core samples in areas of high density and creates clusters around these samples. It also identifies outliers, where an outlier is not within *EPS* distance from any of the resulting clusters [11]. Samples within EPS distance from any core sample are assigned to the respective cluster. For our implementation, we utilized scikit-learn and set the parameter *min samples* as fixed. The min samples parameter describes the minimum amount of data points in proximity of a data point to be considered as a core sample [24]. This means that only a single algorithm specific parameter that needs to be adjusted by the user is required. This parameter was the previously mentioned distance EPS, which describes the maximum distance between two points to be considered as neighbors. For the clustering of time series data, we split each time series into multiple windows, whereby each window is clustered and outliers (i.e. that cannot be assigned to a cluster) are returned as possible error candidates. Therefore, the user is also required to select a fitting window size for the specific use-case. For multivariate time series, we concatenate the n-th window of each dimension into a single time series and let DBSCAN cluster these concatenated windows.

In conclusion, the user only needs to set two numerical parameters via the user interface. Hereby, the window size describes the accuracy of the labels: larger windows result in more inaccurate suggestions, while smaller windows are more precise but also require longer run times. The EPS parameter describes the threshold at which a window is declared as an outlier. When compared to the isolation forest based approach purposed by Zhao et al., we can omit the additional feature engineering through this implementation, which would require more methodical knowledge from the user, and also reduce the run time as we only work on distance and not on multiple features.

We receive error candidates from our unsupervised outlier detection that we now need to match to the correct error classes (e.g. small crack or large crack for our first exemplary use case). In order to determine the classes for each candidate, we construct patterns from the previously labeled segments of each error class and assign our candidates to the most similar pattern. For each error class c, we calculate a reference pattern $ref_c(x)$ as the average series $ref_c(x) = \frac{f_1(x) + f_n(x)}{n}$ of all $1..n$ time series segments $f_n(x)$ that have been labeled with this class. We calculate the similarity of a candidate to each reference pattern with an optimized version of *Dynamic Time Warping (DTW)*. We adopt two commonly used optimizations for the algorithm: First, we reduce the search space for the shortest warping path using the Sakoe-Chiba band [27] and, second, implement early stopping [26], which stops the similarity calculation if the processed pattern can no longer be better than the best pattern found so far. As the calculated patterns improve with each labeled segment also our suggestions become more accurate over time.

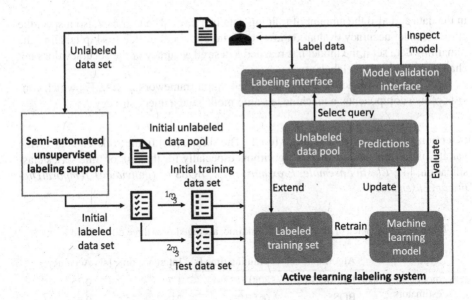

Fig. 6. Overview of our extended label support system using active learning.

5.2 Active Learning Labeling Support

We extend our semi-automated labeling approach with an active learning based approach. Active learning is a machine learning method in which a machine learning model is trained and iteratively queries the user to label new data samples. The general idea behind the extended approach is to use our original semi-automated labeling approach to let a user efficiently create an initial data set for the active learning model and then continue labeling with the active learning based approach until the model reaches a desired accuracy or the accuracy does not improve with addition labels anymore. This eliminates the need to label the entire data set and provides the user with a trained model at the end of the process. Figure 6 shows an overview of the combined approach. A user uses the semi-automated unsupervised labeling support system to quickly create an initial labeled data set. The data set is then split into an initial training data set for the first iteration of the machine learning model and a test data set that is used to evaluate the model after each iteration. The rest of the unlabeled data from the whole set serves as initial pool of unlabeled data. For each iteration, the model is trained on the set of labeled data and evaluated on the test data. Afterwards, it performs predictions for all unlabeled data from the pool. Then the system selects a predefined number of samples from the pool, that the model is most uncertain of, passes them to the user for labeling. This is done with the intention to guide the user through different regions of the data, which contains data patterns, unfamiliar to the user. The labeled sample is then added to the pool of labeled data for the next iteration. Just as with our semi-automated system, time series data that are not pre-segmented are split into windows with each window either belonging to the training data, if it was already labeled or the unlabeled data pool, if not labeled. By repeating this process the user learns a variety of patterns in the data and also the amount of training data increases. The user can also inspect the current model accuracy evaluation after each iteration and decide to skip labeling the remaining data set if the model has reached a desired accuracy or the accuracy does not change with additional labels.

We implement an active learner with the Python framework *modAL* [7] which consists of two components: a machine learning model and a query strategy.

Selection of a Machine Learning Model. The sktime [20] framework features a wide variety of current state-of-the-art algorithms especially for the task of time series classification, like *ElasticEnsemble*, *ProximityForest*, *BOSS* or *(individual) TemporalDictionaryEnsemble*.

Table 2. Performance of estimators from sktime, averaged over three different data sets.

Type	Algorithm	Training time [s]	Inference time [s]	Accuracy
dictionary based estimators	Ind. BOSS	2.50	1.79	0.75
	BOSS	665.10	21.12	0.91
	Con. BOSS	44.31	27.45	0.89
	WEASEL	27.59	19.23	0.91
	MUSE	46.32	43.87	0.93
	Ind. TDE	1.27	2.42	0.80
	TDE	174.97	92.27	0.91
distance based estimators	KNeighbors	0.71	83.13	0.75
	ElasticEnsemble	10037.29	44.31	0.78
	ProximityForest	2919.97	1555.92	0.21
	ProximityTree	1226.62	399.59	0.26
	ProximityStump	0.65	139.45	0.27
interval based estimators	TSFC	3.16	2.15	0.79
	STSF	43.50	5.38	0.80
	CIF	1291.87	175.31	0.86
	DrCif	2002.84	132.35	0.89
	RISE	10.48	10.27	0.87

For our approach of using an active learner to guide the user, we need to select a classifier that has a good balance between accuracy, training time and inference time. In order to find the best model for our use case, we train all dictionary-, interval- and distance-based estimators from sktime on three different pre-made data sets, also provided by the sktime framework. All models were then evaluated in regards to: the average time it took to train the model, the average time it took the model to predict unknown data and the accuracy of the classifier. Table 2 shows the averaged performance of each algorithm, over the *ArrowHead*, *GunPoint* and *ACSF1* data sets. Most of the algorithms, like BOSS or ElasticEnsemble have a good accuracy in our test but take up to more than two and a half hours to fit on even small data sets.

Therefore, we take the top 3 performing models (Ind. BOSS, Ind. TDE and TSFC) and evaluate them again on the real use-case data from our deep drawing use case, to see which of the model performs best on large volumes of data. We found that Ind. BOSS and Ind. TDE scale worse, since the run time of the training and query cycles increases significantly as data volume increases. With TimeSeriesForestClassifier, on the other hand, the run time remains almost constant, so that we decide to use this model in our first approach.

Query Strategy. For the query function, modAL offers several built-in strategies in order to select data samples for labeling. We use ranked batch-mode sampling as we expect to have relatively long training and inference times because of the large size of our data sets and this strategy allows to query multiple samples per iteration. Ranked batch-mode queries were developed by Cardoso et al. [5] and are defined as:

$$score = \alpha(1 - \Phi(x, X_{labeled})) + (1 - \alpha)U(x), \tag{2}$$

where $\alpha = \frac{|X_{unlabeled}|}{|X_{unlabeled}| + |X_{labeled}|}$ and $X_{labeled}$ is the labeled data set, $U(x)$ the uncertainty of predictions for x with $U(x) = 1 - P(\hat{x} \mid x)$, where $\hat{x}$ is the probability of the models prediction for x, and Φ is a so-called similarity function, for instance cosine similarity.

6   Evaluation

We evaluate multiple factors of our tool which we also presented in part in [18]. First, we measure integration and loading times of the deep drawing exemplary use case presented in Sect. 2.1 to see if it can deal with large data sets. Then, we conduct a qualitative user study to evaluate the efficiency and usability of our tool. We compare the time it takes to label with our tool and the quality of the resulting labels with traditional labeling without a special tool to see if it provides a significant advantage in practical use. We finally evaluate the improvement of the labeling process by our extended approach using active learning.

6.1   Performance

To measure the performance of our tool in relation to the amount of data, we create test data sets from the entire data set of 3,400 samples (4.4 GB) of the data from our deep drawing use case. Table 3 shows an overview of those data sets.

Table 3. An overview of our test data sets as presented in [18]. We create subsets from the deep drawing data set presented in Sect. 2.1.

Samples	Data points (approx. mil.)	Data size (approx. MB)
500	28	650
1000	56	1315
2000	112	2637
3400	190	4400

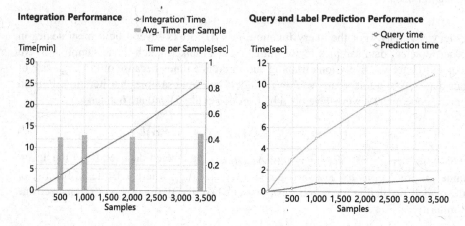

Fig. 7. Results of performance test: integration performance (left) and query and label prediction performance of our semi-automated label support system (right).

We integrate all of the data sets on a notebook with an Intel Core i9-9980HK CPU and 64 GB of RAM main memory and cross-reference all data in the overview.

An overview of measured times is shown in Fig. 7. The left chart shows number of samples (x-axis) and the corresponding integration times (y-axis). The integration time increases linearly with the number of samples from 3 min and 26 s for 500 samples to 25 min and 12 s for 3,400 samples, i.e. the complete data set. As the overall time increases, the integration time per sample remains almost constant at just over 0.4 s. In the chart on the right, we measure the query times and run time of our labeling support system for each test data set. We load the overview chart for each data set and record the time that it takes for the request to complete in the browser. The times range from 300 ms for 500 samples to 1.2 s for the complete data set. Run times of our semi-automated support system range from 3 s to 11 s for the test data sets.

Our performance tests indicate that our system is well suited for exploring and labeling large data volumes. While integration takes much longer relative to query time, it increases only linearly with data volume which means that it scales well with increasingly large data sets. By caching and aggregating views with many data points, the response times of Gideon-TS stays within 2 to 4 s for common tasks (i.e., visualizations) and 8 to 12 s for more complex tasks (i.e., label prediction) for interactive systems [29].

6.2 User Study

We first collected a reference data set and labeling time from a process expert who we asked to label time intervals in a randomly selected subset of one hundred consecutive segments of the deep drawing data set with the tools of his choice. This process took him about one and a half hours, whereby he first spend 30 min exploring and visualizing the data and then one hour actually labeling the data. The final data set contained labels of two small and six large cracks. To compare this process, we carried out a user study with twelve participants with a STEM background to test usability and efficiency of our tool. At the beginning, participants had time to familiarize themselves with our tool and were given a short introduction to the user interface and the functionalities of our tool. We then asked all participants to label time intervals within the same data set as the expert using our tool at their own discretion. We collected time, labeled data set and comments on the general usability and workflow of the tool from the participants.

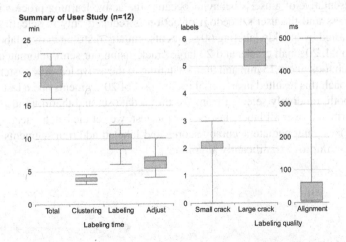

Fig. 8. Summary of our user study results performed with 12 participants as presented in [18]: labeling speed (left) and labeling quality (right).

We give an overview of the result regarding labeling time and quality in Fig. 8. All participants proceeded in the same way by first tuning the clustering algorithm of the labeling support system, then quickly going over the generated label predictions and accepting predictions that they felt were correct, and finally spending time refining the range of labels for error cases. Overall, participants took a median of 20 min to label the complete data set. These were distributed in median on four minutes tuning of the support system, nine and a half minutes labeling of all samples and six minutes refinement of label ranges. They labeled a median of two small cracks and 5.5 large cracks.

The alignment of the labels with our reference data set was at a median difference of 60 ms and at a maximum difference of 500 ms. With a sampling rate of two milliseconds and 7,000 values per sample, this corresponds to a maximum difference of 3.5% within the sample.

In summary, the time required to label the data set was significantly reduced compared to our reference. The labels were mostly assigned in the same way, but there were also deviations in the number of identified small and large cracks. On the one hand, we attribute this to the fact that the distinction between small and large cracks is subjective. On the other hand, it also became apparent that the participants relied on the labeling support system, which also led to cracks being overlooked in individual cases if the parameterization was not optimal. The assigned labels did not deviate much from our reference labels, so we conclude that our system predicts labels that assist users in producing qualitatively comparable results after minor adjustments.

6.3   Active Learning

To measure the performance of our active learning system, we performed an exemplary run with one user labeling the full deep drawing data set. We argue that this exemplary run is representative of a user's behavior because the active learning process is a fully guided process and the user's freedom of action is reduced only to selecting a label class. We started by quickly labeling 60 segments, equally distributed over label classes with 20 normal, 20 small cracks and 20 large crack, using our semi-automated support system which took about 15 min and then switching to the active learning system. Based on our approach this resulted in an initial training set of 20 segments and a test set of 40 segments, both, randomly selected from the initial data set and therefore also roughly equally distributed over all label classes. For our test, we set the batch query size to 10 to receive many intermediate accuracy scores and labeled addition segments until the model did not improve significantly anymore.

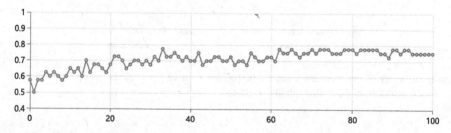

Fig. 9. Summary of accuracy results for our exemplary active learning run over 100 training and query iterations labeling 10 segments per iteration.

Figure 9 shows an overview of the accuracy results of our exemplary run. Over 100 iterations of labeling 10 segments the model improved its accuracy score from approx. 0.5% to approx. 0.7%. We then decided to stop labeling additional segments as the model did not improve for 25 iterations. Overall resulting in 1060 labels (60 unsupervised plus 1000 active learning) corresponding to about one third of the complete data set. The active learning part of our system exceeds the reference time windows for interactive systems by Shneiderman [29] in our exemplary test for performing a training and query cycle. However, active learning is not a method that is designed to be highly interactive all the time, but an iterative approach that reduces interaction times with the labeling system overall and allows the user to stop labeling the data set early. In our example, we would stop labeling after 100 iterations because the model achieves a good accuracy score. During the actual labeling phase, the user performs only labeling interactions which have a near instant response time.

6.4 Discussion

In this work, we extended our existing labeling support with an active learning component. It turns out that there is added value in combining both systems and using the unsupervised approach to quickly create an initial data set and then switching to the active learning system. This results overall in less time spend labeling data, since active learning selects samples with a predicted high knowledge gain. However, the total time required to create a labeled data set increases due to the addition of training and query cycle times.

Another advantage of active learning is that after performing several cycles, a relatively accurate model for the respective use case is obtained. On the other hand, the model is not tuned to the specific use case and the data, so there is still potential. At this point, one could think about automatically tuning the model after reaching certain accuracy scores and then continue with the labeling. It could also be worthwhile to compare different models, as we did in the development of the deep drawing use case.

An additional factor that we found is that active learning requires more data science knowledge, as the whole process is strongly dependent on the quality of the initial data set, which has to be large enough and evenly weighted between label classes to deliver accurate validation results afterwards. To this end, assessment metrics could also be further used to better guide domain experts. In addition, it might be useful to support the selection and parameterization of an appropriate model. For example, a library of models used in similar use cases could be used for this purpose.

7 Conclusion and Outlook

With the increase in supervised machine learning algorithms in the manufacturing industry, the need for labeled data sets is also growing. We found that current tools are not well suited for the requirements of the manufacturing industry. Based on this observation, we have developed our own labeling tool that supports these aspects. In this extended paper, we therefore build on our previously developed tool Gideon-TS, which supports the integration of arbitrary time series formats, is able to visualize large amounts of data and to efficiently label them interactively with a support system. Based on a second use case, we have further developed the tool to include small time series and extended the labeling support system with an active learning component. Our evaluation indicates that our extended tool further reduces human time effort for labeling of industrial sensor data.

With our tool we support efficient visualization and labeling of industrial sensor data. However, labeling requires a lot of domain knowledge and the design of the exploration process is currently completely up to the user. We therefore see potential in the introduction of a guidance system that guides users through the most important insights in the data set and thus enables a faster start of the labeling process [16].

References

1. Adi, E., Anwar, A., Baig, Z., Zeadally, S.: Machine learning and data analytics for the IoT. Neural Comput. Appl. **32**(20), 16205–16233 (2020). https://doi.org/10.1007/s00521-020-04874-y
2. von Ahn, L., Dabbish, L.: Labeling images with a computer game. In: Proceedings of the SIGCHI Conference on Human Factors in Computing Systems, CHI 2004, pp. 319–326. Association for Computing Machinery, New York (2004). https://doi.org/10.1145/985692.985733
3. Angular CDK scrolling (2021). https://material.angular.io/cdk/scrolling/overview. Accessed 01 Nov 2021
4. Bernard, J., Zeppelzauer, M., Sedlmair, M., Aigner, W.: VIAL: a unified process for visual interactive labeling. Vis. Comput. **34**(9), 1189–1207 (2018). https://doi.org/10.1007/s00371-018-1500-3
5. Cardoso, T.N., Silva, R.M., Canuto, S., Moro, M.M., Gonçalves, M.A.: Ranked batch-mode active learning. Inf. Sci. **379**, 313–337 (2017). https://doi.org/10.1016/j.ins.2016.10.037
6. baidu/curve (2021). https://github.com/baidu/Curve. Accessed 01 Nov 2021
7. Danka, T., Horvath, P.: modAL: a modular active learning framework for Python (2018). https://github.com/cosmic-cortex/modAL
8. DIN 8584-3: Manufacturing processes forming under combination of tensile and compressive conditions - part 3: deep drawing; classification, subdivision, terms and definitions. Beuth Verlag, Berlin (2003)
9. Dudley, J.J., Kristensson, P.O.: A review of user interface design for interactive machine learning. ACM Trans. Interact. Intell. Syst. **8**(2), 1–37 (2018). https://doi.org/10.1145/3185517
10. Eirich, J., et al.: IRVINE: a design study on analyzing correlation patterns of electrical engines. IEEE Trans. Vis. Comput. Graph. **28**(1), 11–21 (2021). https://doi.org/10.1109/TVCG.2021.3114797
11. Ester, M., Kriegel, H.P., Sander, J., Xu, X.: A density-based algorithm for discovering clusters in large spatial databases with noise. In: Proceedings of the Second International Conference on Knowledge Discovery and Data Mining, KDD 1996, pp. 226–231. AAAI Press (1996)
12. Heimerl, F., Koch, S., Bosch, H., Ertl, T.: Visual classifier training for text document retrieval. IEEE Trans. Vis. Comput. Graph. **18**(12), 2839–2848 (2012). https://doi.org/10.1109/TVCG.2012.277
13. Hu, B., Chen, Y., Keogh, E.: Classification of streaming time series under more realistic assumptions. Data Min. Knowl. Disc. **30**(2), 403–437 (2016). https://doi.org/10.1007/s10618-015-0415-0
14. Kah, P., Suoranta, R., Martikainen, J.: Advanced gas metal arc welding processes. Int. J. Adv. Manuf. Technol. **67**(1), 655–674 (2013). https://doi.org/10.1007/s00170-012-4513-5

15. Keogh, E., Chu, S., Hart, D., Pazzani, M.: An online algorithm for segmenting time series. In: Proceedings 2001 IEEE International Conference on Data Mining, pp. 289–296 (2001). https://doi.org/10.1109/ICDM.2001.989531

16. Langer, T., Meisen, T.: System design to utilize domain expertise for visual exploratory data analysis. Information **12**(4), 140 (2021). https://doi.org/10.3390/info12040140

17. Langer, T., Meisen, T.: Visual analytics for industrial sensor data analysis. In: Proceedings of the 23rd International Conference on Enterprise Information Systems - Volume 1: ICEIS, pp. 584–593. INSTICC, SciTePress (2021). https://doi.org/10.5220/0010399705840593

18. Langer, T., Welbers, V., Meisen, T.: Gideon-TS: efficient exploration and labeling of multivariate industrial sensor data. In: Proceedings of the 24th International Conference on Enterprise Information Systems - Volume 1: ICEIS, pp. 321–331. INSTICC, SciTePress (2022). https://doi.org/10.5220/0011037200003179

19. Lasi, H., Fettke, P., Kemper, H.-G., Feld, T., Hoffmann, M.: Industry 4.0. Bus. Inf. Syst. Eng. **6**(4), 239–242 (2014). https://doi.org/10.1007/s12599-014-0334-4

20. Löning, M., Bagnall, A., Ganesh, S., Kazakov, V., Lines, J., Király, F.J.: sktime: a unified interface for machine learning with time series. In: Workshop on Systems for ML at NeurIPS 2019 (2019)

21. Madrid, F., Singh, S., Chesnais, Q., Mauck, K., Keogh, E.: Matrix profile XVI: efficient and effective labeling of massive time series archives. In: 2019 IEEE International Conference on Data Science and Advanced Analytics (DSAA), pp. 463–472 (2019). https://doi.org/10.1109/DSAA.2019.00061

22. MDN web Docs (2022). https://developer.mozilla.org/en-US/docs/Web/JavaScript/Reference/Global_Objects/Date. Accessed 24 Aug 2022

23. Meyes, R., Donauer, J., Schmeing, A., Meisen, T.: A recurrent neural network architecture for failure prediction in deep drawing sensory time series data. Procedia Manuf. **34**, 789–797 (2019). https://doi.org/10.1016/j.promfg.2019.06.205

24. Pedregosa, F., et al.: Scikit-learn: machine learning in Python. J. Mach. Learn. Res. **12**, 2825–2830 (2011)

25. Peng, F., Luo, Q., Ni, L.M.: ACTS: an active learning method for time series classification. In: 2017 IEEE 33rd International Conference on Data Engineering (ICDE), pp. 175–178 (2017). https://doi.org/10.1109/ICDE.2017.68

26. Rakthanmanon, T., et al.: Searching and mining trillions of time series subsequences under dynamic time warping. In: Proceedings of the 18th ACM SIGKDD International Conference on Knowledge Discovery and Data Mining, KDD 2012, pp. 262–270. Association for Computing Machinery, New York (2012). https://doi.org/10.1145/2339530.2339576

27. Sakoe, H., Chiba, S.: Dynamic programming algorithm optimization for spoken word recognition. IEEE Trans. Acoust. Speech Sig. Process. **26**(1), 43–49 (1978). https://doi.org/10.1109/TASSP.1978.1163055

28. Saund, E., Lin, J., Sarkar, P.: PixLabeler: user interface for pixel-level labeling of elements in document images. In: 2009 10th International Conference on Document Analysis and Recognition, pp. 646–650 (2009). https://doi.org/10.1109/ICDAR.2009.250

29. Shneiderman, B., Plaisant, C., Cohen, M.S., Jacobs, S., Elmqvist, N., Diakopoulos, N.: Designing the User Interface: Strategies for Effective Human-Computer Interaction. Pearson (2016)

30. Souza, V.M., Rossi, R.G., Batista, G.E., Rezende, S.O.: Unsupervised active learning techniques for labeling training sets: an experimental evaluation on sequential data. Intell. Data Anal. **21**(5), 1061–1095 (2017). https://doi.org/10.3233/IDA-163075

31. microsoft/taganomaly (2021). https://github.com/Microsoft/TagAnomaly. Accessed 01 Nov 2021

32. Tercan, H., Meisen, T.: Machine learning and deep learning based predictive quality in manufacturing: a systematic review. J. Intell. Manuf. (2022). https://doi.org/10.1007/s10845-022-01963-8

33. Timescaledb (2021). https://docs.timescale.com/. Accessed 01 Nov 2021

34. Tkachenko, M., Malyuk, M., Shevchenko, N., Holmanyuk, A., Liubimov, N.: Label studio: data labeling software (2020–2021). https://github.com/heartexlabs/label-studio

35. Walker, J.S., et al.: TimeClassifier: a visual analytic system for the classification of multi-dimensional time series data. Vis. Comput. **31**(4), 1067–1078 (2015). https://doi.org/10.1007/s00371-015-1112-0

36. Zhao, N., Zhu, J., Liu, R., Liu, D., Zhang, M., Pei, D.: Label-Less: a semi-automatic labelling tool for KPI anomalies. In: IEEE INFOCOM 2019 - IEEE Conference on Computer Communications, pp. 1882–1890 (2019). https://doi.org/10.1109/INFOCOM.2019.8737429

Resource Planning in Workflow Nets Based on a Symbolic Time Constraint Propagation Mechanism

Lorena Rodrigues Bruno[✉] and Stéphane Julia

Federal University of Uberlândia, Uberlândia, MG, Brazil
lorenarb94@gmail.com, stephane@ufu.br

Abstract. The main proposal of this work is to present an approach based on the formalism of Petri nets in order to describe the time constraints of resources over the activities of Workflow Management Systems. Using Colored Petri nets, it is possible to separate the process model from the resource model and formally define the communication mechanisms between the two models. Each token from the Colored Petri net will then be able to carry the time information of each case, such as the start and end time of each activity the case will have to perform to complete the Workflow process. Such temporal data can be represented as a set of colors from the CPN model and will be used to monitor the execution of the process in order to find the right amount of resources involved in the execution of activities. The simulation of the CPN with time information then allows estimating the percentage of cases that respect the time constraints of the process (deadline delivery dates). In addition with the CPN simulation model proposed in this work for the improvement of resource planning in Workflow Management System, a time constraint propagation mechanism based on the sequent calculus of Linear Logic and on symbolic dates is proposed. Such propagation mechanism allows the instantaneous calculation of the time constraints that each new case must respect when entering the Workflow process.

Keywords: Time petri net · Symbolic expression · Workflow net · Linear logic · Forward propagation · Backward propagation

1 Introduction

The purpose of Workflow Management Systems is to execute Workflow processes, according to [24]. Workflow processes represent the sequence of activities which have to be executed within an organization to treat specific cases and to reach a well defined goal, while respecting some conditions that determine their order of execution.

Many papers have already considered some of the most important features of Petri net which are well adapted to model Workflow Management Systems. They are considered well adapted too to model Real Time Systems because they allow for the treatment of critical resources which have to be used for specific activities in real time.

Many analysis techniques can be considered in a workflow process; in particular, two major approaches are presented in [24]: qualitative and quantitative. Qualitative

© The Author(s), under exclusive license to Springer Nature Switzerland AG 2023
J. Filipe et al. (Eds.): ICEIS 2022, LNBIP 487, pp. 109–132, 2023.
https://doi.org/10.1007/978-3-031-39386-0_6

analysis is concerned with the logical correctness of the workflow process, while quantitative analysis is concerned with performance and capacity requirements [24].

The Soundness property is an important criterion to be satisfied when treating Workflow processes, and the proof of Soundness is related to the qualitative analysis.

The proposed approach is based on the sequent calculus of Linear Logic to verify qualitative and quantitative properties of Workflow nets. The qualitative analysis is based on the proof of Soundness criterion defined for Workflow nets. The quantitative analysis is based on the concern with the resource planning of each activity of a process mapped into Timed Workflow nets.

In [3], the proposed work presented and approach based on Linear Logic to calculate symbolic formulas based on $(max, +)$ or $(min, -)$ operators that express the possible visibility intervals in which a token (a case of a Workflow process) should start and finishes its activities. Such symbolic expressions were used to define the numerical time constraints associated to specific cases treated by the Workflow process. In this work, the results obtained in [3] are considered to model, simulate and monitor the Workflow process using a Hierarchical Colored Petri net and the CPN Tools. The main is then to estimate, given a certain amount of resources, the number of cases that respect the time constraint of the process imposed by the specified visibility intervals.

In Sect. 2 some important definitions for the understanding of this work are given. In Sect. 3 the explanation about the time constraint propagation mechanism is presented. In Sect. 4 the CPN implementation of the model and the simulation experimentation are produced. Finally, the last section concludes this work.

2 Theoretical Foundations

2.1 Workflow Net

A Petri net that models a Workflow process is called a Workflow net [1, 24]. A Workflow net satisfies the following properties:

- It has only one source place named $Start$ and only one sink place named End. These are special places, such that the place $Start$ has only outgoing arcs and the place End has only incoming arcs.
- A token in $Start$ represents a case that needs to be handled and a token in End represents a case that has been handled.
- Every task t (transition) and condition p (place) should be in a path from place $Start$ to place End.

2.2 Soundness

The correctness criterion of Soundness was defined in the context of Workflow nets. A Workflow net is Sound if and only if the three following requirements are satisfied [24]:

- For each token put in the place $Start$, it is possible to reach place End.
- If there is a token in place End for a specific case, it means that all other places are empty for this case.
- For each transition (task), it is possible to move from the initial state to a state in which that transition is enabled, i.e. there are no dead transitions.

2.3 Process

A process defines which tasks need to be executed and in which order [24]. Modeling a Workflow process in terms of a Workflow net is rather straightforward: transitions are active components and models the tasks, places are passive components and model conditions (pre and post), and tokens model cases [1,24]. In order to illustrate the mapping of a process into a Workflow net, the process for handling complaints, shown in [24] can be understood as follows: an incoming complaint is first recorded. Then the client who has complained along with the department affected by the complaint are contacted. The client is approached for more information. The department is informed of the complaint and may be asked for its initial reaction. These two tasks may be performed in parallel, i.e. simultaneously or in any order. After that, data is gathered and a decision is made. Depending upon the decision, either a compensation payment is made or a letter is sent. Finally, the complaint is filed. In Fig. 1, a Workflow net that correctly model this process is shown.

2.4 Time Petri Net

Time and Timed Petri nets can associate timestamps to places (p-time and p-timed Petri net) or transitions (t-time and t-timed Petri nets). In the p-timed Petri net case, according to [22], time constraints are represented by durations (positive rational numbers) associated with places. In the t-timed Petri net case, according to [19], time constraints are represented by durations (positive real numbers) associated with transitions. In the t-time Petri net case, time constraints are represented by an interval $[\delta_{min}, \delta_{max}]$ associated with each transition; the time interval associated with a specific transition corresponds to an imprecise enabling duration, according to [13,14]. For example, the

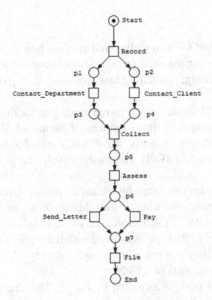

Fig. 1. Handle Complaint Process, from [3].

interval $[8, 12]$ associated with a transition indicates that the transition will be fired at least eight time units after it has been enabled and, at the most, twelve time units after its enabling instant.

2.5   Colored Petri Net

Petri nets are traditionally divides into low-level Petri nets and high-level Petri nets. Low-level Petri nets are characterized by simple tokens (natural number associated to places) that generally indicate the active state of a system or the availability of a resource. High-level Petri nets are aimed at practical use, in particular because they allow the construction of compact and parameterized models.

Classic Petri nets belong to the class of low-level Petri nets. They allow the representation of parallelism and synchronization, thus they are appropriate for the modeling of distributed systems. However, when classic Petri nets are used for the modeling of process, the size of very large and complex systems has become an issue of major complication. In that way, many extensions of basic Petri net models arose from the need to represent these complex systems. One of them is the Colored Petri Net (CPN) [2].

The idea of CPN is to put together the ability to represent synchronization and competition of Petri nets with the expressive power of programming languages with their data types and concept of time. Colored Petri Nets (CPNs) belong then to the class of high-level Petri nets, and they are characterized by the combination of Petri nets and a functional programming language [15], called CPN ML. Thus, the formalism of Petri nets is well suited for describing concurrent and synchronizing actions in distributed system, whereas the functional programming language can be used to define data types and manipulation of data [8].

2.6   CPN Tools

The practical application of CPN modeling and analysis heavily relies on the existence of computer tools, supporting the creation and manipulation of CPN models. CPN Tools [7] is a tool suited for editing, simulating and providing state space analysis of CP-net models.

In this paper, the CPN Tools is used to represent graphically and analyze the proposed models, and to simulate the timed versions of the model. By performing analysis and simulation of the proposed models, it will make possible to investigate different scenarios and to explore qualitatively and quantitatively the global behavior of CPN model.

The states of a CPN are presented by means of places. Every place is associated with a type which determines the kind of data that the place may contain. Each place will contain a varying number of tokens too. Every token has a data value, that is known as a color [2], and belongs to the type associated with the place. These colors do not just mean colors or patterns, they can represent complex data types [8]. Figure 2 illustrates an example of the basic elements of a CPN.

The CPN in Fig. 2 has two places called $p1$ and $p2$. The places have the type (color set) $TOKEN$, as well as the variable x of the same type, associated to the arcs of

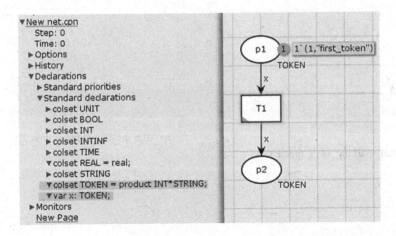

Fig. 2. Elements of a Colored Petri Net.

the net. The $TOKEN$ type is formed by the cartesian product of the color sets INT (integers) and $STRING$, as it can be seen in the left part of the figure, highlighted in yellow. The places accept only tokens of this type. In that way, the inscription $1'(1,$ "$first\_token$") corresponds to one token whose attributes are given by the integer 1 and the string "$first\_token$".

In the example of Fig. 2 the transition $T1$ will be enabled only if there is at least one token in place $p1$ (input place of $T1$). During the firing of a transition in a CPN model, the variables of its input arcs will be replaced with the token value. After firing $T1$, in the example of Fig. 2, a new token is produced in place $p2$.

CPN models allow to add timing information to investigate the performance of systems. For this, a global clock is introduced. The clock value may either be discrete or continuous [7]. In a timed CPN model it is the token that carry a time value, called timestamp. To calculate the timestamp to be given to a token it is necessary to use time delay inscriptions attached to the transitions or to individual output arcs [8]. A time inscription on a transition applies a time delay to all output tokens created by that transition. On the other hand, a time inscription on an output arc applies a time delay only to tokens created at that arc [8].

To exemplify a firing transition in a timed CPN, consider the Fig. 3a. Transition $T1$ represents an operation which takes 5 time units. Thus, $T1$ creates timestamps for its output tokens by using time delay inscriptions attached to the transition (inscription @ + 5). Every time $T1$ is fired, its output token timestamp will be increased by 5 time units. Figure 3b illustrates the state of the marking after the firing of $T1$. The ingoing arc to $p1$ has a time delay expression @ + 3 as it can be seen in Fig. 3b. Thus, the timestamp given to the tokens created on this input arc is the sum of the value of the global clock and the result of evaluating the time delay inscription of the arc (in this case, $5 + 3 = 8$), shown in Fig. 3c.

Another advantage of CPN model is that it can be structured into different related modules. The concept of module in CPN is based on a hierarchical structuring mechanism which supports bottom-up as well as top-down working style. The basic idea

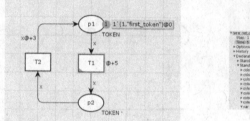

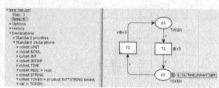

(b) After the firing of transition $T1$.

(a) Initial marking of a timed CPN model.

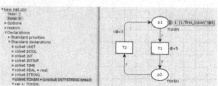

(c) After the firing of transition $T2$.

Fig. 3. Timed Colored Petri Net.

behind hierarchical CPN is to allow the modeler to construct a large model by combining a number of small CPNs [7]. According to [2], it facilitates the modeling of large and complex systems, such as information systems and business processes.

One of the builders of hierarchical CPN language is substitution transitions. This builder allows the construction of a more complex hierarchical CPN, by composing many less complex CPNs. The aim of substitution transition is to allow a relationship between a transition and its arcs with a more complex CPN, that usually provides a more detailed description. A set of CPNs is called page. Subpage is a page that contains a detailed description of the functionality modeled by the corresponding substitution transition.

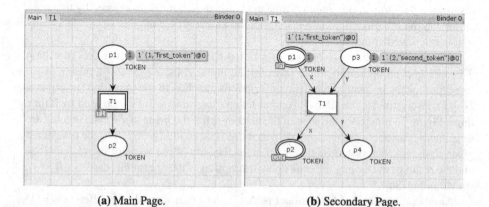

(a) Main Page.　　　　　　　　　　(b) Secondary Page.

Fig. 4. Hierarchical CPNs - Substitution transition.

In the example shown in Fig. 4a the highest-level page titled $Main$ has the transitions $T1$, that can be replaced by the subpage $T1$, shown in Fig. 4b. It should be noted that $p1$ and $p2$, which are respectively input and output of $T1$, appears both in the main page and the secondary page. In the main page, this places are called $sockets$, and in the secondary page, they are called $ports$ and are labeled, respectively, with the labels In and Out.

2.7 Resource Allocation Mechanism

Reference [24] defines some concepts related to Workflow nets. The main objective of a Workflow system is to deal with $cases$. A case can be modeled by a token in a Petri net. Each case has a unique identity and has a limited lifetime. An example of a case is an insurance claim (represented by a token in the place start in Fig. 1), which begins at the moment when the claim is submitted and ends when the processing of the claim has been completed. Another important concept is $task$. A task is a logical unit of work that is indivisible; so if something goes wrong during the execution of a task, the task must be re-initiated. Register a complaint, contact the client, contact the department and send a letter are some examples of tasks. A task refers to a generic piece of work and not to the performance of an activity for one specific case. To avoid confusion between these concepts, the following terms are used: $work$ $item$ and $activity$. A work item is the combination of a case and a task that is about to be executed and, as soon as a work begins upon the work item, it becomes an activity. So, an activity can be understood as the performance of a work item. The work item is a concrete peace of work, which is enabled for a specific case. This can be understood as the combination of a task, a case and a trigger (optional) [1].

A $trigger$ is an external condition that leads to the execution of an enabled task [1]. There are four types of triggering [1]:

- Automatic: at the moment it is enabled, the task is triggered.
- User: a human participant triggers a task.
- Message: an external event is responsible for triggering an enabled task.
- Time: a clock triggers an enabled task instance.

In the handle complaint process, illustrated in Fig. 1, it is possible to identify eight tasks, of which three are triggered automatically ($Record$, $Collect$ and $File$) and five are triggered by a user ($Contact\_Department$, $Contact\_Client$, $Assess$, Pay and $Send\_Letter$).

The allocation of resources to work items is the decision of which resource is allowed to execute a work item [1]. There exist two main mechanisms for allocating resources:

- Push control: the decision mechanism decides which user will execute which work item from the Workflow Management System; for each work item, a resource is chosen.

– Pull control: the Workflow Management System sends a copy of each work item available to each available user; when a user selects one specific work item, all other copies of this work item disappear for the other users.

Formalization of resource allocation mechanisms using the formalism of Petri net was proposed in [9].

2.8   Linear Logic

Linear Logic [5] was proposed by Jean-Yves Girard in 1987. In Linear Logic, the propositions are considered as resources which are consumed and produced in each change of state, which is different from classic logic where propositions have Boolean values (true or false) [18]. The Linear Logic introduces seven new connectors divided into three groups:

– Multiplicative connective: "times"($\otimes$), "par"($\otimes$) and "linear implies"($\multimap$).
– Additive connective: "plus"($\oplus$) and "with"($\&$).
– Exponential connective: "of course"($!$) and "why not"($?$).

In this paper, just two of these connectives will be used: $times$, denoted by $\otimes$ and $linear\ implies$, denoted by $\multimap$. The first symbol represents the simultaneous availability of resources; for example, $A \otimes B$ represents simultaneous availability of resources A and B. The second symbol represents a possible change of state; for instance $A \multimap B$ means that B is produced when A is consumed.

The translation of a Petri net into formulas of Linear Logic is performed as presented in [20]. A marking M is a monomial in $\otimes$, which is represented by $M = A1 \otimes A2 \otimes ... \otimes Ak$ in which Ai are place names. In Fig. 1, the initial marking is just $Start$ because of the token in place $Start$. A transition is an expression of the form $M1 \multimap M2$ where $M1$ and $M2$ are markings. For instance, the transition $Record$ of the Workflow net in Fig. 1 can be represented as $Record = Start \multimap p1 \otimes p2$.

A sequent represents the triggering of a transition (or a sequence of possible transitions). The sequent $M, t_i \vdash M'$ represents a scenario where M is the initial marking, M' is the final marking, and t_i is a set of non-ordered transitions. To prove the correctness of a sequent, a proof tree should be built by applying the rules of sequent calculus. According to [6] the proof of a sequent in Linear Logic is equivalent to the corresponding reachability problem in the Petri net theory.

In this paper, only some of the Linear Logic rules will be considered. These rules will be used to build the proof trees in the context of the reachability problem of the Petri net theory. For such, F, G and H are considered as formulas and, Γ and Δ are blocks of formulas. The following rules will be used in this paper:

– The $\multimap_L$ rule, $\frac{\Gamma \vdash F \quad \Delta, G \vdash H}{\Gamma, \Delta, F \multimap G \vdash H}$ $\multimap_L$ expresses a transition firing and generates two sequents. The left sequent represents the tokens consumed by the transition firing and the right sequent represents the subsequent that still needs to be proved.
– The $\otimes_L$ rule, $\frac{\Gamma, F, G \vdash H}{\Gamma, F \otimes G \vdash H} \otimes_L$, transforms a marking into an atom list.
– The $\otimes_R$ rule, $\frac{\Gamma \vdash F \quad \Delta \vdash G}{\Delta, \Gamma \vdash F \otimes G} \otimes_R$, transforms a sequent such as $A, B \vdash A \otimes B$ into two identity sequents $A \vdash A$ and $B \vdash B$.

To exemplify these above mentioned rules, considering the transition $Record = Start \multimap p1 \otimes p2$ of the Workflow net in Fig. 1, by applying the rule $\multimap L$, two sequents are generated: $Start \vdash Start$ representing the token consumed by this transition firing, and $p1 \otimes p2$ is part of the sequent that still needs to be proved, $\frac{Start \vdash Start \quad p1 \otimes p2, \ldots \vdash End}{Start, Start \multimap p1 \otimes p2, \ldots \vdash End} \multimap L$. Using the marking $p1 \otimes p2$ produced as a result of the application of the rule $\multimap L$ in the transition $Record$, one can apply the rule $\otimes L$ to transform it into a list of atoms $p1, p2$, $\frac{p1, p2, \ldots \vdash End}{p1 \otimes p2, \ldots \vdash End} \otimes L$. The rule $\otimes R$ transforms a sequent such as $A, B \vdash A \otimes B$ into to identity sequents $A \vdash A$ and $B \vdash B$. For example, using another transition, $Collect = p3 \otimes p4 \multimap p5$ of the Workflow net in Fig. 1, the sequent that represents the tokens consumed by this transition firing, $p3, p4 \vdash p3 \otimes p4$, also needs to be proved, using the $\otimes R$ rule, $\frac{p3 \vdash p3 \quad p4 \vdash p4}{p3, p4 \vdash p3 \otimes p4} \otimes R$.

A Linear Logic proof tree is read from bottom-up. The proof stops when there is no longer any rule that can be applied, or when all the leaves of the proof tree are identity sequent.

2.9   Linear Logic for Soundness Verification

The proof of Soundness is a way to ensure that a Workflow process was correctly mapped by a Workflow net. First, it is important to define a scenario based on the context of this paper. A scenario is a set of partially ordered transition firings passing from initial marking to the final marking. For example, one possible scenario S_{c1} of the Workflow net represented in Fig. 1, could be defined by the sequent $Start, t1, t2, t3, t4, t5, t6, t8 \vdash End$.

Using Linear Logic in the study of the reachability problem of Petri nets allows us to consider partial markings of the net or sub-processes and not only mechanisms with global markings of the whole model. The global Workflow process is the result of integration of its sub-processes. Therefore, if a problem occurs in one of these sub-processes, a new study will only be necessary in its corresponding process. Another advantage of using Linear Logic is that information related to task execution time interval can be extracted directly from the proof trees used to prove the Soundness criterion of a Workflow net.

The first step toward proving the Soundness correction criterion is to represent the Workflow net through formulas of Linear Logic [17]. If the Workflow net in question has more than one scenario, each of these must produce its corresponding linear sequent [17]. In this approach, each linear sequent has only one atom $Start$, which represents the initial marking of the Workflow net. To prove the Soundness criterion, just one token in place $Start$ is necessary and sufficient. Necessary because without it the proof of the linear sequent cannot begin, and sufficient because if there is more than one token in place $Start$, only one will be used in the construction of the proof tree for the Soundness property verification [17].

After the construction of the proof tree, some important aspects needs to be analyzed and respected:

1. For each proof tree:
   - If just one atom End was produced in the proof tree (this is represented in the proof tree by the identity $End \vdash End$), then the first requirement of Soundness is proved.

- The second requirement of Soundness is proved if there is no more available atoms in the proof tree to be consumed in any of the places of the Workflow net.
- Finally, the third requirement corresponds to the fact that there is no transition formula available in the proof tree that has not been fired.

2. Considering the scenarios $S_{c1}, S_{c2}, ..., S_{cn}$ for the Workflow net analyzed, each transition $t \in T$ needs to appear in at least one scenario. This proves that all transitions were fired (i.e., there is no dead transitions), i.e. the third requirement for Soundness is verified.

If the conditions 1 and 2 above are satisfied, the Workflow net is Sound.

3   Time Constraint Propagation Mechanism

Typically, the time required to execute an activity in a Workflow process is non-deterministic. According to [20], explicit time constraints, which exist in systems with real-time characteristics, can be formally specified using a static time interval associated to each task (transition) of the model. The dynamic behavior of the corresponding time Petri net thus depends on the marking of the network, as well as on the tokens temporal situation that is given by a visibility interval [21]. A visibility interval $[(\delta_p)_{min}, (\delta_p)_{max}]$ associated with a token in a place p of a time Petri net specified the minimum date $(\delta_p)_{min}$ at which a token is available in p to trigger an output transition of p (earliest start date of an activity), and the maximum date after which the token becomes unavailable (dead) and cannot be used to trigger any transition (latest start date of the corresponding activity).

In a Workflow Management System, the visibility interval depends on a global clock connected to the entire net which calculates the passage of time from date $\delta = 0$, which corresponds to the start of system operations. In particular, the existing waiting time between sequential activities will be represented by a visibility interval, for which the minimum and maximum bounds will depend on the earliest and latest delivery of a case process by the Workflow net. Through knowledge of the beginning date and the maximum duration of a case, it should be possible to calculate the visibility intervals associated with the tokens in the waiting places.

The calculus of the visibility intervals associated with the tokens in the waiting places will be realized in this proposal by considering the different kinds of routs that can exist in a Workflow process, and which can be expressed through the sequents of Linear Logic. An iterative rout can be substituted by a global activity as shown in [23] and will not be considered in this work.

The constraint propagation technique proposed in [3] and reinforced in this article is based on two different approaches: a forward reasoning to produce the minimum bounds of the visibility intervals (earliest dates to initiate the activities of the process), and a backward reasoning to produce the maximum bounds of the visibility intervals (latest dates to initiate the activities of the process). The produced visibility intervals will be given through symbolic date expressions instead of numerical ones. The main advantage of using symbolic dates is that once the expressions have been calculated for a specific Workflow process, these can be used for any case that will be handled by the same Workflow process.

In this paper, D_i will denote a date and d_i a duration associated with a transition (t_i) firing. A pair (D_p, D_c) will be associated with each token of the proof tree and represents respectively the production and consumption date of a token. As transition firings are instantaneous ones in t-time nets (there is no token reservation), the following definition can be produced:

Definition 1. *The production date D_p of a token is equal to the firing date of the transition which has produced it and the consumption date D_c is equal to the firing date of the transition which has consumed it.*

For each triggered activity on the t-time Workflow net analyzed, the dates of production D_p and consumption D_c of the atoms that represents the preconditions of the activity should be extracted. The date of production of the atom D_p, corresponds to the start execution of the activity associated with the transition and the date of consumption, D_c, corresponds to its conclusion. Thus, an interval of dates $[D_p, D_c]$ will be generated, and the resource that will handle the referred activity will be able to be requested within the calculated interval.

Since production and consumption dates depend on d_i (duration associated to a transition firing) and take there values within time intervals $\Delta_i = [\delta_{imin}, \delta_{imax}]$, many tasks execution intervals can be considered according to a strategic planning. For instance, the execution interval $I_{Exec} = [D_{pmin}, D_{cmax}]$ considers that the allocation of resources to handle the task can occur between the earliest beginning and the latest conclusion of this activity. To illustrate the proposal of this work, the Workflow net shown in Fig. 1 is considered.

3.1 Forward Propagation

The forward propagation consider the Workflow net in its normal flow, which is a case represented by a token that begins in place $Start$ and finishes in place End, while following the ordered places and transitions.

The date computation in the canonical proof for the Workflow net is then the following one:

- Assign a production date D_i to the initial token in place $Start$ (initial marking of the t-time Petri net model).
- For each $\multimap L$, compute the firing date of the corresponding transition: it is equal to the maximum of the production dates of the consumed atoms, increased by the enabling duration d_j associated with the considered transition.
- Update all the temporal stamps of the atoms which have been consumed and produced.

As in t-time Petri net model [14], an enabling duration d_i takes its values from within a time interval $\Delta_i = [\delta_{imin}, \delta_{imax}]$. The computed symbolic dates depend on d_i and its domains will depend on time interval.

Considering Fig. 5, to verify the soundness criterion, two linear sequents must be proved, where each one represents a different scenario. Such linear sequents can be

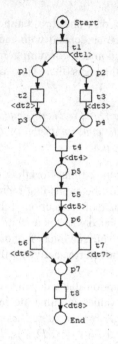

Fig. 5. t-Time Workflow net - Forward Propagation, from [3].

obtained automatically applying a t-invariant algorithm (linear equation solving) as proposed in [16].

For scenario S_{c1}, the following sequent has to be proved:

$$Start, t1, t2, t3, t4, t5, t6, t8 \vdash End \tag{1}$$

For scenario S_{c2}, the following sequent has to be proved:

$$Start, t1, t2, t3, t4, t5, t7, t8 \vdash End \tag{2}$$

The transitions of the Workflow net are represented by the following formulas of Linear Logic:

$$t_1 = Start \multimap p1 \otimes p2,$$
$$t_2 = p1 \multimap p3,$$
$$t_3 = p2 \multimap p4,$$
$$t_4 = p3 \otimes p4 \multimap p5,$$
$$t_5 = p5 \multimap p6,$$
$$t_6 = p6 \multimap p7,$$
$$t_7 = p6 \multimap p7,$$
$$t_8 = p7 \multimap End.$$

Considering $Seq = D_S + d_{t1} + max\{d_{t2}, d_{t3}\} + d_{t4}$, the proof tree with dates corresponding to scenario S_{c1} is as follow:

$$\cfrac{\cfrac{\cfrac{\cfrac{\cfrac{\cfrac{\cfrac{\cfrac{\cfrac{p7(Seq+d_{t5}+d_{t6},Seq+d_{t5}+d_{t6}+d_{t8})\vdash p7 \quad End(Seq+d_{t5}+d_{t6}+d_{t8},.)\vdash End}{p6(Seq+d_{t5},Seq+d_{t5}+d_{t6})\vdash p6 \quad p7(Seq+d_{t5}+d_{t6},.),p7\multimap End\vdash End}{}_{\multimap L}}{p5(Seq,Seq+d_{t5})\vdash p5 \quad p6(Seq+d_{t5},.),p6\multimap p7,t8\vdash End}{}_{\multimap L}}{\cfrac{p3(D_S+d_{t1}+d_{t2},Seq)\vdash p3 \quad p4(D_S+d_{t1}+d_{t3},Seq)\vdash p4}{p3(D_S+d_{t1}+d_{t2},Seq),p4(D_S+d_{t1}+d_{t3},Seq)\vdash p3\otimes p4}{}_{\otimes R} \quad p5(Seq,.),p5\multimap p6,t6,t8\vdash End}{}_{\multimap L}}{p2(D_S+d_{t1},D_S+d_{t1}+d_{t3})\vdash p2 \quad p3(D_S+d_{t1}+d_{t2},.),p4(D_S+d_{t1}+d_{t3},.),p3\otimes p4\multimap p5,t5,t6,t8\vdash End}{}_{\multimap L}}{p1(D_S+d_{t1},D_S+d_{t2})\vdash p1 \quad p2(D_S+d_{t1},.),p3(D_S+d_{t1}+d_{t2},.),p2\multimap p4,t4,t5,t6,t8\vdash End}{}_{\multimap L}}{p1(D_S+d_{t1},.),p2(D_S+d_{t1},.),p1\multimap p3,t3,t4,t5,t6,t8\vdash End}{}_{\otimes L}}{Start(D_S,D_S+d_{t1})\vdash Start \quad p1(D_S+d_{t1},.)\otimes p2(D_S+d_{t1},.),p1\multimap p3,t3,t4,t5,t6,t8\vdash End}{}_{\multimap L}}{Start(D_S,.),Start\multimap p1\otimes p2,t2,t3,t4,t5,t6,t8\vdash End}$$

$$(3)$$

The proof tree with dates for scenario S_{c2} is similar to the proof tree of S_{c1}, the only difference is that in the first lines of the proof, instead of d_{t6} (duration of the transition firing $t6$), it is d_{t7} (duration of the transition firing $t7$) which is considered. The entire proof can be found in [23].

$$\cfrac{\cfrac{p7(Seq+d_{t5}+d_{t6},Seq+d_{t5}+d_{t7}+d_{t8})\vdash p7 \quad End(Seq+d_{t5}+d_{t7}+d_{t8},.)\vdash End}{p6(Seq+d_{t5},Seq+d_{t5}+d_{t7})\vdash p6 \quad p7(Seq+d_{t5}+d_{t7},.),p7\multimap End\vdash End}{}_{\multimap L}}{}_{\multimap L}$$

$$(4)$$

$$\dots$$

$$Start(D_S,.),Start\multimap p1\otimes p2,t2,t3,t4,t5,t7,t8\vdash End$$

Considering the proof trees for scenarios S_{c1} and S_{c2}, the Workflow net shown in Fig. 5 is Sound because conditions 1 and 2 are satisfied.

The symbolic dates obtained from the proof tree of forward propagation mechanism are shown in the Table 1.

Table 1. Symbolic date intervals for forward propagation in scenario S_{c1}, from [3].

Tasks	Production Date	Consumption Date
$Start$	D_s	$D_s + d_{t1}$
$p1$	$D_s + d_{t1}$	$D_s + d_{t1} + d_{t2}$
$p2$	$D_s + d_{t1}$	$D_s + d_{t1} + d_{t3}$
$p3$	$D_s + d_{t1} + d_{t2}$	$D_s + d_{t1} + max(d_{t2}, d_{t3}) + d_{t4}$
$p4$	$D_s + d_{t1} + d_{t3}$	$D_s + d_{t1} + max(d_{t2}, d_{t3}) + d_{t4}$
$p5$	$D_s + d_{t1} + max(d_{t2}, d_{t3}) + d_{t4}$	$D_s + d_{t1} + max(d_{t2}, d_{t3}) + d_{t4} + d_{t5}$
$p6$	$D_s + d_{t1} + max(d_{t2}, d_{t3}) + d_{t4} + d_{t5}$	$D_s + d_{t1} + max(d_{t2}, d_{t3}) + d_{t4} + d_{t5} + d_{t6}$
$p7$	$D_s + d_{t1} + max(d_{t2}, d_{t3}) + d_{t4} + d_{t5} + d_{t6}$	$D_s + d_{t1} + max(d_{t2}, d_{t3}) + d_{t4} + d_{t5} + d_{t6} + d_{t8}$
End	$D_s + d_{t1} + max(d_{t2}, d_{t3}) + d_{t4} + d_{t5} + d_{t6} + d_{t8}$	D_{End}

3.2 Backward Propagation

From the proof tree obtained by the forward propagation model, considering the start date D_S, it is possible to determine part of the visibility intervals to trigger the tasks (represented by a token in P_i). As in [23], the forward propagation technique allows us to find the minimum bounds of the visibility intervals.

Backward propagation is used to calculate the maximum bounds of the same visibility interval. To accomplish this goal, the approach is based on an inverted model as those presented in [11, 16]. In [11], a backward reasoning was applied on a Petri net model with all the arcs reversed in order to identify feared scenarios in mechatronic systems. Paper [16] also proposed a backward reasoning applied on a Workflow net with all arcs inverted in order to diagnose the causes of deadlock situations in Service-Oriented Architectures.

For the backward propagation technique, the arcs of the Workflow net are then inverted in such a way that the initial marking is initiated in place End and reaches the $Start$ place at the end of the execution of the inverted Workflow net, as illustrated in Fig. 6.

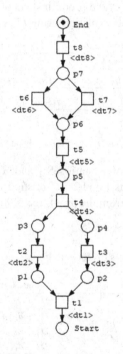

Fig. 6. t-Time Workflow net - Backward Propagation, from [3].

The proof of a sequent on the inverted Workflow net is obtained using the same proof algorithm as in [23] for the applied rules of Linear Logic, but with different computation of temporal stamps. We denote D_i for dates and d_i for durations.

When the consumption date is not fully computed, some pairs are denoted $(D_p, .)$ will appear in the proof tree. Identity sequents generated on the left side of the tree will produce pair stamps (D_p, D_c) completely defined. These final leaves stamps will correspond to the main temporal results (maximum bound of visibility intervals) of the backward propagation mechanism.

The date computation in the canonical proof tree for the inverted Workflow net is then the following one:

- Assign a production date D_i to the initial token in place End (initial marking of the inverted net).
- For each $\multimap L$, compute the firing date of the corresponding transition: it is equal to the minimum of the production dates of the consumed atoms, decreased by the enabling duration d_j associated with the considered transition.
- Update all the temporal stamps of the atoms which have been consumed and produced.

The rest of the process to prove the sequent is the same as the one presented in the forward propagation mechanism.

Considering the Workflow net in Fig. 6, the proof of the Soundness criterion on this inverted model corresponds to two different scenarios given by the following sequents: For scenario S_{c1}:

$$End, t8, t6, t5, t4, t3, t2, t1 \vdash Start \qquad (5)$$

and for scenario S_{c2}:

$$End, t8, t7, t5, t4, t3, t2, t1 \vdash Start \qquad (6)$$

The transitions of the Workflow net are represented by the following formulas of Linear Logic:

$$t_8 = End \multimap p7,$$
$$t_7 = p7 \multimap p6,$$
$$t_6 = p7 \multimap p6,$$
$$t_5 = p6 \multimap p5,$$
$$t_4 = p5 \multimap p3 \otimes p4,$$
$$t_3 = p4 \multimap p2,$$
$$t_2 = p3 \multimap p1,$$
$$t_1 = p1 \otimes p2 \multimap Start.$$

Considering $Seq_1 = D_{End} - d_{t8} - d_{t6} - d_{t5} - d_{t4}$ and $Seq_2 = D_{End} - d_{t8} - d_{t7} - d_{t5} - d_{t4}$, the proof tree with dates for S_{c1} is the following:

$$ (7) $$

The proof tree with dates for S_{c2} is:

$$\cfrac{\cfrac{p1(Seq_2-d_{t2},Seq_2-min\{d_{t2},d_{t3}\}-d_{t1})\vdash p1 \quad p2(Seq_2-d_{t3},Seq_2-min\{d_{t2},d_{t3}\}-d_{t1})\vdash p2}{p1(Seq_2-d_{t2},Seq_2-min\{d_{t2},d_{t3}\}-d_{t1}),p2(Seq_2-d_{t3},Seq_2-min\{d_{t2},d_{t3}\}-d_{t1})\vdash p1\otimes p2}\otimes_R \quad Start(Seq_2-min\{d_{t2},d_{t3}\}-d_{t1},..)\vdash Start}{\cfrac{p3(Seq_2,Seq_2-d_{t2})\vdash p3 \quad p1(Seq_2-d_{t2},.),p2(Seq_2-d_{t3},.),p1\otimes p2\multimap Start\vdash Start}{\cfrac{p4(Seq_2,Seq_2-d_{t3})\vdash p4 \quad p3(Seq_2,.),p2(Seq_2-d_{t3},.),p3\multimap op1,t1\vdash Start}{\cfrac{p4(Seq_2,.),p3(Seq_2,.),p4\multimap op2,t2,t1\vdash Start}{\cfrac{p5(D_{End}-d_{t8}-d_{t7}-d_{t5},D_{End}-d_{t8}-d_{t7}-d_{t5}-d_{t4})\vdash p5 \quad p3\otimes p4,p4\multimap op2,t2,t1\vdash Start}{\cfrac{p6(D_{End}-d_{t8}-d_{t7},D_{End}-d_{t8}-d_{t7}-d_{t5})\vdash p6 \quad p5(D_{End}-d_{t8}-d_{t7}-d_{t5},.),p5\multimap op3\otimes p4,t3,t2,t1\vdash Start}{\cfrac{p7(D_{End}-d_{t8},D_{End}-d_{t8}-d_{t7})\vdash p7 \quad p6(D_{End}-d_{t8}-d_{t7},.),p6\multimap op5,t4,t3,t2,t1\vdash Start}{\cfrac{End(D_{End},D_{End}-d_{t8})\vdash End \quad p7(D_{End}-d_{t8},..),p7\multimap op6,t5,t4,t3,t2,t1\vdash Start}{End(D_{End},.),End\multimap op7,t7,t5,t4,t3,t2,t1\vdash Start}\multimap_L}\multimap_L}\multimap_L}\otimes_L}\multimap_L}\multimap_L}\multimap_L}\multimap_L}$$

$$(8)$$

Considering the proof trees for scenarios S_{c1} and S_{c2}, the Workflow net shown in Fig. 6 is Sound because conditions 1 and 2 are satisfied.

The symbolic dates obtained from the proof tree of backward propagation mechanism are shown in the Table 2.

Table 2. Symbolic date intervals for backward propagation in scenario S_{c1}, from [3].

Tasks	Production Date	Consumption Date
End	D_{End}	$D_{End} - d_{t8}$
p7	$D_{End} - d_{t8}$	$D_{End} - d_{t8} - d_{t6}$
p6	$D_{End} - d_{t8} - d_{t6}$	$D_{End} - d_{t8} - d_{t6} - d_{t5}$
p5	$D_{End} - d_{t8} - d_{t6} - d_{t5}$	$D_{End} - d_{t8} - d_{t6} - d_{t5} - d_{t4}$
p4	$D_{End} - d_{t8} - d_{t6} - d_{t5} - d_{t4}$	$D_{End} - d_{t8} - d_{t6} - d_{t5} - d_{t4} - d_{t3}$
p3	$D_{End} - d_{t8} - d_{t6} - d_{t5} - d_{t4}$	$D_{End} - d_{t8} - d_{t6} - d_{t5} - d_{t4} - d_{t2}$
p2	$D_{End} - d_{t8} - d_{t6} - d_{t5} - d_{t4} - d_{t3}$	$D_{End} - d_{t8} - d_{t6} - d_{t5} - d_{t4} - min(d_{t2},d_{t3}) - d_{t1}$
p1	$D_{End} - d_{t8} - d_{t6} - d_{t5} - d_{t4} - d_{t2}$	$D_{End} - d_{t8} - d_{t6} - d_{t5} - d_{t4} - min(d_{t2},d_{t3}) - d_{t1}$
Start	$D_{End} - d_{t8} - d_{t6} - d_{t5} - d_{t4} - min(d_{t2},d_{t3}) - d_{t1}$	D_s

Finally, symbolic visibility intervals can be produced for the tokens in each places of the Workflow net of Fig. 5. The minimum bounds correspond to the production dates D_p of Table 1, and the maximum bounds correspond to the production dates of Table 2.

Table 3. Symbolic visibility intervals for task execution in scenario S_{c1}.

Tasks	Minimum Bound of Visibility Interval	Maximum Bound of Visibility Interval
Start	D_s	$D_{End} - d_{t8} - d_{t6} - d_{t5} - d_{t4} - min(d_{t2},d_{t3}) - d_{t1}$
p1	$D_s + d_{t1}$	$D_{End} - d_{t8} - d_{t6} - d_{t5} - d_{t4} - d_{t2}$
p2	$D_s + d_{t1}$	$D_{End} - d_{t8} - d_{t6} - d_{t5} - d_{t4} - d_{t3}$
p3	$D_s + d_{t1} + d_{t2}$	$D_{End} - d_{t8} - d_{t6} - d_{t5} - d_{t4}$
p4	$D_s + d_{t1} + d_{t3}$	$D_{End} - d_{t8} - d_{t6} - d_{t5} - d_{t4}$
p5	$D_s + d_{t1} + max(d_{t2},d_{t3}) + d_{t4}$	$D_{End} - d_{t8} - d_{t6} - d_{t5}$
p6	$D_s + d_{t1} + max(d_{t2},d_{t3}) + d_{t4} + d_{t5}$	$D_{End} - d_{t8} - d_{t6}$
p7	$D_s + d_{t1} + max(d_{t2},d_{t3}) + d_{t4} + d_{t5} + d_{t6}$	$D_{End} - d_{t8}$
End	$D_s + d_{t1} + max(d_{t2},d_{t3}) + d_{t4} + d_{t5} + d_{t6} + d_{t8}$	D_{End}

4   CPN Implementation of the Workflow Net with Resources

In order to show the modeling and implementation of the visibility intervals calculated using the Linear Logic and the resource allocation models in the execution of the process activities, the CPN Tools will be used to model the "Complaints Handling Process" of Fig. 1. In the modeling proposal of this work, the process representation is separated from the resource allocation mechanism, using the hierarchical possibilities of the CPN Tools. To define the duration of each activity, a random time based on the uniform distribution is used. The total expected (planned) duration of a case should be 105 units of time for this process.

4.1   CPN Model

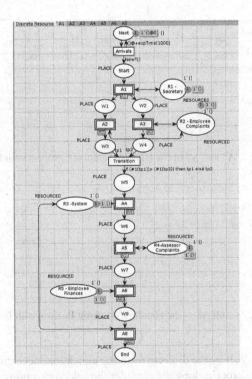

Fig. 7. Complaint Handling Process with Discrete Resources.

Figure 7 shows the implementation for the Complaint Handling Process defined in [24] with explicit inclusion of discrete resource mechanisms and time duration intervals. Such a model uses the hierarchical aspect of the CPN Tools where each activity corresponding to a specific transition can be transformed into a new CPN model that describes in detail the corresponding activity. An exponential distribution rate is used to simulate the arrival of the cases into the Workflow process.

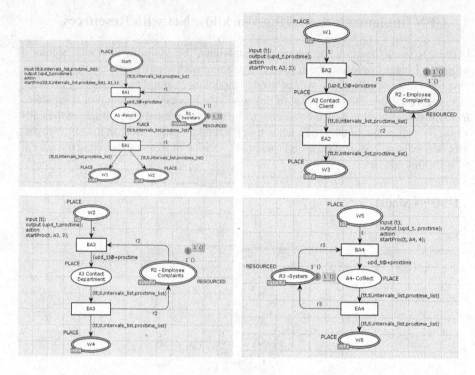

Fig. 8. Activities A1, A2, A3 and A4 of the Handle Complaint Process with Discrete Resources.

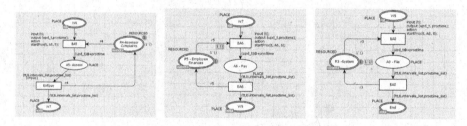

Fig. 9. Activities A5, A6 and A8 of the Handle Complaint Process with Discrete Resources.

Figure 8 and Fig. 9 show the discrete resource allocation mechanisms.

This model of the process does not consider the selective routing between tasks $A6$ and $A7$. Instead, in order to simplify the simulation of the visibility intervals calculated by the symbolic formulas, only the activity $A6$ is considered.

According to Table 4 some colsets should be highlighted:

- Tp: represents current time;
- t1: represents the initial time of a time interval;
- t2: represents the final time of a time interval;
- W: represents a time interval of an activity;
- INTERVALSlist: represents a list of time intervals;

Table 4. Colset Declarations for the CPN Model.

colset Tp = INT;
colset t1 = REAL;
colset t2 = REAL;
colset W = product t1 * t2;
colset INTERVALSlist = list W;
colset transition = INT;
colset T = product transition*Tp;
colset PROCTIMElist = list T;
colset PLACE = product Tp*Tp*INTERVALSlist*PROCTIMElist timed;

- transition: represents a transition id;
- T: represents a pair of data: the transition id and the firing date of the transition;
- PROCTIMElist: represents a list of values of type T;
- PLACE: represents the information that each token (case) carries: date of criation of the case, current time, list of visibility time intervals, and list of processing times (firing dates of the transition corresponding to the activities process).

The process start time is set when a new token is produced in place *Start*. The current time is the application clock time. The visibility intervals are calculated for each token, based on the results presented in Table 3, using a function implemented in Standard ML whose purpose is to transform the symbolic dates of the visibility intervals into numeric ones considering the time of creation of the case.

Some monitors were added to the CPN model. These monitors generate .txt files with the temporal data. The produced data is then processed by a Python Script which extract automatically the percentage of cases that respect the time constraints given by the corresponding visibility intervals. These information will then be used to understand which activities are causing the most delay in the process.

The Python Script considers in particular the visibility intervals and processing time for each case and based on that, calculates for each activity, if the process time respect the visibility interval for that activity.

4.2 Simulation Results

Untimed statistics					
Name	Count	Sum	Avrg	Min	Max
Duration_A2	625	15687	25.099200	20	30
Duration_A3	625	18759	30.014400	25	35
Duration_A5	625	12393	19.828800	15	25
Duration_A6	625	6290	10.064000	5	15
Durtion_A1	625	4697	7.515200	5	10
Number_of_cases	625	625	1.000000	1	1
Total_duration	625	59314	94.902400	65	189

Fig. 10. CPN Tools Output Result for Simulation Considering 1 Resources in $R1$, $R2$, $R3$, $R4$ and $R5$.

A simulation of the process was done considering 10000 steps (transitions firings). Some simulation results, obtained from monitors added to the CPN net, are presented in Fig. 10, that indicate that 625 cases completed the whole process. We can see in particular the average time of each activity.

Looking at the results produced by the Python Script shown in Fig. 11 it is possible to note that 8.32% of the cases finishes the process with delay (the ending date of the case at the end of the process is superior to the maximum duration allowed to complete the process) and 45.17% of the activities are triggered with delay which means that the token (case) started the activity (trigger the transition related to the activity) after the maximum time of the visibility interval defined to that activity. Another information that can be seen on Fig. 11, considering the number of times each activity is triggered with delay, is that the greater accumulation of cases that do not respect the maximum constraint given by their visibility intervals occurs on the activities $A2$, $A5$ and $A6$. Adding a resource in $R2$ and $R4$ should improve the respect of the process deadlines for most of the cases. With 2 resources in $R2$ and $R4$, and 1 resource in $R1$, $R3$ and $R5$, the simulation results are shown in Fig. 12 and the Python Script result is presented in Fig. 13. It can be seen that the average values shown in Fig. 10 and Fig. 12 are very similar. But the difference between the delay analysis in Fig. 11 and Fig. 13 shows that the new resource allocation was able to decrease the delay rate in the process as a whole from 8.32% to 0.00% and decrease the delay rate per activity from 45.17% to 2.72%.

```
8.32% of the cases finish the process with delay (> 105)
45.17% of the activities are triggered with delay
Number of times each activity is triggered with delay
A1: 4 cases
A2: 310 cases
A3: 100 cases
A4: 201 cases
A5: 625 cases
A6: 407 cases
A7: 0 cases
A8: 47 cases
```

Fig. 11. Python Code Analysis About Delay Considering 2 Resources in $R1$, $R2$, $R3$, $R4$ and $R5$.

Untimed statistics					
Name	**Count**	**Sum**	**Avrg**	**Min**	**Max**
Duration_A2	625	15741	25.185600	20	30
Duration_A3	625	18747	29.995200	25	35
Duration_A5	625	12540	20.064000	15	25
Duration_A6	625	6260	10.016000	5	15
Durtion_A1	625	4657	7.451200	5	10
Number_of_cases	625	625	1.000000	1	1
Total_duration	625	42625	68.200000	53	98

Fig. 12. CPN Tools Output Result for Simulation Considering 2 Resources in $R2$ and $R4$ and 1 Resource in $R1$, $R3$ and $R5$.

```
0.00% of the cases finish the process with delay (> 105)
2.72% of the activities are triggered with delay
Number of times each activity is triggered with delay
A1: 0 cases
A2: 74 cases
A3: 0 cases
A4: 0 cases
A5: 24 cases
A6: 4 cases
A7: 0 cases
A8: 0 cases
```

Fig. 13. Python Code Analysis About Delay Considering 2 Resources in $R2$ and $R4$ and 1 Resource in $R1$, $R3$ and $R5$.

5   Related Works

Several works have already dealt with time constraint propagation in Workflow nets for the Planning/Scheduling of shared resources in Workflow Management Systems.

In [10], forward and backward reasoning are presented to produce visibility intervals of the cases in a Workflow process modeled by a time Petri net, but such an approach considers only numerical information. Consequently, the forward and backward propagation mechanisms must be applied each time a new case appears at the beginning of the Workflow process.

In [12], an energetic reasoning is used for applying a kind of filter to the time constraints of a Workflow process given by visibility intervals. The forward and backward mechanisms are implemented in a constraint programming language; these however need to be applied to each new case coming into the process due to the numerical values of the intervals associated with task durations.

In [4], fuzzy forward and backward propagation techniques are presented. The advantage is being able to consider more flexible time constraints given by fuzzy intervals, however, the task durations are also given by fuzzy numerical values and the propagation technique must be applied each time a new case is dealt with the Workflow process.

In [23], the sequent calculus of Linear Logic was used on a time Workflow net, which produced time symbolic constraints, in a manner very similar to that one presented in this article. However, only the forward propagation technic was presented. No backward propagation mechanism applied to an inverted time model was considered. The time constraints given by visibility intervals of the processed cases were not so precise and as such, not suitably adapted for use in the practical Panning/Scheduling problem of the resources used for the implementation of the activities in the Workflow process in real time.

Finally, the calculus of the visibility intervals were introduced in [3], but no implementation or exploitation of these results in a simulation model were presented.

6   Conclusions

The specification of a Workflow process is not a trivial task, making it necessary to use diagrams and charts to visually specify sets of requirements. Thus, this work presented a study based on Timed Workflow nets for the definition of the time constraints that exist in business processes. Discrete resource allocation mechanism modeled by Petri nets were also used to model the execution of activities in Workflow Management Systems.

Forward and backward mechanisms based on $(max, +)$ and $(min, -)$ operators were presented to produce symbolic formulas that indicate the earliest and latest dates authorized for the beginning of the activities of the cases treated by the Workflow process. Such propagation mechanisms were treated by the Workflow process. Such propagation mechanisms were derived from the proof trees of sequent calculus of Linear Logic. In particular, the computation of symbolic dates produced data that can be used for the planning of the resources involved in the activities of the Workflow process, since the computed dates are symbolic instead of numerical.

Using a Hierarchical Colored Petri net model, it was possible to implement in a separate way the Workflow process and the Discrete Resource allocation mechanism used in the execution of activities. Temporal data of each case were then specified on the tokens of the CPN and, using the CPN Tools, some simulations were performed to monitor each activity execution when considering the numerical temporal constraints attached to each case and automatically implemented by the specific Standard ML function whose main argument corresponds to the symbolic formulas. The main result of the simulation was the identification of the percentage of cases that do not respect the time constraints (delivery deadline dates).

As a future study, the authors intend to implement several classical scheduling policies as the ones presented in [9] in order to exploit in the best possible way the time constraints of each case. The future CPN model could then be seen as a kind of prototyping tool for the evaluation of scheduling strategies in Workflow Management Systems.

Acknowledgement. The authors would like to thank FAPEMIG, CNPq and CAPES for the financial support provided.

References

1. Van der Aalst, W.M.: The application of Petri nets to workflow management. J. Circuits Syst. Comput. **8**(01), 21–66 (1998)
2. Van der Aalst, W.M.P.: Timed coloured Petri nets and their application to logistics (1992)
3. Bruno, L.R., Julia, S.: A symbolic time constraint propagation mechanism proposal for workflow nets. In: ICEIS (1), pp. 537–544 (2022)
4. de Freitas, J.C.J., Julia, S.: Fuzzy time constraint propagation mechanism for workflow nets. In: 2015 12th International Conference on Information Technology-New Generations, pp. 367–372. IEEE (2015)
5. Girard, J.Y.: Linear logic. Theor. Comput. Sci. **50**(1), 1–101 (1987)
6. Girault, F., Pradier-Chezalviel, B., Valette, R.: A logic for Petri nets. Journal européen des systèmes automatisés **31**(3), 525–542 (1997)
7. Jensen, K.: An introduction to the practical use of coloured Petri nets. In: Reisig, W., Rozenberg, G. (eds.) ACPN 1996. LNCS, vol. 1492, pp. 237–292. Springer, Heidelberg (1998). https://doi.org/10.1007/3-540-65307-4_50
8. Jensen, K., Kristensen, L.M.: Coloured Petri Nets: Modelling and Validation of Concurrent Systems. Springer, Heidelberg (2009). https://doi.org/10.1007/b95112
9. Jeske, J.C., Julia, S., Valette, R.: Fuzzy continuous resource allocation mechanisms in workflow management systems. In: 2009 XXIII Brazilian Symposium on Software Engineering, pp. 236–251. IEEE (2009)
10. Julia, S., de Oliveira, F.F., Valette, R.: Real time scheduling of workflow management systems based on a p-time Petri net model with hybrid resources. Simul. Model. Pract. Theory **16**(4), 462–482 (2008)
11. Khalfhoui, S., Demmou, H., Guilhem, E., Valette, R.: An algorithm for deriving critical scenarios in mechatronic systems. In: IEEE International Conference on Systems, Man and Cybernetics, vol. 3, pp. 6-pp. IEEE (2002)
12. Medeiros, F.F., Julia, S.: Constraint analysis based on energetic reasoning applied to the problem of real time scheduling of workflow management systems. In: ICEIS (3), pp. 373–380 (2017)

13. Menasche, M.: Analyse des réseaux de Petri temporisés et application aux systèmes distribués (1982)
14. Merlin, P.M.: A study of the recoverability of computing systems. University of California, Irvine (1974)
15. Milner, R., Tofte, M., Harper, R., MacQueen, D.: The Definition of Standard ML: Revised. MIT Press, Cambridge (1997)
16. Oliveira, K.S., Julia, S.: Detection and removal of negative requirements of deadlock-type in service-oriented architectures. In: 2020 International Conference on Computational Science and Computational Intelligence (2020)
17. Passos, L.M.S., Julia, S.: Qualitative analysis of workflow nets using linear logic: soundness verification. In: 2009 IEEE International Conference on Systems, Man and Cybernetics, pp. 2843–2847. IEEE (2009)
18. Pradin-Chézalviel, B., Valette, R., Kunzle, L.A.: Scenario durations characterization of t-timed Petri nets using linear logic. In: Proceedings 8th International Workshop on Petri Nets and Performance Models (Cat. No. PR00331), pp. 208–217. IEEE (1999)
19. Ramamoorthy, C., Ho, G.S.: Performance evaluation of asynchronous concurrent systems using Petri nets. IEEE Trans. Softw. Eng. 5, 440–449 (1980)
20. Riviere, N., Pradin-Chezalviel, B., Valette, R.: Reachability and temporal conflicts in t-time Petri nets. In: Proceedings 9th International Workshop on Petri Nets and Performance Models, pp. 229–238. IEEE (2001)
21. dos Santos Soares, M., Julia, S., Vrancken, J.: Real-time scheduling of batch systems using Petri nets and linear logic. J. Syst. Softw. 81(11), 1983–1996 (2008)
22. Sifakis, J.: Use of Petri nets for performance evaluation. Acta Cybernet. 4(2), 185–202 (1979)
23. Soares Passos, L.M., Julia, S.: Linear logic as a tool for qualitative and quantitative analysis of workow processes. Int. J. Artif. Intell. Tools 25(03), 1650008 (2016)
24. Van Der Aalst, W., Van Hee, K.M., van Hee, K.: Workflow Management: Models, Methods, and Systems. MIT Press, Cambridge (2004)

Implementation of the Maintenance Cost Optimization Function in Manufacturing Execution Systems

Andrzej Chmielowiec[1]([✉]) [iD], Leszek Klich[1] [iD], Weronika Woś[1] [iD],
and Adam Błachowicz[2] [iD]

[1] Faculty of Mechanics and Technology, Rzeszow University of Technology,
Kwiatkowskiego 4, Stalowa Wola, Poland
{achmie,l.klich,weronikawos}@prz.edu.pl
[2] Pilkington Automotive Poland, Strefowa 17, Chmielów, Poland
adam.blachowicz@pl.nsg.com

Abstract. In this chapter, we consider the global problem of spare parts operating costs optimization from an implementation perspective. So far, the problem of the optimal replacement time for a spare part has been extensively studied in the context of a single part. The available analytical models allow for a very accurate estimation of further service life and savings related to the replacement. Nevertheless, the simultaneous optimization problem of the operating costs for all spare parts used in the factory is beyond the scope of the developed analytical methods. This chapter presents a proposal of a method allowing for global cost optimization. This is one of many approaches that can be used. However, we want to show what kind of problems the global optimization task boils down to in this case. Our goal is to propose modern architecture for IT systems supporting the maintenance department. We believe that the methods presented in this chapter will be an important foundation for this type of solution.

Keywords: Maintenance · Cost optimization · Weibull distribution ·
Exploitation costs · Manufacturing execution system · Knapsack problem

1 Introduction

The development of information and measurement technologies makes modern production companies more and more involved in implementation of innovative techniques for monitoring the state of machine functioning. Despite the installation of an increasing number of sensors and monitoring the degree of wear of many systems and their subassemblies, there is still a large number of spare parts which usability status is determined in a binary manner: functional/non-functional. An example is an oil hose for which the performance level on the fly cannot be determined. Perhaps in the future, advanced vision systems will be able to automatically detect microcracks on its surface and track wear levels. At present, however, the operation of such cases is purely reactive. The exchange is made at a strictly defined moment or the failure is removed when it occurs. From the point of view of maintenance personel, this is a rather uncomfortable situation. However, it should be remembered that it is included in the activities of the enterprise. Too high operational reliability always leads to disproportionately high

© The Author(s), under exclusive license to Springer Nature Switzerland AG 2023
J. Filipe et al. (Eds.): ICEIS 2022, LNBIP 487, pp. 133–154, 2023.
https://doi.org/10.1007/978-3-031-39386-0_7

maintenance costs. For this reason, each plant determines the acceptable level of failure, which is included in its operating costs. However, for each such spare part on the production line, we can determine the optimum working time. Optimum time is understood here as the time that minimizes the costs of servicing. Intuition suggests that if the removal of a failure is much more expensive than a scheduled replacement, the standard working time of such an spare part should be shorter than in a situation where the costs of removing a failure and scheduled replacement are similar. However, the question is: when should the replacement be made so that the long-term costs are as low as possible? This problem will be discussed in this chapter. In the following parts, it will be shown how modern IT tools combined with the classical theory of reliability can support the decision-making process in a manufacturing company. Our goal is to propose modern architecture for IT systems supporting the maintenance department. We will show how classical statistical analysis, data from sensors and artificial intelligence methods can contribute to the optimization of production maintenance costs.

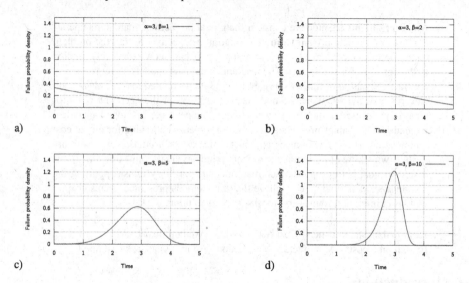

Fig. 1. Probability density for the Weibull distribution with different parameters α and β.

The Fig. 1 shows the Weibull distribution with different parameters α and β. This distributions have been chosen as an example to illustrate the reliability of aging components. In practice, for each spare part, the α and β parameters should be estimated based on historical data (which will be presented in the following sections) or information from the manufacturer. The presented graphs clearly show that they can cover a fairly large range of possible life cycles of an spare part. From a maintenance standpoint, the higher the β value, the better. This is due to the fact that distributions with a large value of this parameter are more concentrated around the mean value, and this in turn means that it is possible to predict the failure time more precisely. In this respect, the distributions for which the value of the beta parameter is equal to or slightly greater than one, are the worst. In these kinds of situations, the optimal replacement time is either infinite ($\beta = 1$) or of such a great value that in practice the best strategy for such a case is waiting for a failure.

The Fig. 1 shows that the systematic replacement of such an spare part every one time unit practically eliminates the problem of failure. Therefore, the manufacturer may recommend that users regularly replace them at a specific time period. However, is this approach economically justified? Unfortunately, there is no clear answer to this problem. The solution depends on the cost of replacing this spare part during standard service activities, and the cost of replacement in the event of a failure. It should be emphasized that the costs related to downtime should be added to the cost of the service replacement as well as to the cost of the replacement during a failure. This is a very important cost component that can cause the same spare part used at different points in the production line to be treated differently. In fact, any finite resource at the company's disposal that has any value can be considered as a cost. This can be time, money, energy, etc. In principle, cost should be understood to mean all the necessary resources that are consumed during replacement (both service replacement and failure replacement).

The problem of age replacement policy is the subject of many authors publications. The common feature of all these articles is the single spare part focused optimization approach. The developed analytical methods give a very good picture of the viability of specific system components, but do not allow to approach them globally. In this chapter we will demonstrate that such a holistic approach is possible. However, it requires the use of completely different methods. We will demonstrate how the problem of optimizing the operation time of many spare parts fits into the methods of discrete optimization. Before we get down to that, we will present a literature review related to policies for optimizing the working time of individual spare parts. First [5] and [6] have analyzed the optimization of the expected operating costs and the impact of the replacement policy on these costs. Moreover, for several common failure distributions methods have been developed by [20]. On the other hand, [17] and [35] consider the optimal replacement time, taking into account discounting - this issue is especially important in the case of rapidly changing prices. The publications [12, 13, 36] and [38] present a generalized approach to the optimization of operating costs and preventive replacement. The confidence intervals with unknown failure distributions for the problem of minimizing the expected replacement costs were also quite intensively investigated, which can be found in the articles [18, 22, 27]. Next, [7] compares an age-related exchange policy with exchange policy at strictly defined periods. The articles [11] and [3] analyze the asymptotic operating costs assuming the replacement policy according to the age of the spare part. A very extensive bibliography on the problem of optimization of maintenance costs was presented by [30]. In the review article [44] an up-to-date summary of the state of knowledge in this field and perspectives for further research directions can be found. It should be emphasized that the problem of the replacement policy as a result of aging is also valid today, as evidenced by publications from recent years [14, 19, 24, 39, 45]. As mentioned before, one problem in the practical use of the methods developed so far is focusing attention on a single spare part or spare part. This chapter presents a global approach based on a conference publication [9]. In the following sections analytical methods will be combined with discrete optimization methods reducing the problem to a two- or one-dimensional knapsack problem.

The rest of this chapter is organized as follows. Section 2 provides a standardized description of the analytical theory for determining the optimal life of non-repairable components. The formulas contained therein are used to calculate the relative cost of operation per unit of time, assuming the knowledge of the failure probability distribution. Section 3 presents methods of determining parameters for specific spare parts based on their failure history and costs generated during their replacement. This section also presents IT tools that can be used to perform the necessary calculations. Finally, Sect. 4 presents the global problem of cost optimization during component replacement. This section also presents the reduction of said problem to a two-dimensional knapsack problem.

2 Minimal Operating Cost for Single Spare Part

In this section, an analytical approach to optimizing the cost of ownership of a single spare part will be presented. Most of the concepts presented here are well established in the area of reliability. However, we are introducing them in a formal way to clarify the approach and unify the optimization problem.

Let C_F be the cost of removing the failure of a given spare part, and C_M the cost of the scheduled service replacement. Moreover, let $a = C_F/C_M$. In addition, it is assumed that $C_F = C_M + \tau C_D$, where C_D is the downtime cost per unit of time and τ is the downtime required during a failure. It follows from the assumptions that the coefficient $a > 1$. It is then assumed that there will be regular maintenance replacements for this item at exactly T units of time since the previous replacement. If $F(t)$ is the failure probability at time t, and $G(t) = 1 - F(t)$ is the reliability at time t, then the expected cost of service activities for the given replacement time T is estimated as

$$C_T = C_F F(T) + C_M G(T). \tag{1}$$

In the formula above, the cost C_T is simply the weighted average cost of failure C_F that occurs with probability $F(T)$ and service cost C_M that occurs with probability $G(T) = 1 - F(T)$. If we consider that $C_F = a \cdot C_M$, we get the following equality

$$C_T = C_M \big(1 + (a - 1)F(T)\big). \tag{2}$$

The calculated cost C_T is not enough. It does not yet take into account the frequency with which such changes will be made. There may be a situation in which the average cost of C_T is small, but the frequency with which it will be borne is so high that it is not an optimal solution. So let μ_T denote the mean time between replacements of a given spare part, both due to the failure and the planned service replacement. Then the value μ_T^{-1} determines the frequency of replacement of the spare part and allows to define the ε_T function which expresses the relative cost of maintaining the efficiency of the spare part per unit of time

$$\varepsilon_T = \frac{C_T}{C_M} \cdot \mu_T^{-1}. \tag{3}$$

Minimizing the ε_T function is our main task. Before we get down to that, it is necessary to define the formula for the expected replacement time μ_T. It is given by the following formula [31]

$$\mu_T = \int_0^T t \, dF(t) + TG(T) = \int_0^T G(t) \, dt. \tag{4}$$

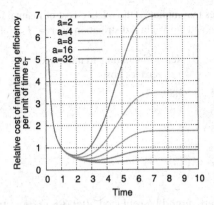

Fig. 2. The functions of the relative cost of maintaining efficiency in the time unit ε_T for example values of the a parameter and the Weibull distribution with the parameters $\alpha = \beta = 5$ [9].

Thus, the ε_T function defining the relative cost of maintaining efficiency per unit of time is expressed as follows

$$\varepsilon_T = \frac{1 + (a - 1)F(T)}{\displaystyle\int_0^T G(t)\,\mathrm{d}t}. \tag{5}$$

Nakagawa [30] showed that T minimizing the value of ε_T meets one of the following conditions:

1. is finite and is given uniquely,
2. is infinite.

It should be noted that in the absence of a finite value of T, all service replacements are unprofitable. In such cases, the only economically sensible strategy is to wait for a failure. It should be noted that the integral appearing in the formula (5) may not be determinable by analytical methods. Therefore, there are no general formulas for the minimum value of the ε_T function. In practice, it means that the minimum value of this function must be numerically computed for almost every distribution.

   Now let's see what the example plot of the ε_T function looks like for different values of the a parameter. The introduced concepts will be illustrated on an exemplary Weibull distribution. The Fig. 2 shows the graphs of the ε_T function for a few example values of a and the Weibull failure probability

$$F(t) = 1 - e^{-(t/\alpha)^\beta}, \tag{6}$$

with the following parameters: $\alpha = \beta = 5$. It can be seen from these graphs that the function of the relative cost of maintaining efficiency takes its minimum the sooner the higher the value of the a coefficient. This is in line with the intuition that the higher the cost of failure compared to the cost of service replacement, the shorter the time should be the component used. If the cost of removing the failure is twice as much as the cost

Table 1. Replacement times that minimize the relative cost of maintaining the device in service for the specified values of a and the Weibull distribution of $\alpha = \beta = 5$ [9].

a	α	β	Replacement time to minimize maintenance costs
2	5	5	3.80
4	5	5	3.05
8	5	5	2.57
16	5	5	2.21
32	5	5	1.91

of servicing, the spare part should be replaced after 3.8 units of time. When the cost of removing a failure exceeds four times the cost of regular service, the spare part should be replaced after 3.05 time units. A summary of all values for the above-mentioned graph is given in the Table 1.

3   Determining the Optimal Cost of a Single Spare Part

In order to determine the optimal service life of each spare part, it is necessary to have a basic set of data about them. The first is historical data on the service life of a given spare part. In other words, these are the operating times from the moment of assembly

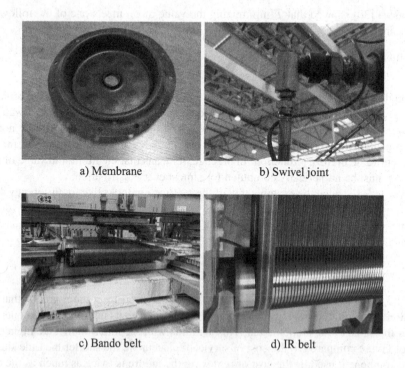

a) Membrane   b) Swivel joint

c) Bando belt   d) IR belt

Fig. 3. Industrial spare parts used as examples for maintenance cost optimisation [9].

Table 2. Uptime in days for each industrial spare part [9].

Spare part	Uptime (days)
Membrane	1, 40, 45, 16, 6, 29, 6, 76, 14, 34, 58, 91, 49, 42, 156, 28, 35, 88, 96, 106, 13, 17
Swivel joint	72, 208, 51, 153, 245, 119, 41, 692, 79, 235, 99, 342
Bando belt	14, 55, 107, 44, 146, 63, 18, 78, 42, 41, 40, 28, 59, 7, 106, 48, 24, 66, 23, 50, 3, 67, 7, 43, 72, 1, 75, 37, 27, 49, 34, 49, 6, 74, 91
IR belt	130, 373, 172, 178, 272, 138, 303, 85

to their failure. The second important information is the cost of service replacement, line downtime costs and the time needed to replace a given spare part. The calculation of the necessary parameters will be presented in this section in detail using four real examples of spare parts in operation at the factory. For the purposes of this chapter, certain parts have been selected. They are: membrane, swivel joint, bando belt and IR belt. This parts are shown in Fig. 3. The data on the failure-free operation time of individual spare parts can be found in the Table 2. They were used to determine the parameters for the Weibull distribution using the variable linearization method and the least squares method [25, 26]. The approximation results were illustrated with probability graphs for each of the analyzed spare parts. The `reliability` library available in Python was used to calculate and visualize the results. Listing 1.1 shows sample code for setting parameters and visualizing the fit.

```python
# Import libraries.
from reliability.Distributions import Weibull_Distribution
from reliability.Fitters import Fit_Weibull_2P
import matplotlib.pyplot as plt

# Set uptime data.
data = [85,130,138,172,178,272,303,373]
# Prepare Probability Plot using least
# square method. It also determine Weibull
# distribution parameters.
fit = Fit_Weibull_2P(failures=data, method='LS', linewidth=2.5)

# To get details about determined distribution
# use properties and functions of fit.distribution
# object. It is possible to get both plots and
# numeric data.

# ... Add plot options and show ...
plt.show()
```

Listing 1.1. The program that determines the best fit for the Weibull distribution.

For more information on using Python libraries to analyze and parameterize the Weibull distribution, see the article [8]. The obtained results will be the basis for determining the optimal operating times for each part. They can be considered as the first level to support the parts replacement management process.

It should be noted that the choice of the Weibull distribution as a distribution modeling the life cycle of spare parts used in industry is not accidental. Since the introduction [42], this distribution is widely used in numerous industrial issues: strength of glass [23], pitting corrosion [37], adhesive wear of metals [34], failure rate of shells [2], failure frequency of brittle materials [16], failure rate of composite materials [33], wear of concrete elements [28], fatigue life of aluminum alloys [21], fatigue life of Al-Si castings [1], strength of polyethylene terephthalate fibers [43], failure of joints under shear [4] or viscosity-temperature relation for damping fluids [10].

On the basis of financial data, the optimal times for replacing individual spare parts were also determined. Analytical determination of optimal replacement times is generally impossible. This is due to the fact that the integral appearing in the formula (5) is usually indeterminate. Nevertheless, the numerical determination of the minimum value of ε_T poses no major problems. For the purposes of this chapter, a simple spreadsheet was created that allows to determine the searched the value. This worksheet is available from the public https://github.com/achmie/maintenance repository. This is of course only an overview tool. For the needs of the company, it is worth creating the appropriate software connected with the database (Fig. 4).

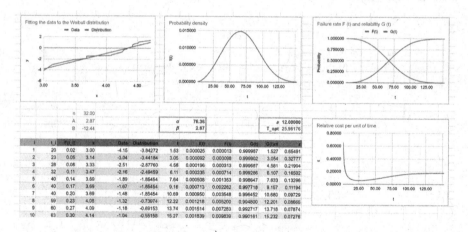

Fig. 4. A screenshot of the tool used to determine the optimal time of spare part replacement.

Although the determined times optimize the operating costs of each part separately, they do not allow to look at the problem from a global point of view – there are hundreds of such elements functioning in a manufacturing company. Therefore, in the next Section the use of rectangular knapsack problem algorithms to support the management process of maintenance costs will be discussed.

3.1 Real Industrial Spare Parts

Membrane. The membrane is one of the key components of the pneumatic valve actuator, which is used to create a vacuum in an electric press. The press and the valve mounted in it operate in cycles of approximately 20 s. During this time, the membrane is subjected to forces causing its systematic wear. A rupture in the membrane causes the production line to stop completely for $\tau_1 = 4$ h. The cost of replacing an spare part is $C_{M,1} = 200$ EUR, and the cost of stopping a line is approximately $C_{D,1} = 650$ EUR h. Therefore, the cost of an emergency replacement of this spare part should be estimated at $C_{F,1} = C_{M,1} + \tau_1 \, C_{D,1} = 2\,800$ EUR. This means the a ratio for the membrane is 14.0.

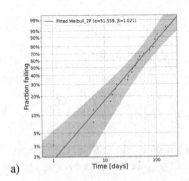
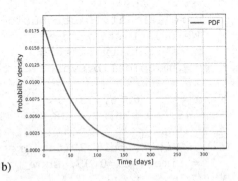

a)   b)

Fig. 5. Membrane Weibull probability plot determined by coordinate linearization and less square method and Weibull PDF for $\alpha = 51.559$ and $\beta = 1.021$.

Using the data on the failure rate of the spare part, it is possible to estimate the parameters of the distribution. The Fig. 5(a) shows the method of determining the parameters using the least squares method. The parameters obtained by this way have the value $\alpha_1 = 51.559$ and $\beta_1 = 1.021$. The code presented in Listing 1.2 is used to generate a plot showing the best fit using the least squares method.

```
1 # Import libraries.
2 from reliability.Distributions import
      Weibull_Distribution
3 from reliability.Fitters import Fit_Weibull_2P
4 import matplotlib.pyplot as plt
5 # Times of failure-free operation period.
6 data = [1, 6, 6, 13, 14, 16, 17, 28, 29, 34, 35, 40, 42,
      45, 49, 58, 76, 88, 91, 96, 106, 156]
7 # Finding best fit using LS (Least Squares) method.
8 fit = Fit_Weibull_2P(failures=data, method='LS', color='
      #CC2529', linewidth=2.5)
9 # Save plot in SVG format.
10 plt.title("")
```

```
11 plt.xticks(size=16)
12 plt.yticks(size=16)
13 plt.ylabel("Fraction failing", fontsize=24)
14 plt.xlabel("Time [days]", fontsize=24)
15 plt.legend(fontsize=16)
16 plt.savefig('MEM_Wb_PP.svg', format='svg', pad_inches
      =0.1, bbox_inches="tight")
```

Listing 1.2. The program that determines fit of Weibull parameters for membrane.

The Weibull distribution density plot for such parameters is presented in Fig. 5(b). The code presented in Listing 1.3 is used to generate the plot.

```
1 # Import libraries.
2 from reliability.Distributions import
      Weibull_Distribution
3 from reliability.Fitters import Fit_Weibull_2P
4 import matplotlib.pyplot as plt
5 # Times of failure-free operation period.
6 data = [1, 6, 6, 13, 14, 16, 17, 28, 29, 34, 35, 40, 42,
      45, 49, 58, 76, 88, 91, 96, 106, 156]
7 # Finding best fit using LS (Least Squares) method.
8 fit = Fit_Weibull_2P(failures=data, method='LS', CI_type
      ='time', CI=0.95, show_probability_plot=False)
9 # Finding probability density function.
10 fit.distribution.PDF(label='PDF', color='#CC2529',
      linewidth=2.5)
11 # Save plot in SVG format.
12 plt.grid()
13 plt.title("")
14 plt.xticks(size=8)
15 plt.yticks(size=8)
16 plt.ylabel("Probability density", fontsize=12)
17 plt.xlabel("Time [days]", fontsize=12)
18 plt.legend(fontsize=12, loc='upper right')
19 plt.savefig('MEM_Wb_PDF.svg', format='svg', pad_inches
      =0.1, bbox_inches="tight")
```

Listing 1.3. The program that determines probability density function for membrane.

Determining the optimal replacement time as a value that minimizes the ε_T function given by the Eq. (5), we get $T_1 = 980$ days. Observing the historical data, it can be concluded that in practice it means replacement only in the event of a failure. This is due to the fact that the β parameter value for the membrane is very close to 1, and for this value, preventive replacements are unprofitable at all [30].

Swivel Joint on an Industrial Robot. The swivel joint is an spare part installed on the head of an industrial robot that applies a seal to the car window. It connects the material

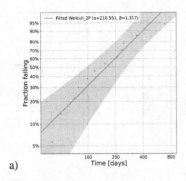
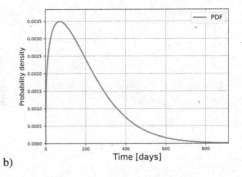

a)

b)

Fig. 6. Swivel joint Weibull probability plot determined by coordinate linearization and less square method and Weibull PDF for $\alpha = 210.551$ and $\beta = 1.317$.

feeding hose (polyurethane mass) with the movable robot head. The tool on which the swivel is installed travels around the three edges of the glass approximately every 70 s. During this time, the joint rotates 270° and returns to the zero position at the end of the robot cycle. Swivel joint wear occurs as a result of repeated rotation, and its sign is loss of tightness or blockage of movement. Damage of the swivel causes the production line to stop completely for $\tau_2 = 1$ h. The cost of replacing it is $C_{M,2} = 60$ EUR, and the cost of stopping the line is the same as membrane, approximately $C_{D,2} = 650$ EUR h. Therefore, the cost of an emergency replacement for this item should be estimated at $C_{F,2} = C_{M,2} + \tau_2\, C_{D,2} = 710$ EUR. This means the a factor for a swivel joint is 11.833.

Using the data on the failure rate of the swivel joint, it is possible to estimate the parameters of the distribution. The Fig. 6(a) shows the process of determining the parameters using the least squares method. The parameters calculated with this method have the value $\alpha_2 = 210.551$ and $\beta_2 = 1.317$. The Weibull distribution density plot for the determined parameters is presented in Fig. 6(b). Determining the optimal replacement time for this item, we get $T_2 = 87$ days. After this time, the swivel joint should therefore be serviced.

Bando Belt. Bando belt is an spare part that transports glass waste after the breaking operation. It works in a fast cycle, every 16 s, starting and braking with great torque. The belt works at ambient temperature and is exposed to high tensile forces. The symptom of failure is partial or complete breaking of the belt. Damage to the Bando belt causes a complete production line stop at $\tau_3 = 4$ h. The cost of replacing it is $C_{M,3} = 1\,000$ EUR, and the cost of stopping the line is the same as the membrane and the swivel joint, approximately $C_{D,3} = 650$ EUR h. Therefore, the cost of an emergency replacement of this spare part should be estimated at $C_{F,3} = C_{M,3} + \tau_3\, C_{D,3} = 3\,600$ EUR. This means that the a ratio for a Bando belt is 3.6.

Using the Bando belt failure rate data it is possible to estimate distribution parameters. The Fig. 7(a) shows the process of determining the parameters using the least

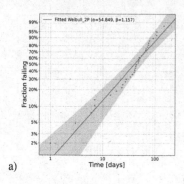
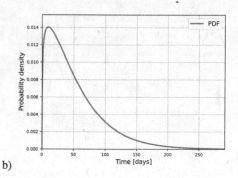

a)   b)

Fig. 7. Bando belt Weibull probability plot determined by coordinate linearization and less square method and Weibull PDF for $\alpha = 54.849$ and $\beta = 1.157$.

squares method. The parameters obtained with this method have the value $\alpha_3 = 54.849$ and $\beta_3 = 1.157$. The Weibull distribution density plot for the determined parameters is presented in Fig. 7(b). Determining the optimal replacement time for this item, we get $T_3 = 237$ days. After this time, the Bando belt should be serviced.

IR Belt. The IR belt is a transport mesh that transports the glass through the paint drying zone after the screen printing process. The belt works with constant dynamics (speed and acceleration), but in an environment of increased temperature - 180 °C. Belt failure usually manifests as belt stretching beyond the adjustment value (critical length) or breaking completely. Damage to the IR belt causes the production line to stop completely for $\tau_4 = 8$ h. The cost of replacing it is $C_{M,4} = 5\ 200$ EUR, and the cost of stopping the line is the same as for all other spare parts, and is approximately $C_{D,4} = 650$ EUR h. Therefore, the cost of an emergency replacement of this spare part should be estimated at $C_{F,4} = C_{M,4} + \tau_4\ C_{D,4} = 10\ 400$ EUR. This means that the a ratio for a Bando belt is 2.0.

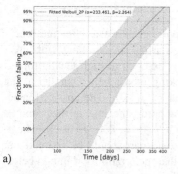
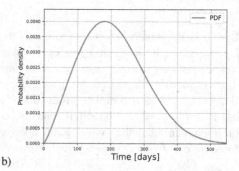

a)   b)

Fig. 8. IR belt Weibull probability plot determined by coordinate linearization and less square method and Weibull PDF for $\alpha = 233.461$ and $\beta = 2.264$.

Using the IR belt failure rate data it is possible to estimate distribution parameters. The Fig. 8(a) shows the process of determining the parameters using the least squares method. The parameters obtained by this method have the value $\alpha_4 = 233.461$ and $\beta_4 = 2.264$. The Weibull distribution density plot for the determined parameters is presented in Fig. 8(b). Determining the optimal replacement time for this item, we get $T_4 = 222$ days. After this time, service replacement of the IR belt should take place.

3.2 Spare Part Controlled by Sensors and AI System

Note that the real examples have one quite important feature – they all have a relatively low value of the β parameter. In practice, this means that the error made when determining the average lifetime of a given spare part is relatively large. This results in the ε_T function having a virtually imperceptible minimum. This means that for some components, the optimization process may not have any significant impact on cost reduction. The only way to change this is to use additional information from sensors supported by an artificial intelligence and machine learning system.

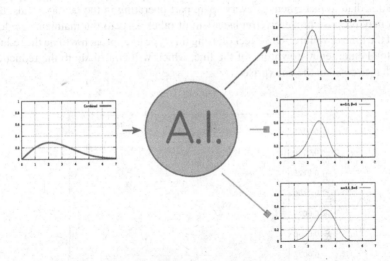

Fig. 9. Diagram of an artificial intelligence system selecting from a cumulative distribution a case corresponding to a given spare part based on data from sensors.

Imagine you are dealing with a situation that is illustrated in Fig. 9. The analysis covers an spare part whose failure probability distribution resulting from the historical analysis is strongly flattened. The risk of erroneous estimation of the expected operating time of such an spare part is very high. Nevertheless, during its operation, various types of signals related to the correct operation of the spare part are measured – e.g. vibrations. At this point, the machine learning system may try to classify the spare part and refine the failure probability distribution characterizing that spare part.

Initially, we treat an spare part as one of many belonging to the same group. We assign it parameters of the failure probability distribution in accordance with the collected historical data. Nevertheless, during its operation, we begin to compare the signal provided by the sensors with the signals stored in the database. After catching certain regularities, we are able to limit the behavior of an spare part to a specific subset. In this way, we reduce the uncertainty of determining its properties, and the failure probability distribution for this spare part becomes more cumulative (if it is the Weibull distribution, then β becomes larger and larger).

4   Implementation of the Spare Part Cost Optimization Algorithm in MES

The examples presented in the previous section clearly show that the optimal replacement times vary greatly between the different parts, as it is shown in Fig. 10. Therefore, it becomes practically impossible to optimize the replacement time for all spare parts, which may be even several hundred in the enterprise. However, the question arises whether the downtime caused by a failure cannot be used to replace other parts as well, and not only the damaged one. Computer aided calculations can estimate the profitability of immediate replacement of every spare part operating in the factory. Thus, the IT system may propose preventive replacement of other parts to the maintenance department at the time of failure. The effect of using this type of approach will be the reduction of the total emergency downtime of the line, which will contribute to the reduction of maintenance costs in the enterprise.

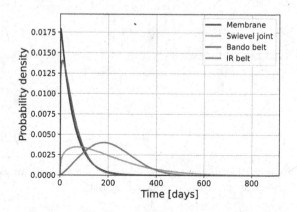

Fig. 10. Common Weibull PDF plot for all presented parts [9].

The presented proposal is a definitely new approach to the problem of optimization of operating costs in an enterprise. The classical methods are limited to the use of analytical methods of the probability theory and reliability theory. This article presents the global optimization method, which combines analytical and discrete methods. The cost reduction method presented here is one of many possible approaches to optimizing the performance of maintenance teams. However, it fits perfectly into modern trends in predictive maintenance for the needs of large-scale production of automotive industry companies. It also opens up wide possibilities of using machine learning mechanisms to better estimate the failure rate function $F(t)$ and to determine global minima for the operating cost function.

In order to create an algorithm to support the decision-making process in the maintenance system, it is necessary to define a common measure that determines the condition of the production line. Such a measure will be the expected cost of keeping the line operational from now until the next planned downtime. Suppose the line uses n spares $U_1, U_2, \ldots, U_n$. Each of the U_i parts has a set of parameters that define it. It can therefore be said that

$$U_i = (t_i, \alpha_i, \beta_i, F_i, G_i, \mu_i(t), C_{M,i}, C_{D,i}, \tau_i, \sigma_i), \tag{7}$$

where t_i is the current time of failure-free operation of the spare part, α_i, β_i are the Weibull distribution parameters defining the failure rate of F_i and the reliability of G_i, $\mu_i(t)$ is the expected uptime of the spare part after working t time units, $C_{M,i}$ is the cost of component replacement during planned downtime, $C_{D,i}$ is the cost of production line downtime, τ_i is the downtime required to replace the component, and σ_i is the part of the human resources that is used during the replacement of a given item. Below, in Listing 1.4, an example implementation of the U_i structure is presented. The code aims to show what the spare part is from the optimization process point of view.

```
1  # Import mathematical libraries.
2  import math
3  import numpy as np
4  import scipy.integrate as integrate
5  # Import reliability libraries.
6  from reliability.Distributions import
       Weibull_Distribution
7  from reliability.Fitters import Fit_Weibull_2P
8  # Import plot libraries.
9  import matplotlib.pyplot as plt
10
11 # Spare part class.
12 class SparePart:
13    def __init__(self, name, imageFile = '', workTime = 0,
          maintenanceCost = 0, downtimeCost = 0,
          replacementTime = 0, failedTimes = [], runningTimes
          = []):
14       # Name of the spare part.
15       self.name = name
16       # File with the spare part image.
17       self.imageFileName = imageFile
18       # Spare part work time.
19       self.workTime = workTime
20       # Spare part maintenance cost (cost of the planned
          replacement).
```

```
21    self.cM = maintenanceCost
22    # Production line downtime cost per hour.
23    self.cD = downtimeCost
24    # Spare part replacement time (from line failure to
      line start).
25    self.rT = replacementTime
26    # Spare part total failure cost (maintenance +
      downtime costs).
27    self.cF = self.cM + self.rT*self.cD
28    # Spare part failure history (working time from
      beginning to failure).
29    self.failedTimes = failedTimes
30    # Spare part replacement history (working time from
      beginning to replacement).
31    self.runningTimes = runningTimes
32    # Determination of Weibull parameters based on spare
      failure and replacement part history.
33    if len(self.failedTimes) > 1:
34      self.fit = Fit_Weibull_2P(
35        failures = self.failedTimes,
36        right_censored = self.runningTimes,
37        method = 'LS',
38        CI_type = 'time',
39        CI = 0.95,
40        print_results = False,
41        show_probability_plot = False
42      )
43    # Weibull alpha parameter.
44    self.a = self.fit.alpha
45    # Weibull beta parameter.
46    self.b = self.fit.beta
47    # Expected lifetime for the spare part from
      beginning.
48    self.u0 = self.u()
49
50  # Conditional reliability function (take into
    consideration workTime).
51  def R(self, t):
52    return math.exp(-math.pow((self.workTime + t) / self
      .a, self.b) + math.pow(self.workTime / self.a, self.
      b))
53
54  # Conditional failure function (take into
    consideration workTime).
55  def F(self, t):
```

```
56      return 1 - self.R(t)

57

58   # Conditional hazard function (take into consideration
         workTime).
59   def h(self, t):
60      return (self.b / self.a) * math.pow((self.workTime +
         t) / self.a, self.b - 1)

61

62   # Expected lifetime.
63   def u(self, t = float('NaN')):
64      def absoluteR(t):
65         return math.exp(-math.pow(t / self.a, self.b))
66      if np.isnan(t):
67         t = self.workTime
68      return integrate.quad(absoluteR, t, np.inf)[0] /
         absoluteR(t)
```

Listing 1.4. Example implementation of spare part structure U_i.

If it is assumed that the next scheduled stoppage will take place for T units of time, then the expected cost of operating a part of U_i at that time is

$$C(U_i, T) = \bar{k}_i C_{F,i} + G_i(t'_i)C_{M,i}, \tag{8}$$

where $C_{F,i} = C_{M,i} + \tau_i C_{D,i}$ is the cost of an emergency replacement part, $\bar{k}_i = 1 + k_i + F_i(t'_i)$ is the expected number of failures in time to the scheduled downtime, and $G_i(t'_i)$ is the probability of failure-free operation of the part from the time of the last forecast failure to the scheduled downtime. The expected number of failures can be obtained by assuming the following values for k_i and t'_i

$$k_i = \left\lfloor \frac{T - \mu_i(t_i)}{\mu_i(0)} \right\rfloor, t'_i = \min\left(T - \mu_i(t_i) - k_i\mu_i(0), T\right). \tag{9}$$

Note that if $T < \mu_i(t_i)$, then $k_i = -1$ and the expected number of failures is $\bar{k}_i = F_i(T)$. The interpretation of k_i and t'_i is shown in Fig. 11.

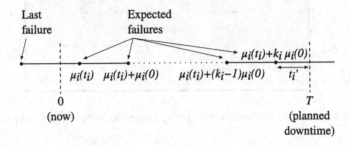

Fig. 11. Expected failures in the case of not carry out preventive replacement of the spare part [9].

Additionally, we define the alternative expected cost of operating the U_i part

$$C'(U_i, T) = \bar{s}_i C_{F,i} + (1 + G_i(t_i''))C_{M,i}, \tag{10}$$

for which $\bar{s}_i = s_i + F_i(t_i'')$, $s_i = \lfloor T/\mu_i(0) \rfloor$ and $t_i'' = T - s_i\mu_i(0)$. This function expresses the expected alternative costs that will be incurred in the event that part of U_i has been replaced at present on the occasion of downtime due to failure of another part. The interpretation of s_i and t_i'' is shown in Fig. 12.

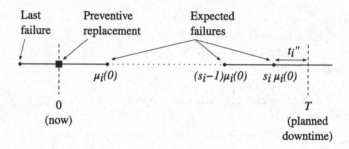

Fig. 12. Expected failures in the event of a preventive replacement of an spare part [9].

For the functions C and C' defined in this way, we create a sequence of values $V_i = C(U_i, T) - C'(U_i, T)$, which express the expected profit on the replacement of U_i during the emergency downtime. If we assume that the part with index j failed, then the optimization problem can be reduced to a two-dimensional knapsack problem. It should be assumed that the knapsack has the shape of a rectangle with a width of τ_j and a height of $1 - \sigma_j$, and the spare parts arranged in it are rectangles of width τ_i, height σ_i and values V_i, where $i \neq j$. It should also be emphasized that the spare parts placed in the knapsack cannot be twisted, as time and human resources are not interchangeable. Figure 13 illustrates the problem described above.

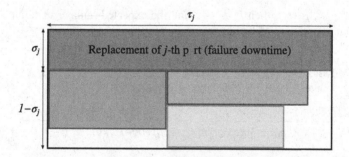

Fig. 13. Optimum use of downtime for the preventive replacement of additional parts [9].

The literature devoted to this problem is very rich. The best review article to date is [32]. During a systematic review of bibliographic items, the authors distinguished four basic approaches to the construction of algorithms for solving the two-dimensional rectangular knapsack problem: approximation, exact, heuristic and hybrid. The article presents a very in-depth bibliometric analysis indicating which publications have the most significant impact on the development of algorithms solving this problem. The review includes over one hundred publications issued by 2020. The intensity of research in this area is also evidenced by numerous publications issued in the last two years: [15,29,41,46] and [40].

Finally, it is worth considering a certain simplification of the optimization problem posed. From a practical point of view, it is often the case that a manufacturing company divides maintenance workers into groups. Teams formed in this way are usually the same number and for safety reasons their division is not foreseen. If, for example, there are three such units in a factory, the overall two-dimensional knapsack problem can be reduced as shown in Fig. 14. As a result of such reduction, optimization is reduced to a one-dimensional multi-knapsack problem, which is an issue with lower computational complexity. If we assume that the available time period τ_j can be quantized with an accuracy of, for example, 5 min, we get a fairly reasonable framework for creating a dedicated optimization algorithm.

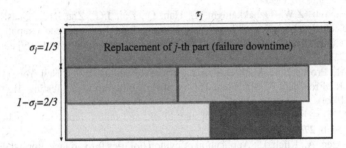

Fig. 14. Easy case of optimum use of downtime for the preventive replacement of additional parts.

5   Conclusions

The chapter is an extension of the conference article [9]. Its main goal is to approach the previously presented topic from the computational and implementation side. The included code snippets are intended to illustrate how to go from a purely theoretical description to a practical implementation.

We believe that the MES (Manufacturing Execution System) equipped with this type of algorithms can keep the maintenance manager informed about the probability of failure of a given spare part and the expected costs of its removal. It provides the current knowledge of what events should be prepared for and what to expect in the near future. On the other hand, the Eqs. (8) and (10) introduced in the fourth part constitute the basis for estimating the actual value of the service activity.

Particular attention should be paid to the simplifications of the knapsack problem suggested at the end of the fourth section. They result from a practical approach to the problem of removing failures in a manufacturing company. Nevertheless, they reduce the overall two-dimensional knapsack problem to the one-dimensional multi-knapsack problem.

It can be said that bringing the presented method to the production stage requires solving two fundamental problems. The first is such a simplification of the optimization problem that will allow the operator of the MES system to indicate possible steps in the event of a failure within a dozen or so seconds. On the other hand, the second issue is the way of using the methods of artificial intelligence and machine learning to modify the failure probability distributions.

References

1. Aigner, R., Leitner, M., Stoschka, M.: Fatigue strength characterization of AL-Si cast material incorporating statistical size effect. In: Henaff, G. (ed.) 12th International Fatigue Congress (FATIGUE 2018). MATEC Web of Conferences, vol. 165 (2018). https://doi.org/10.1051/matecconf/201816514002
2. Almeida, J.: Application of Weilbull statistics to the failure of coatings. J. Mater. Process. Technol. **93**, 257–263 (1999)
3. Ansell, J., Bendell, A., Humble, S.: Age replacement under alternative cost criteria. Manag. Sci. **30**(3), 358–367 (1984)
4. Bai, Y.L., Yan, Z.W., Ozbakkaloglu, T., Han, Q., Dai, J.G., Zhu, D.J.: Quasi-static and dynamic tensile properties of large-rupture-strain (LRS) polyethylene terephthalate fiber bundle. Constr. Build. Mater. **232**, 117241 (2020). https://doi.org/10.1016/j.conbuildmat.2019.117241
5. Barlow, R., Proschan, F.: Mathematical Theory of Reliability. Wiley, New York (1965)
6. Barlow, R., Proschan, F.: Statistical Theory of Reliability and Life Testing. Holt Rinehart, Austin (1975)
7. Berg, M., Epstein, B.: Comparison of age, block, and failure replacement policies. IEEE Trans. Reliab. **27**(1), 25–29 (1978)
8. Chmielowiec, A., Klich, L.: Application of python libraries for variance, normal distribution and Weibull distribution analysis in diagnosing and operating production systems. Diagnostyka **22**(4), 89–105 (2021). https://doi.org/10.29354/diag/144479
9. Chmielowiec, A., Klich, L., Woś, W., Błachowicz, A.: Global spare parts exploitation costs optymization and its reduction to rectangular knapsack problem. In: Proceedings of the 24th International Conference on Enterprise Information Systems - Volume 1: ICEIS, pp. 364–373. INSTICC, SciTePress (2022). https://doi.org/10.5220/0011067400003179
10. Chmielowiec, A., Woś, W., Gumieniak, J.: Viscosity approximation of PDMS using Weibull function. Materials **14**(20), 6060 (2021). https://doi.org/10.3390/ma14206060
11. Christer, A.: Refined asymptotic costs for renewal reward processes. J. Oper. Res. Soc. **29**(6), 577–583 (1978)
12. Cléroux, R., Dubuc, S., Tilquin, C.: The age replacement problem with minimal repair and random repair costs. Oper. Res. **27**(6), 1158–1167 (1979)
13. Cléroux, R., Hanscom, M.: Age replacement with adjustment and depreciation costs and interest charge. Technometrics **16**(2), 235–239 (1974)
14. Finkelstein, M., Cha, J., Levitin, G.: On a new age-replacement policy for items with observed stochastic degradation. Qual. Reliab. Eng. Int. **36**(3), 1132–1143 (2020). https://doi.org/10.1002/qre.2619
15. Fırat, H., Alpaslan, N.: An effective approach to the two-dimensional rectangular packing problem in the manufacturing industry. Comput. Ind. Eng. **148**, 106687 (2020). https://doi.org/10.1016/j.cie.2020.106687

16. Fok, S., Mitchell, B., Smart, J., Marsden, B.: A numerical study on the application of the Weibull theory to brittle materials. Eng. Fract. Mech. **68**, 1171–1179 (2001)
17. Fox, B.: Letter to the editor - age replacement with discounting. Oper. Res. **14**(3), 533–537 (1966)
18. Frees, E., Ruppert, D.: Sequential nonparametric age replacement policies. Ann. Stat. **13**(2), 650–662 (1985)
19. Fu, Y., Yuan, T., Zhu, X.: Optimum periodic component reallocation and system replacement maintenance. IEEE Trans. Reliab. **68**(2), 753–763 (2018). https://doi.org/10.1109/TR.2018. 2874187
20. Glasser, G.: The age replacement problem. Technometrics **9**(1), 83–91 (1967)
21. Hemphill, M., et al.: Fatigue behavior of $Al_0.5CoCrCuFeNi$ high entropy alloys. Acta Materialia **60**(16), 5723–5734 (2012). https://doi.org/10.1016/j.actamat.2012.06.046
22. Ingram, C., Scheaffer, R.: On consistent estimation of age replacement intervals. Technometrics **18**(2), 213–219 (1976)
23. Keshevan, K., Sargent, G., Conrad, H.: Statistical analysis of the Hertzian fracture of pyrex glass using the Weibull distribution function. J. Mater. Sci. **15**, 839–844 (1980)
24. Knopik, L., Migawa, K.: Optimal age-replacement policy for non-repairable technical objects with warranty. Eksploatacja i Niezawodnosc Maint. Reliab. **19**(2), 172–178 (2017). https://doi.org/10.17531/ein.2017.2.4
25. Lai, C.: Generalized Weibull Distributions. Springer, Heidelberg (2014). https://doi.org/10. 1007/978-3-642-39106-4
26. Lai, C., Murthy, D., Xie, M.: Weibull distributions and their applications. In: Pham, H. (ed.) Springer Handbook of Engineering Statistics. SHB, pp. 63–78. Springer, London (2006). https://doi.org/10.1007/978-1-84628-288-1_3
27. Léger, C., Cléroux, R.: Nonparametric age replacement: bootstrap confidence intervals for the optimal cost. Oper. Res. **40**(6), 1062–1073 (1992)
28. Li, Q., Fang, J., Liu, D., Tang, J.: Failure probability prediction of concrete components. Cem. Concr. Res. **33**, 1631–1636 (2003)
29. Li, Y.B., Sang, H.B., Xiong, X., Li, Y.R.: An improved adaptive genetic algorithm for two-dimensional rectangular packing problem. Appl. Sci. **11**(1), 413 (2021). https://doi.org/10. 3390/app11010413
30. Nakagawa, T.: Maintenance and optimum policy. In: Pham, H. (ed.) Handbook of Reliability Engineering, pp. 367–395. Springer, London (2003). https://doi.org/10.1007/1-85233-841-5_20
31. Nakagawa, T.: Statistical models on maintenance. In: Pham, H. (ed.) Springer Handbook of Engineering Statistics. SHB, pp. 835–848. Springer, London (2006). https://doi.org/10.1007/ 978-1-84628-288-1_46
32. Neuenfeldt, A., Jr., Silva, E., Francescatto, M., Rosa, C., Siluk, J.: The rectangular two-dimensional strip packing problem real-life practical constraints: a bibliometric overview. Comput. Oper. Res. **137**, 105521 (2022). https://doi.org/10.1016/j.cor.2021.105521
33. Newell, J., Kurzeja, T., Spence, M., Lynch, M.: Analysis of recoil compressive failure in high performance polymers using two-, four-parameter Weibull models. High Perform. Polym. **14**, 425–434 (2002)
34. Queeshi, F., Sheikh, A.: Probabilistic characterization of adhesive wear in metals. IEEE Trans. Reliab. **46**, 38–44 (1997)
35. Ran, A., Rosenlund, S.: Age replacement with discounting for a continuous maintenance cost model. Technometrics **18**(4), 459–465 (1976)
36. Scmeaffer, R.: Optimum age replacement policies with an Ikreasing cost factor. Technometrics **13**(1), 139–144 (1971)
37. Sheikh, A., Boah, J., Hansen, D.: Statistical modelling of pitting corrosion and pipeline reliability. Corrosion **46**, 190–196 (1990)

38. Subramanian, R., Wolff, M.: Age replacement in simple systems with increasing loss functions. IEEE Trans. Reliab. **25**(1), 32–34 (1976)
39. Wang, J., Qiu, Q., Wang, H.: Joint optimization of condition-based and age-based replacement policy and inventory policy for a two-unit series system. Reliab. Eng. Syst. Saf. **205**, 107251 (2021). https://doi.org/10.1016/j.ress.2020.107251
40. Wang, T., Hu, Q., Lim, A.: An exact algorithm for two-dimensional vector packing problem with volumetric weight and general costs. Eur. J. Oper. Res. (2021). https://doi.org/10.1016/j.ejor.2021.10.011
41. Wei, L., Lai, M., Lim, A., Hu, Q.: A branch-and-price algorithm for the two-dimensional vector packing problem. Eur. J. Oper. Res. **281**(1), 25–35 (2020). https://doi.org/10.1016/j.ejor.2019.08.024
42. Weibull, W.: A statistical theory of the strength of material. Ingeniors Vetenskaps Akademiens Handligar **151**, 5–45 (1939)
43. Xie, S., et al.: A statistical damage constitutive model considering whole joint shear deformation. Int. J. Damage Mech **29**(6), 988–1008 (2020). https://doi.org/10.1177/1056789519900778
44. Zhao, X., Al-Khalifa, K., Hamouda, A., Nakagawa, T.: Age replacement models: a summary with new perspectives and methods. Reliab. Eng. Syst. Saf. **161**, 95–105 (2017). https://doi.org/10.1016/j.ress.2017.01.011
45. Zhao, X., Li, B., Mizutani, S., Nakagawa, T.: A revisit of age-based replacement models with exponential failure distributions. IEEE Trans. Reliab. (2021). https://doi.org/10.1109/TR.2021.3111682
46. Zhou, S., Li, X., Zhang, K., Du, N.: Two-dimensional knapsack-block packing problem. Appl. Math. Model. **73**, 1–18 (2019). https://doi.org/10.1016/j.apm.2019.03.039

Quantitative Comparison of Translation by Transformers-Based Neural Network Models

Alexander Smirnov[1], Nikolay Teslya[1](✉) [ID], Nikolay Shilov[1] [ID], Diethard Frank[2], Elena Minina[2], and Martin Kovacs[2]

[1] SPIIRAS, SPC RAS, 14th line 39, St. Petersburg, Russia
{smir,teslya,nick}@iias.spb.su
[2] Festo SE & Co. KG, Ruiter Str. 82, Esslingen, Germany
{diethard.frank,elena.minina,martin.kovacs}@festo.com

Abstract. One of the major tasks in producer-customer relations is a processing of customers' feedback. Since for the international companies provide their products for a lot of countries the feedback could be provided with various languages. That can produce a number of difficulties during automatic text processing. One of the solutions is to use one common language to process feedback and automatically translate feedbacks from the various languages to the common one. Then the translated feedback could be processed with the configured pipeline. This paper compares existing open models for automatic text translation. Only language models with Transformer architecture were considered due to the best results of translation over other existing approaches. The models are: M2M100, mBART, OPUS-MT (Helsinki NLP). Own test data was built due to requirement of translation for texts specific to the subject area. To create dataset Microsoft Azure Translation was chosen as the reference translation with manual translation verification for grammar. Translations produced by each model were compared with the reference translation using two metrics: BLEU and METEOR. The possibility of fast fine-tuning of models was also investigated to improve the quality of address the translation on specific lexicon of the problem area. Among the reviewed models, M2M100 turned out to be the best in terms of translation quality, but it is also the most difficult to fine-tune it.

Keywords: Machine Translation · DNN Translation · Comparison · Training · Transformers · Fine-Tuning

1 Introduction

Machine learning has made a significant progress in models variety and data processing quality recently and has become an efficient tool for automation in various domains [1, 2]. It is used un the wide range of problem domains including natural language processing, and machine translation (MT, automatic translation of text from one natural language into another) in particular [3, 4].

Statistical and rule-based MT dominated for decades and mainly relies on various count-based models. Most novel models uses concept of neural machine translation

J. Filipe et al. (Eds.): ICEIS 2022, LNBIP 487, pp. 155–174, 2023.
https://doi.org/10.1007/978-3-031-39386-0_8

(NMT) based on the language modelling with a neural network [5]. With the increasing translation quality such models have been widely used both as third party services [6, 7] or internally hosted services for text translation.

Each of the solutions has some advantages and disadvantages. Third-party translation services provide enough computational power to run relatively heavy neural networks, but they are very limited with possibilities for fine-tuning of pre-deployed language models to specific terminology, which might be critical in certain scenarios. However infrastructure by developers of the custom services for neural machine translation hosting (either on own hardware or using MaaS (Machine-as-a-Service) could provide such possibilities for training but require significant financial investment to operate translation service.

Automated translation can be used by international industries to automate the processing of requests from customers. Such companies provide their service in many countries selling and supporting products. However, for the development of the industries in the context of digital business transformation, it is important to collect customer feedback and provide centralized analysis within a single NLP platform. It reduces load on the human operators and provide faster analysis and response to the feedback. To process messages in different languages within such platform, it is proposed to automatically translate them into common language using neural machine translation models. Due to the most models use English as a core language model it is proposed to also use it as a common language to translate customer feedback from any language.

The paper extends previous work by the authors [8] with detailed description of the dataset and more experiment results on models fine tuning and presents an analysis of different state-of-the-art neural machine translation models for texts from a specific domain as well as their abilities for fine tuning based on examples from specific terminology in the area of industrial automation.

The paper is structured as follows. Section 2 presents information about models, metrics and concepts related to machine translation and translation quality assessment. Section 3 contains a description of the experimental methodology, the prepared data set, and the results of the translation experiment. Section 4 reports the results of the experiment on fine-tuning models. Section 5 summarizes the results of the comparison with a discussion and concludes on the applicability of the considered models.

2 Related Works

This section provides an overview of the state-of-the-art research in the area of machine translation and related topics such as translation quality metrics and machine translation model fine-tuning.

The use of machine learning for automatic translation of texts is an actively developed area of research with significant progress during last decade. Early statistical models based on analysis of word usage frequency show a result that poorly reflects the meaning of the original phrase. However, the development of artificial neural networks has significantly improved the quality of translation. Over the past years, the architecture of RNN networks has been used most often for machine translation [9–11]. However, such networks are not very good in keeping track of the context, and the longer the text, the worse they do, even with using of LSTM [9].

The quality of translation has been significantly improved as a result of the development of the Transformers architecture [12–14]. In this architecture, the main focus is not on dependency tracking, but on the so-called attention mechanism in dependency calculation. Like RNN, the Transformers architecture is designed to transform one sequence into another, however, it does this not sequentially over the elements of the sequence, but over all the sequences at once.

Currently, most machine translation models are based on the Transformers architecture. For this work, the following models are considered: M2M100 [15], mBART [16], and OPUS-MT (Helsinki NLP) [17].

M2M100 is a multilingual encoder-decoder (seq-to-seq) model trained for Many-to-Many multilingual translation [15]. The model is provided in two types: with 1.2 bullion parameters and 418 million parameters. Both support 100 languages in total and allow for automatic translation between any pairs of languages.

Helsinki NLP is based on the MarianMT project that is essentially an add-on over the Marian C++ library for fast learning and translation. Using this library, a number of bilingual models have been trained by Helsinki University as well as two large models for translation from several languages into English and from English into several languages. List of supported languages contains 277 entries, excluding different alphabet for the same languages.

mBART is a sequence-to-sequence denoising auto-encoder pre-trained on large-scale monolingual corpora in many languages using the BART objective. mBART is one of the first methods for pre-training a complete sequence-to-sequence model by denoising full texts in multiple languages, while previous approaches have focused only on the encoder, decoder, or reconstructing parts of the text.

There are two versions of MBart: (1) mbart-large-cc25 (with support for 25 languages) and (2) mbart-large-50 (MBart-50, with support for 50 languages). MBart-50 is created using the original mbart-large-cc25 checkpoint by extending its embedding layers with randomly initialized vectors for an extra set of 25 language tokens and then pre-trained on 50 languages.

Fine-tuning of multilingual transformer models is a common practice. It is usually done to improve the models' performance in a given domain or for a given language. However, unlike, for example, fine-tuning of convolutional networks when most of the layers are remained unchanged, fine-tuning of transformer networks usually assumes model training without freezing any layers, e.g. [22, 23].

Though normally, fine-tuning is done using relatively large training datasets (thousands of samples), in this particular research we will consider fine-tuning on small datasets. The reason for that is the absence of such a dataset or lack of time for its creation, when one needs to fine-tune the model only for several specific terms. The results of this study are presented in Sect. 4.

Since machine translation is primarily aimed at translating massive volumes of texts, manual evaluation of their quality is time-consuming and basically impractical [18]. As a result, number of metrics have been developed for this purpose.

One of the most popular metrics is the BLEU score [19] based on the comparison of n-grams of the one or several ground truth translations and the candidate (evaluated) translation. It is a universal language-independent metric.

Another popular metric is METEOR [20]. It is also based on n-gram comparison, however unlike BLEU it takes into account language-specific information such as synonyms, paraphrasing, and stemmed forms of words. As a result, though in general the METEOR metric correlates with the BLEU metric, it is less strict and often closer to the human judgement. However, it is language-dependent and requires additional language resources for translation evaluation.

One more metric that is worth to mention is NIST [21]. It is also an n-gram based metric, however, it has a different n-gram scoring mechanism than the BLUE metric aimed for better correlation with human assessment. However, it has not got such a popularity as the previous two metrics.

Since the BLEU and METEOR metrics are the most common, they were used in this research for the translation quality evaluation.

3   Analysis Methodology

3.1   Dataset Preparation

The dataset was formed from 210,910 domain-specific texts presented in 32 languages, including English. Since the texts had contained some language independent items like e-mails, numbers, etc. the first step was cleaning of the dataset. The following items were removed:

1. Blank lines and extra spaces;
2. References to objects attached to the source text;
3. E-mails, records marked CONFIDENTIAL, and hyperlinks;
4. All lines shorter than 20 characters.
5. All numbers were replaced by 1

After cleaning, the dataset contains 147866 texts in 32 languages. The distribution of the lengths of all texts is shown in Fig. 1.

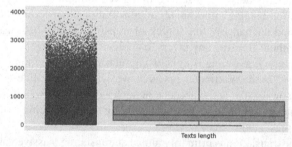

Fig. 1. Distribution of texts length in dataset before filtering [8].

3.2 Language Detection

Since the goal of the process is to translate texts into English, then only non-English texts should be left in the dataset. For this, four utilities were used to determine the text language: langdetect, langid, fasttext, and cld3. Each text was processed by all of four utilities (Fig. 2). Since the most of text samples has the same languages from all language detection tools, some of them has a conflict. The conflict here is a situation when at least one of the utilities gives a different result from the others. There were 24393 text samples in the dataset with conflict. For example, for some of the texts it was a situation, when the used utilities provide absolutely different results for detected languages (i.e., set('en', 'de', 'fr', 'nl'). In this case it is impossible to detect language, since only few models provide their confidence levels. If at least two models identify the same languages (i.e., set ('en','en', 'de', 'fr') then this language is chosen as the language text ('en' in this example). For the subsequent work, the rows of the dataset were selected, for which all four utilities determined the same language. Since translation into English requires text in a language other than English, additional filtering was carried out. This was done using a filter over the data frame to select rows that do not contain English language detected by the consensus of all four models.

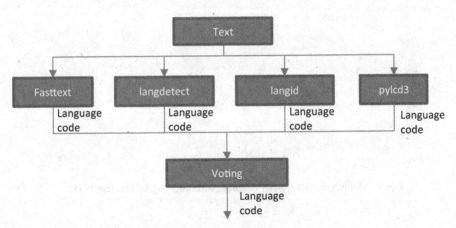

Fig. 2. Language detection procedure.

According to the results of the mask selection, 90219 texts remained. The distribution of texts by languages other than English is shown in Fig. 3.

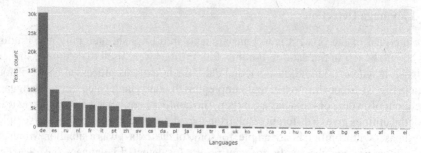

Fig. 3. Dataset distribution by language after filtering by language.

To improve the quality of the translation, it was decided to select texts over 500 characters long. This is justified by the need to choose the language for the text and to study the influence of the context on the accuracy of the translation. After filtering texts by length, the distribution by language turned out as follows (Fig. 4). Distribution by text length after filtering is shown in Fig. 5.

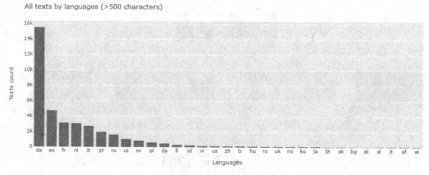

Fig. 4. Dataset distribution by language after filtering by text length [8].

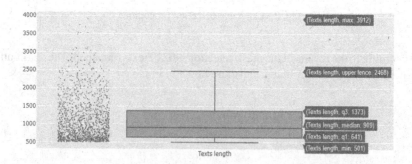

Fig. 5. Distribution of texts length in dataset after filtering.

3.3   Automatic Translation

The system with Intel(R) Core(TM) i9-10900X CPU @ 3.70 GHz and 64 Gb DDR4 was used as the test bed for the neural network translation model analysis. All services were launched in a virtual environment based on Docker container technology with no restrictions on the amount of resources used.

The following models are considered in test of automatic translation process: M2M100-1.2B, M2M100-418M, mBART-50 and HelsinkiNLP mul2en. The process is organized as sequential translation of the data set from the original language into English. The original language is determined automatically with language annotation tools described below in the next subsection. The Hugginface Transformers library for Python is used to run the models. The result of the translation is saved as a separate column in the data set. After the translation is completed, the received text on the English language are evaluated with the reference translation using the BLEU and METEOR metrics and visualization is carried out for their manual analysis.

An experiment was carried out in two directions. The first one is the direct translation of text with preliminary language detection for the whole text and the second one is independent translation of paragraphs with a preliminary definition of the language for the each one. The idea of the second experiment can be defined by the specifics of the problem area. Each text can contain parts in different languages due to the customers from all over the world. In case of whole text analysis, the automatic language detection identifies the most frequently encountered language, which can significantly worsen the translation result. To circumvent this situation, a solution was proposed with the division of the text into paragraphs and detecting language for each paragraph in independent manner.

3.4   Translation with MS Azure Cloud

Due to the great size of the dataset and 32 languages for texts it is highly difficult to create reference translation manually. Therefore, it is needed to find solution for automation of this process. After the comparison of online translation tools, the Translation service from the Microsoft Azure Cognitive Services, was chosen to translate texts from the dataset to English and use this translation as a reference for benchmark metric calculation. It should be noted that Azure translation is also not ideal, and a better choice would be to compare with text translated by professionals. Therefore, this experiment should be interpreted solely as a comparison of the translation quality of the selected models between each other relating to the translation of MS Azure Translator.

A free Tier of MS Azure Translation service was used, which provides the ability to translate up to two million characters per month. At the same time, there is an additional limitation on the frequency and size of request to the server, which should not exceed 33300 characters per minute.

Since there is a limit of two million characters for using the translation service, the limit of 70 texts per language was set. This gave a total of 1,658 examples with a total of 1,819,898 characters.

In order to meet the limit on the number of characters per minute, a delay was set in the cycle in which the translation takes place, depending on the length of the text that was submitted for translation. The delay duration is calculated as the nearest integer when the text length is divided by 550 (maximum number of characters per second).

Metric Evaluation. The general BLEU metric is shown in Fig. 6 using the box plot. The figure shows that the translation by the model M2M100-1.2B (BLEU - 0.51) is the closest to the reference text. HelsinkiNLP has the lowest rating (BLEU - 0.27).

For the METEOR metric, the relative results were the same (Fig. 7). Model M2M100-1.2B is the leader with METEOR = 0.74 and HelsinkiNLP has the lowest score with METEOR = 0.56. The absolute results of the METEOR metrics are higher due to the use of synonyms when comparing texts.

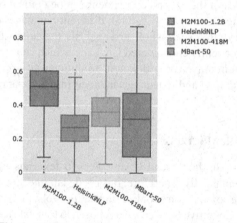

Fig. 6. BLEU metrics comparison by model [8].

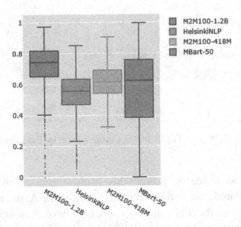

Overall METEOR metric

Fig. 7. METEOR metrics comparison by model [8].

Additionally, the models were compared by language. When analyzing models for each language, it was taken into account that Mbart-50 does not support some of the languages from the dataset. In Figs. 8 and 9, this can be seen from the values of the metrics close to zero. It can also be seen that the highest median metrics values are for translations from Dutch (nl), Spanish (es), French (fr) (see Table 1). The Afrikaans language (af) also has high metric values, but one should note extremely small number of examples for this language (2 pieces in the dataset).

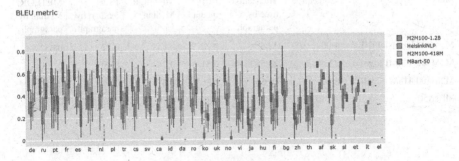

BLEU metric

Fig. 8. BLEU metrics comparison by language.

For translation by paragraphs, the BLEU and METEOR metrics were also calculated (Fig. 10, 11, 12 and 13). The results were noticeably lower than when translating whole texts. Presumably this is due to several reasons. First, for shorter texts, the accuracy of determining the language may be lower than for large texts. Because of this, the translation quality may be lower, since many models need to specify the source language. This assumption is justified by the fact that in some cases (2651 out of 23928) the language defined by the models did not match the language defined by the MS Azure

METEOR metric

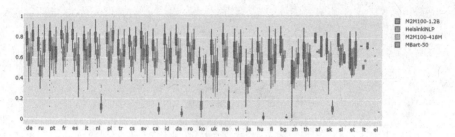

Fig. 9. METEOR metrics comparison by language.

service. Another possible reason is that when translating a multilingual text, part of it may remain untranslated, both when translating with MS Azure and when translating with the models under study. Since the text is not translated, it is simply transferred to the translation result, which is why it is possible that the same parts of the text appear, which is perceived by the metrics as a correct translation and the estimate is overestimated.

In the case of paragraph translation, a smaller division of the text occurs, and the language is determined for each separate part, which will subsequently be translated. Because of this, the score may be lower.

Table 1 shows the estimates of the models in terms of execution time and average metrics of translation quality.

Table 1. Comparison of model indicators [8].

Model	Translation time	Translation time by paragraph	BLEU median	METEOR Median	BLEU median by paragraph	METEOR Median by paragraph
MarianMT/HelsinkiNLP	4:09	4:55	0,27	0,56	0,22	0,65
M2M100 1.2B (Large)	22:50	23:58	0,51	0,74	0,39	0,73
M2M100 418M (small)	8:47	10:56	0,36	0,61	0,35	0,71
MBart50	10:40	11:29	0,30	0,60	0.00	0,57

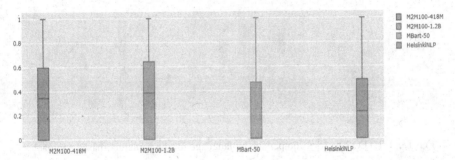

Fig. 10. BLEU metrics comparison by model for the paragraph-based approach [8].

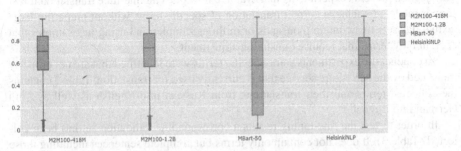

Fig. 11. METEOR metrics comparison by model for the paragraph-based approach [8].

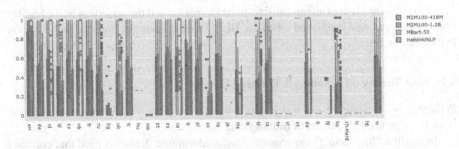

Fig. 12. BLEU metrics comparison by language for the paragraph-based approach.

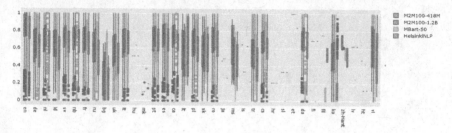

Fig. 13. METEOR metrics comparison by language for the paragraph-based approach.

4 Fine-Tuning

This section presents experiments that evaluate how one can fine-tune translation transformer models to achieve correct translation of specific terms without preparing large textual datasets that contain thousands or millions samples and using just samples that contain only terms that require translation adjustment.

To conduct the experiments some specific terms were identified that are not properly translated by the MS Azure service that is currently used for translation. Table 2 contains the collected terms and their translations from Russian into English as well as from German into English.

In order to test the fine-tuning results a corresponding validation set has been collected (Table 3). It does not contain only terms but complete sentences including those with wordings similar to the terms in the training dataset (but still different). The reason for that is to be sure that fine-tuning does not only forces the translation models to translate the required terms in the right way but also does not corrupt the translation of texts, which should not be affected by the fine-tuning.

In the presented experiments the fine-tuning is performed the same way as the regular training process, and it finished when the BLEU metric does not improve any more on the validation set.

4.1 Fine-Tuning on a Reduced Training Dataset

Presented below study research if it is possible to fine-tune models' neural translation models using limited amount of training data.

In the first "fine-tuning" experiment a reduced training dataset is used that contains only Russian translations of English terms in the singular form (lines 10–14 from Table 2). However, the evaluation of the fine-tuning is done via applying the resulting model not only to Russian-to-English translation but also to German-to-English translation. This is done with the goal to learn if model fine-tuning for a language would affect the translation results for other languages. Figures 14, 15, 16 and 17 present the results of the experiment. Below the draw conclusions are presented.

1. Fine-tuning of M2M100 models cannot be performed on such a small dataset. These models require larger datasets that consist of multiple sentences for fine-tuning. It can be seen from Figs. 16–17 that translation quality does not significantly improve when

the models are fine-tuned on limited datasets and in some cases even gets worse. Fine-tuning of M2M100 models on small datasets eventually leads to extremely shortened results. For example "Ein Normzylinder hat zwei Sensornuten." After several epochs is translated as "standard cylinder".

6. Fine-tuning of Helsinki NLP and MBart Large models is generally successful. It can be seen (Figs. 14 and 17) that metrics significantly increase.

7. Helsinki NLP and MBart Large models are capable of transferring the translation used for fine-tuning from one language to another language (or other languages). For example, given training only on Russian singular terms, the translation from German to English also improves:

Text to be translated: "Anfrage der technischen Dokumentation zu MPA Ventilinsel."

Original translation: "Application for technical documentation to MPA Ventilinsel."

Translation after fine-tuning: "request of technical documentation to MPA valve manifold."

Table 2. Training dataset.

Sample #	Singular / plural	Source language	Source text	Target language	Target text (the ground truth)
1	sing	de	die Sensornut	en	sensor slot
2	sing	de	die Endlagendämpfung	en	end-position cushioning
3	sing	de	Normzylinder	en	standard cylinder
4	sing	de	Ventilinsel	en	valve manifold
5	sing	de	der Antrieb	en	actuator
6	plur	de	Sensornuten	en	sensor slots
7	plur	de	der Normzylinder	en	standard cylinders
8	plur	de	Ventilinseln	en	valve manifolds
9	plur	de	Antriebe	en	actuators
10	sing	ru	паз для датчиков	en	sensor slot
11	sing	ru	фиксированное демпфирование	en	end-position cushioning
12	sing	ru	стандартизированный цилиндр	en	standard cylinder
13	sing	ru	пневмоостров	en	valve manifold
14	sing	ru	привод	en	actuator
15	plur	ru	пазы для датчиков	en	sensor slots
16	plur	ru	стандартизированные цилиндры	en	standard cylinders
17	plur	ru	пневмоострова	en	valve manifolds
18	plur	ru	приводы	en	actuators

Table 3. Validation dataset.

Source language	Text to translate	The ground truth
ru	Датчик находится в непраильном пазе.	The sensor is in the wrong slot.
ru	Является ли цена продукта фиксированной?	Is the product price fixed?
ru	Продукт имеет форму цилиндра.	The product has the shape of the cylinder.
ru	Какой паз для датчиков имеет этот пневмоостров?	Which sensor slot does this valve manifold have?
ru	На моих пневмоостровах отсутствуют необходимые пазы для датчиков.	There are no required sensor slots on my valve manifolds.
ru	Я обнаружил, что привод DSBC не имеет фиксированного демпфирования.	I have found that DSBC actuator does not have end-position cushioning.
ru	Вашим приводам требуется фиксированное демпфирование.	Your actuators require end-position cushioning.
ru	Все стандартизированные цилиндры оборудованы пазами для датчиков.	All standard cylinders are equipped with sensor slots.
ru	К какой модели паза для датчиков можно подключить стандартизированные цилиндры типа DSBC?	To which sensor slot model can standard cylinder DSBC be attached?
ru	Функция фиксированного демпфирования не совместима с приводами VTUG.	The end-position cushioning function is not compatible with VTUG actuators.
ru	Сколько приводов можно одновременно подключить к пневмоострову ABCD-23?.	How many actuators can be attached to valve manifold ABCD-23 simultaneously?
ru	Стандартный пневмоостров поддерживает до 10 пазов для датчиков.	Standard valve manifold supports up to 10 sensor slots.
de	Der Sensor sitzt im falschen Steckplatz.	The sensor is in the wrong slot.
de	Das Produkt hat die Form des Zylinders.	The product has the shape of the cylinder.
de	Wie viele Sensornuten hat ein Normzylinder?	How many sensor slots does the standard cylinder have?
de	Ein Normzylinder hat zwei Sensornuten.	The standard cylinder has two sensor slots.
de	Anfrage der technischen Dokumentation zu MPA Ventilinsel.	Request for technical documentation about MPA valve manifold.
de	Ich benötige elektriche Antriebe in der volgenden Konfiguration.	I need electrical actuators in the following configuration.
de	Ich möchte die zusätzlichen CPX und MPA Ventilinseln bestellen.	I would like to order additional CPX and MPA valve manifolds.
de	Welche Type der Endlagendämpfung hat dieses Produkt?	What type of end-position cushioning this product has?

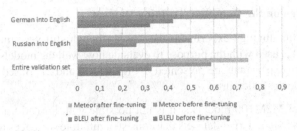

Fig. 14. Results of Helsinki-NLP fine-tuning on a reduced dataset [8].

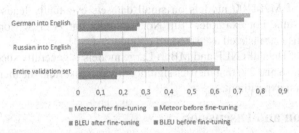

Fig. 15. Results of M2M100 418M fine-tuning on a reduced dataset [8].

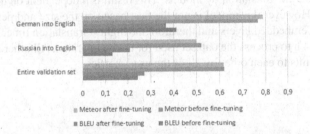

Fig. 16. Results of M2M100 1.2B fine-tuning on a reduced dataset [8].

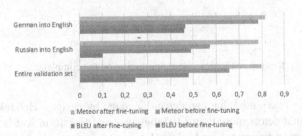

Fig. 17. Results of MBart Large fine-tuning on a reduced dataset [8].

The ground truth translation: "Request for technical documentation about MPA valve manifold."

4.2   Fine-Tuning on the Complete Training Dataset

In the second "fine-tuning" experiment the whole (though still very limited) training dataset (Table 2) is used with the purpose to evaluate how well the models can be tuned. The results of the experiment are presented in Figs. 18, 19 and 21. Based on the figures one can conclude the following:

1. Fine-tuning of M2M100 models still is not possible with limited training datasets. These models require larger datasets that consist of multiple sentences for fine-tuning. It can be seen from Figs. 20–21 that translation quality does not significantly improve when the models are fine-tuned on limited datasets and in some cases even gets worse. Fine-tuning of M2M100 models on small datasets eventually leads to extremely shortened results. For example, "Ein Normzylinder hat zwei Sensornuten." After several epochs is translated as "standard cylinder".
2. Fine-tuning of Helsinki NLP and MBart Large models is generally successful. It can be seen (Figs. 18 and 21) that metrics significantly increase, and the increase is higher than in the first experiment.

5   Conclusion and Discussion

The best translation quality is provided by the M2M100 1.2B model based on the metrics used to compare the translation by models. This result is independent on the translation method used. However this model is not the best based on the size and performance. It has the largest embedded layers and therefore the highest translation time. For instance, it took about 24 h to process the dataset used for the comparison. All other models show quite close results to each other by all of the used metrics.

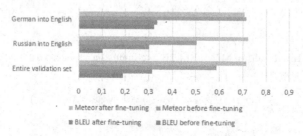

Fig. 18. Results of Helsinki-NLP fine-tuning on the complete dataset [8].

There is only one model that can detect language by design - Helsinki NLP. It has internal layers that detect more than 140 languages and transform words to tokens. As other model it can transform tokens back to text using parameter of target language when translating from many origin languages to one target language. In the translation pipeline it is possible to provide only target language for translation without origin. Other models require to provide both parameter to define translation direction for the tokenizer and translator.

When deciding which translation model should be used in production stage the following factors have to be considered. Two models - MarianMT and HelsinkiNLP

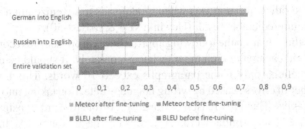

Fig. 19. Results of M2M100 418M fine-tuning on the complete dataset [8].

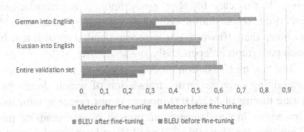

Fig. 20. Results of M2M100 1.2B fine-tuning on the complete dataset [8].

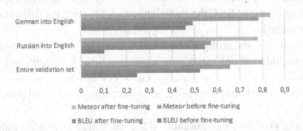

Fig. 21. Results of MBart Large fine-tuning on the complete dataset [8].

have low translation time and acceptable quality. Therefore both of them can be used in situation where time of translation is a critical factor. To increase translation quality a bit text could be split by paragraphs for the separate translation. It translation quality is a critical factor then the best choose is the M2M100 1.2B model. This model also has a lightweight version with lower number of parameters - M2M100 418M. It provides acceptable translation quality with fairly short time and could be good compromise in this two factors among all reviewed models.

It could be a situation sometimes when text contains some words or even sentences in several languages at time. It could cause a lot of problems in tokenization especially in case of use of origin language parameter in the model. The problem is caused by use of wrong tokens for words on other language that is provided as the origin one. Two possible situations were detected in this case. The first situation was when model exclude unknown tokens from the result that is highly unacceptable for translation but in other case it could cause errors in the translation (M2M100-1.2B & M2M100-M418 works in this way). The other situation was when model tries to tokenize unknown word and translate it with unpredictable result (HelsinkiNLP & MBart-50).

To override this situation the following approach was proposed. Text could be divided by paragraphs and each paragraph the translated separately. It should be noted that this approach is beneficial only if the paragraphs exceed 10 words. This is explained by estimated number of errors caused by wrong language detection taking into account large number of examples. The other reason of this result is due to word transferring between languages and use of very similar words in various language groups (like Indo-European language family). In language group it is often difficult to difficult to distinguish what language the considered word belongs to (eg Spanish and Italian languages have a lot of common words). In this case for short paragraphs, it is recommended to use one language for the entire text translation without dividing it by paragraph. Oppositely for long paragraphs (more than 10 words), it is recommended to split text by paragraphs and translated each paragraph independently.

In the part of fine-tuning translation models for correct translation of specific terms using datasets containing only the terms, translation of which should be corrected, it was shown that fine-tuning of M2M100 models is not possible with limited training datasets and they require larger datasets that consist of thousands or millions entries for fine-tuning. However, the Helsinki NLP and MBart Large models showed very good fine-tuning possibilities using limited datasets as well as their ability to adjust translation not only for languages used for fine tuning, but also for other languages, which are not used in the training datasets. Still, the more languages are used for fine-tuning, the better results can be achieved.

It can also be noted that using a large validation dataset is also important for fine tuning. It has to include not only texts with terms that need to be adjusted, but also texts with similar wordings and even completely different texts. The reason for this is to make sure that translation adjustment does not lead to overfitting and translation of other texts is not corrupted.

The fine-tuning process can be organized the same way as regular training process. However, when choosing a metric for early-stopping, one should consider the BLEU metric. This is caused by the fact the METEOR metric is not always capable to perform a valid comparison for terms that require precise translation. On the other hand, the BLEU metric is very strict and sometimes can produce poor results (e.g., 0) for proper translations. As a result, it is suggested to use METEOR for the final evaluation of the fine-tuning quality.

Acknowledgements. The paper is due to the collaboration between SPC RAS and Festo SE & Co. KG. The methodology and experiment setup (Sect. 3) are partially due to the State Research, project number FFZF-2022-0005.

References

1. Usuga Cadavid, J.P., Lamouri, S., Grabot, B., Pellerin, R., Fortin, A.: Machine learning applied in production planning and control: a state-of-the-art in the era of industry 4.0. J. Intell. Manuf. **31**(6), 1531–1558 (2020). https://doi.org/10.1007/s10845-019-01531-7

2. Cioffi, R., Travaglioni, M., Piscitelli, G., Petrillo, A., De Felice, F.: Artificial intelligence and machine learning applications in smart production: progress, trends, and directions. Sustainability **12**, 492 (2020). https://doi.org/10.3390/su12020492

3. Zhang, J., Zong, C.: Neural machine translation: challenges, progress and future. Sci. China Technol. Sci. **63**(10), 2028–2050 (2020). https://doi.org/10.1007/s11431-020-1632-x

4. Jooste, W., Haque, R., Way, A., Koehn, P.: Neural machine translation. Mach. Transl. **35**, 289–299 (2021). https://doi.org/10.1007/s10590-021-09277-x

5. Stahlberg, F.: Neural machine translation: a review. J. Artif. Intell. Res. **69**, 343–418 (2020). https://doi.org/10.1613/jair.1.12007

6. Google: Google Tranlate. https://translate.google.com/. Accessed 2022/01/12

7. Microsoft: Microsoft Bing Translator. https://www.bing.com/translator. Accessed 12 Jan 2022

8. Smirnov, A., Teslya, N., Shilov, A., Frank, D., Minina, E., Kovacs, M.: Comparative analysis of neural translation models based on transformers architecture. In: Proceedings of the 24th International Conference on Enterprise Information Systems, pp. 586–593. SCITEPRESS - Science and Technology Publications (2022). https://doi.org/10.5220/0011083600003179

9. Vathsala, M.K., Holi, G.: RNN based machine translation and transliteration for Twitter data. Int. J. Speech Technol. **23**, 499–504 (2020). https://doi.org/10.1007/s10772-020-09724-9

10. Mahata, S.K., Das, D., Bandyopadhyay, S.: MTIL2017: machine translation using recurrent neural network on statistical machine translation. J. Intell. Syst. **28**, 447–453 (2019). https://doi.org/10.1515/jisys-2018-0016

11. Wang, X., Chen, C., Xing, Z.: Domain-specific machine translation with recurrent neural network for software localization. Empir. Softw. Eng. **24**, 3514–3545 (2019). https://doi.org/10.1007/s10664-019-09702-z

12. Wolf, T., et al.: Hugging face's transformers: state-of-the-art natural language processing. In: Proceedings of the 2020 Conference on Empirical Methods in Natural Language Processing: System Demonstrations. pp. 38–45. Association for Computational Linguistics, Stroudsburg, PA, USA (2020). https://doi.org/10.18653/v1/2020.emnlp-demos.6

13. Wolf, T., et al.: Hugging Face's Transformers: State-of-the-art Natural Language Processing. arXiv Prepr. arXiv1910.03771 (2019)

14. Devlin, J., Chang, M.-W., Lee, K., Toutanova, K.: BERT: pre-training of deep bidirectional transformers for language understanding (2018)

15. Fan, A., et al.: Beyond English-centric multilingual machine translation. J. Mach. Learn. Res. **22**, 1–48 (2021)

16. Liu, Y., et al.: Multilingual denoising pre-training for neural machine translation. Trans. Assoc. Comput. Linguist. **8**, 726–742 (2020). https://doi.org/10.1162/tacl_a_00343

17. Tiedemann, J., Thottingal, S.: OPUS-MT: building open translation services for the world. In: Proceedings of the 22nd Annual Conference of the European Association for Machine Translation, pp. 479–480 (2020)

18. Tan, Z., et al.: Neural machine translation: a review of methods, resources, and tools. AI Open **1**, 5–21 (2020). https://doi.org/10.1016/j.aiopen.2020.11.001

19. Papineni, K., Roukos, S., Ward, T., Zhu, W.-J.: BLEU: a method for automatic evaluation of machine translation. In: Proceedings of the 40th Annual Meeting of the Association for Computational Linguistics (ACL), pp. 311–318. Philadelphia (2002)

20. Denkowski, M., Lavie, A.: Meteor universal: language specific translation evaluation for any target language. In: Proceedings of the Ninth Workshop on Statistical Machine Translation, pp. 376–380. Association for Computational Linguistics, Stroudsburg, PA, USA (2014). https://doi.org/10.3115/v1/W14-3348

21. Doddington, G.: Automatic evaluation of machine translation quality using n-gram co-occurrence statistics. In: HLT '02: Proceedings of the second international conference on

Human Language Technology Research, pp. 138–145. ACM Press (2002). https://doi.org/10.5555/1289189.1289273

22. Chen, B., et al.: Transformer-based language model fine-tuning methods for COVID-19 fake news detection. In: Chakraborty, T., Shu, K., Bernard, H.R., Liu, H., Akhtar, M.S. (eds.) CONSTRAINT 2021. CCIS, vol. 1402, pp. 83–92. Springer, Cham (2021). https://doi.org/10.1007/978-3-030-73696-5_9

23. Mishra, S., Prasad, S., Mishra, S.: Multilingual joint fine-tuning of transformer models for identifying trolling, aggression and cyberbullying at TRAC 2020. In: Proceedings of the Second Workshop on Trolling, Aggression and Cyberbullying at Language Resources and Evaluation Conference (LREC 2020), pp. 120–125. European Language Resources Association (ELRA) (2020)

OptiViTh: A Decision Guidance Framework and System for Design, Analysis and Optimization of Cloud-Manufactured Virtual Things

Xu Han[✉][ID] and Alexander Brodsky[ID]

George Mason University, Fairfax, VA 22030, USA
{xhan21,brodsky}@gmu.edu

Abstract. Entrepreneurs are looking for efficient and effective ways to turn their innovative ideas into marketable products. Due to the lack of accessibility, predictability and agility in the traditional idea-to-market paradigm, stiff hindrances have to be overcome, which is perceived as the manufacturing-entrepreneurship disconnect. Toward bridging this gap, we propose and develop *OptiViTh* - a decision guidance system to recommend Pareto-optimal instantiation and composition of virtual things - parameterized products and services that can be searched, composed and optimized within a hierarchical service network against customer requirements specs. *OptiViTh* is based on an extensible and reusable repository of virtual things and their artifacts, which adhere to a proposed mathematical framework. This mathematical framework formalizes the notions of (1) combined product and service designs, (2) customer-facing specifications, (3) requirements specifications, (4) virtual thing search, as a constraint satisfaction problem, (5) compositions of virtual things that are mutually consistent, Pareto-optimal against requirements specifications, and optimal against customer utility function expressed in terms of customer facing metrics, such as cost, delivery terms and desirable product characteristics. We also demonstrate *OptiViTh* using a case study of cloud-based bicycle production service network combined with its product design.

Keywords: Markets of virtual things · Decision guidance · Formal framework · Cloud manufacturing

1 Introduction

In today's information age new ideas are generated on a daily basis. The ideas reflect the increasing demands of people for desirable conveniences, capabilities and freedom in their lives. Tremendous value can be created by turning these ideas into innovative products and services and bringing them to market in a timely manner. However, in the traditional manufacturing paradigm, due to the rigid processes and limited access to valuable market information, the manufacturers are operating with latency and suboptimal decisions. Thus they cannot realize their full potential to capture the opportunities, as manifested in the manufacturing-entrepreneurship disconnect [2].

Cloud manufacturing [9] is an emerging manufacturing model that enables fast, distributed, collaborative and intelligent manufacturing that capitalizes on the information

J. Filipe et al. (Eds.): ICEIS 2022, LNBIP 487, pp. 175–197, 2023.
https://doi.org/10.1007/978-3-031-39386-0_9

sharing, virtualization, and various computing resources in the cloud environment. Together with recent research advancements in parametric product and process design and optimization [5,6,8,10–12], cloud manufacturing looks promising in providing a solution to enhance the value creation chain. To achieve high agility, accessibility, and predictability in the cloud manufacturing ecosystem, we have proposed the concept of markets of virtual things - parameterized products and services that can be searched, composed and optimized [2], leveraging prior work on decision guidance systems [3,4]. However, the concept paper [2] stopped short of development of a formal mathematical framework and developing a system supporting analysis and optimization of virtual things. There has been work [1] on feasibility of optimization of service networks combined with products' characteristics. However, [1] did not support the concepts of virtual things. Bridging these gaps is exactly the focus of this paper.

More specifically, the contributions of this paper are as follows. First, we develop a formal mathematical framework for *virtual things* or *V-things*: parameterized products and services that can be searched, composed and optimized. The proposed framework formalizes the notions of virtual product and service design and customer-facing specs as well as requirements' specs. Based on these formal concepts, the framework also formalizes the notions of search for and composition of virtual products and services that (1) are mutually consistent with the requirement specs, (2) are Pareto-optimal in terms of customer-facing metrics such as cost, product desirable characteristics and delivery terms; and (3) that are optimal in terms of customer utility function that is expressed in terms of customer facing metrics.

Second, we develop *OptiViTh* - a Decision guidance system to recommend Pareto-optimal instantiation and composition of virtual products and services within a hierarchical service network against customer requirements' specs. *OptiViTh* is based on an extensible and reusable repository of virtual things and their artifacts. Finally, we demonstrate *OptiViTh* using a case study of bicycle production service network combined with product design. This paper is a significant extension of our conference paper [7] with the design and development of *OptiViTh*.

The chapters are organized as follows. In Sect. 2 we review the idea and the intuition for markets of virtual things before we formalize the notions of virtual product and service design and specs in Sect. 3. Then in Sect. 4 we introduce an example of the implementation of the framework as a virtual bike production service network. And we demonstrate one case of virtual service optimization. We conclude with possible directions for follow-up works in Sect. 5.

2 Overview of V-Thing Framework

In the traditional ideation to production value creation chain, the entrepreneurs conceive their ideas of certain products, work with designers to sketch out their ideas, build the prototypes, and then pass the technical specs onto the manufacturers for production. The process requires expertise in various capabilities, and is often inaccurate, slow and rigid. We desire to overcome these defects through the productivity framework of V-things that employs a fundamentally new approach of connecting entrepreneurs with manufacturers in a bootstrapped marketplace of virtual products and services (shown as the V-Things Markets on the right in Fig. 1 [7].

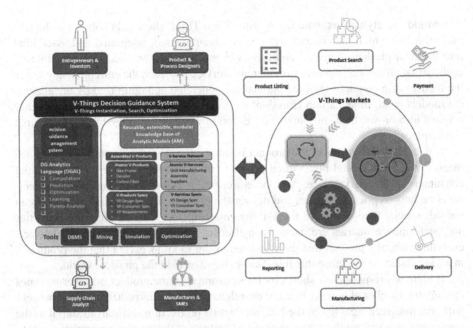

Fig. 1. The V-Things Decision Guidance Ecosystem [7].

Within the marketplace of V-things, the virtual products can be presented by their parameterized CAD design, and searched by their parameters. The virtual services are the parameterized transformations of virtual products (assembly, transportation, etc.). All virtual products and services have their own analytical model respectively that describes their important characteristics [3]. When the entrepreneurs are searching and composing the virtual products in the design environment, they can utilize the AI-powered decision guidance system to reach optimal decisions (shown as the V-Thing Decision Guidance System on the left in Fig. 1). The designers and manufacturers can also proactively participate in the discovery of new market demands within the V-thing marketplace using the V-thing design tools. And insights can be drawn by analyzing the data in the system with the avail of Decision Guidance Analytics Language [4].

Take for example that an entrepreneur wants to make a new product of a special type of bicycle that is non-existent in the real world. She can start with searching for any virtual products that are similar to her idea in the V-things markets. If she finds none, she will consider building her own bike. First, she will initiate her virtual bike project, and start to describe her idea of the bike. She may use the design-by-sketch and design-by-example features to generate a design spec from scratch, or she can specify a more definitive requirements spec of the bike with its components (bike frame, bike chain, tires, etc.). Second, she can search again for the components of the bike as virtual products in the marketplace, and if she cannot find one, she can either design one or solicit for one with a requirements spec. The system will make recommendations and guide her to make the best informed decisions along the way. The entrepreneur can recursively find the components or generate the specs of the components for the bike that would closely approximate her original idea. Third, she would solicit for the services through which the components can be manufactured, integrated and assembled into the final product. After all constraints in the specs are verified to be feasible by each virtual product component provider and service provider, the entrepreneur can see the metrics of manufacturing the bike that are generated and optimized by the analytical models (cost, time, carbon emission, etc.). And she can see if the bike reflects her original idea or not, then she can make a decision whether to list the bike in the virtual marketplace.

If the entrepreneur thinks the product is satisfactory and decides to take the go-to-market move, the special bike virtual product will be made available for the potential consumers in the marketplace. And the entrepreneur can make further business decisions on how to price, advertise, and deliver the product. The potential consumers can see this special type of bike in the V-things marketplace, what its physical properties and performance metrics are, how much it costs, and how long it takes to deliver - everything about a product that the consumers want to know except that the product is in a virtual state in the sense that it is "non-existent-yet" in the physical world.

If there is strong interest shown by the consumers in the product, whether it comes through the number of orders placed or the relevant discussions posted, the entrepreneur will give the green light to put the bike into actual production and start to ship it to the consumers as a real product. She can take advantage of the responsive market reaction as a pilot campaign. After collecting enough user feedback data from the marketing campaign and the product reviews, the entrepreneur can get informed of how well the consumers have received this new product. From there the entrepreneur can initiate a product launch with calculated risk by listing the bike in the real-world online or on-premise markets and sell the product to the targeted user groups.

3   Mathematical Formalization of the V-Thing Framework

3.1   Virtual Products

Intuitively, a virtual product, which is defined formally below, is a parameterized CAD design with an associated analytic model that defines all feasible parametric instantiations along with their consumer-facing metrics. For example, a virtual product can be a final product (a bike), a part of another product (a bike fork or a wheel), or a raw material (steel). The analytic models put the constraints on the virtual products (the dimensions of the bike fork must fit that of the wheels), and describe feasibility of the product based on the parameters and metrics (a bike weighing 1 ton will be infeasible as it violates the tolerance of the wheels).

Definition 1. *A virtual product design spec (or VP for short) is a tuple*

$$VP = <domain, parametersSchema, metricSchema,$$
$$template, analyticModel, feasibilityMetric>$$

where

1. *domain - is a set of all things under consideration. E.g., all components that can be specified using CAD design.*
2. *parametersSchema - is a set*

$$PS = (p_1, D_1), \ldots, (p_n, D_n)$$

of pairs, where for $i = 1, \ldots, n$
   (a) *p_i - a unique parameter name (e.g., geometric dimension, tolerance, density of material) (fixed parameters vs decision variables)*
   (b) *D_i - the domain of p_i*
3. *metricSchema - is a set*

$$MS = \{(m_1, M_1), \ldots, (m_k, M_k)\}$$

of pairs, where for $i = 1, \ldots, k$
   (a) *m_i - a unique metric name (e.g., weight, strength, volume, floatable (Boolean))*
   (b) *M_i - the domain of m_i*
4. *template - is a function*

$$T : D_1 \times \cdots \times D_n \to Domain$$

which gives, for parameter instantiation $(p_1, \ldots, p_n)$ in $D_1 \times \cdots \times D_n$, a specific thing in the Domain of things.
5. *analyticModel - is a function*

$$AM : D_1 \times \cdots \times D_n \to M_1 \times \cdots \times M_n$$

which gives, for parameter instantiation $(p_1, \ldots, p_n)$ in $D_1 \times \cdots \times D_n$, a vector $(m_1, \ldots, m_k)$, where $(m_1, \ldots, m_k)$ are metric values in $M_1 \times \cdots \times M_k$. We will denote resulting metric value m_i, $i = 1, \ldots, n$, using the (.) dot notation as $AM(p_1, \ldots, p_n).m_i$ or $m_i(p_1, \ldots, p_n)$
6. *feasibilityMetric - is a Boolean metric name C in $\{m_1, \ldots, m_n\}$ such that its corresponding domain $D = \{T, F\}$ in metricSchema MS (T to indicate that parameter instantiation is feasible, and F otherwise)*

3.2   Specs of Virtual Products

In order to communicate the information between multiple parties in the V-things marketplace, manufacturers may not want to disclose low-level design parameters (in the parametric schema) but only disclose product metrics (in the metric schema) which we assume contain all product information relevant to consumers. For example, a consumer purchasing a bike may only want to know the high-level commercial and performance information about the bike (such as price, weight, max speed and type) rather than of the more granular supply chain metrics (such as prices of the parts, dimensions of the bike pedals and strength of the steel.) This notion is captured in the concept of VP consumer-facing specs.

Definition 2. *VP consumer-facing spec (or VP CFS for short) is a tuple*

$$\delta = <domain, metricSchema, MFC>$$

where domain, metricSchema are as defined in VP, and MFC: $M_1 \times \cdots \times M_k \rightarrow \{T, F\}$ is a set of feasibility constraints that tells, for every vector of metrics in $M_1 \times \cdots \times M_k$, whether it is feasible or not.

Given a VP design spec, manufacturers may want to extract a VP consumer-facing spec from it, defined next.

Definition 3. *VP consumer-facing spec derived from VP design spec* vp *(denoted CFS(vp) for short). Let*

$$vp = (domain, parametersSchema, metricSchema,$$
$$template, analyticModel, feasibilityMetric)$$

be a VP design spec.

The derived consumer-facing spec CFS(vp) is a tuple (domain, metricSchema, MFC), where MFC is defined as follows: $\forall (m_1, \ldots, m_k) \in M_1 \times \cdots \times M_k$, $MFC(m_1, \ldots, m_k) = \exists (p_1, \ldots, p_n) \in D_1 \times \cdots \times D_n$ s.t. $(m_1, \ldots, m_k) = AM(p_1, \ldots, p_n) \wedge C(p_1, \ldots, p_n) = T$

Note: CFS(vp) defines the set of all feasible metric vectors $\{(m_1, \ldots, m_k) | (m_1, \ldots, m_k) \in M_1 \times \cdots \times M_k \wedge MFC(m_1, \ldots, m_k)\}$

In order to enable the users to search for the VPs, we introduce the VP Requirements spec. The purpose of the Requirements spec is for the users to get conditions for search as they put constraints on the VP space.

Definition 4. *VP (consumer) Requirements Spec (or RS for short) is a tuple RS = <domain, metricSchema, objectiveSchema, objectives, MC, OC> where*

1. objectiveSchema is a set

$$OS = \{(o_1, O_1), \ldots, (o_l, O_l)\}$$

of pairs, where for $i = 1, \ldots, l$
(a) o_i - a unique objective name (e.g., cost, risk, time, etc.)
(b) O_i - the domain of o_i
2. objectives: $M_1 \times \cdots \times M_k \rightarrow O_1 \times \cdots \times O_l$ defines objectives as a function of metrics i.e., objectives$(m_1, \ldots, m_k)$ is a vector of objective values $(O_1, \ldots, O_l)$
3. $MC : M_1 \times \cdots \times M_k \rightarrow \{T, F\}$ define feasibility constraints in the metric space
4. $OC : O_1 \times \cdots \times O_l \rightarrow \{T, F\}$ define feasibility constraints in the objective space

The user search should only render the results that fall in the range of the constraints domains determined by the product characteristics and specified by the user. For example, a user searches for a bike with weight not greater than 10 kg, and price not greater than $200, then the result should only contain products that are feasible within the metrics ranges, and also comply to the user specified objective domains.

Definition 5. *VP Requirements Spec Constraints*

Overall constraints of VP Requirements Spec are the constraints $C : M_1 \times \ldots \times M_k \to \{T, F\}$ defined by: $C(m_1, \ldots, m_n) = MC(m_1, \ldots, m_n) \wedge OC(objectives(m_1, \ldots, m_n))$ Note also that VP Requirements spec defines a set of all feasible metric vectors R: $R = \{(m_1, \ldots, m_n) | (m_1, \ldots, m_n) \in M_1 \times \cdots \times M_n \wedge C(m_1, \ldots, m_n)\}$

To search for a VP in the marketplace, the user specifies the VP Requirements Spec (RS), and the system checks if any VP consumer-facing Spec (CFS) matches the RS.

Definition 6. *Search for VP*

– *Let*

$$VP = <domain, parametersSchema, metricSchema,$$
$$template, analyticModel, feasibilityMetric>$$

be a VP design spec. Out of parameter vector $(p_1, \ldots, p_l, \ldots, p_n)$ in parametersSchema, let $DV = (p_1, \ldots, p_l)$ be designated as decision variables (without loss of generality). We use the term fixed parameters for the vector of remaining parameters $FP = (p_{(l+1)}, \ldots, p_n)$.
– *Let $CFS = <domain, metricSchema, MFC>$ be a VP (consumer-facing) Spec*
– *Let $RS = <domain, metricSchema, objectiveSchema, objectives, MC, OC>$ be a VP Requirements spec*
– *Let $U : O_1 \times \cdots \times O_l \to \mathbb{R}$ be a utility function*

We say that:

– RS and CFS match (or CFS is feasible w.r.t. RS, or CFS and RS are mutually consistent) if
  1. they have identical domains and metric schema $(m_1, M_1), \ldots, (m_k, M_k)$
  2. their joint constraint $C(m_1, \ldots, m_k) = MC(m_1, \ldots, m_n)$ and $OC(objectives(m_1, \ldots, m_n))$ and $MFC(m_1, \ldots, m_k)$ are satisfiable
– A metrics vector $(m_1^*, \ldots, m_k^*) \in M_1 \times \cdots \times M_k$ is Pareto-optimal w.r.t. to RS and CFS if:
  1. the joint constraint $C(m_1^*, \ldots, m_k^*)$ holds
  2. there does not exist a metrics vector $(m_1', \ldots, m_k')$ that
     • satisfies the joint constraint C
     • for objective vectors $(o_1^*, \ldots, o_l^*) = objective(m_1^*, \ldots, m_k^*)$ and $(o_1', \ldots, o_l') = objective(m_1', \ldots, m_k')$, $(\forall i = 1, \ldots, l, o_i' \geq o_i^*)$ and $(\exists j, 1 \leq j \leq l)$ such that $o_j' > o_j^*$

– A metrics vector $(m_1^*, \ldots, m_k^*) \in M_1 \times \cdots \times M_k$ is optimal w.r.t. RS, CFS and U if:

$$(m_1^*, \ldots, m_k^*) \in argmax(m_1, \ldots, m_k) \in M_1 \times \cdots \times M_k(U(objective(m_1, \ldots, m_k)))$$

subject to $C(m_1, \ldots, m_k)$

- Given a vector $FP = (v_{(l+1)}, \ldots, v_n)$ of fixed parameters, parameter vector $(p_1^*, \ldots, p_n^*) \in P_1 \times \cdots \times P_n$ and the corresponding product $p = template(p_1^*, \ldots, p_n^*)$ are Pareto-optimal w.r.t. RS, VP and FP if:
    1. the constraint $C(p_1^*, \ldots, p_n^*) = MFC(AM(p_1^*, \ldots, p_n^*))$ and $\forall j = l+1, \ldots, n, p_j^* = p_j$
    2. there does not exist a parameter vector $(p_1', \ldots, p_n')$ that
        - satisfies the constraint C, and
        - for objective vectors $(o_1^*, \ldots, o_l^*) = objective(AM(p_1^*, \ldots, p_n^*))$ and $(o_1', \ldots, o_l') = objective(AM(p_1', \ldots, p_n'))$, $(\forall i = 1, \ldots, l) o_i' \geq o_i^*)$ and $(\exists j, 1 \leq j \leq l)$ such that $o_j' > o_j^*$
- Given a vector $FP = (v_{(l+1)}, \ldots, v_n)$ of fixed parameters, a parameter vector $(p_1^*, \ldots, p_n^*) \in P_1 \times \cdots \times P_n$ and the corresponding product $p = template(p_1^*, \ldots, p_n^*)$ are optimal w.r.t. RS, VP, U and FP if:
    - $(p_1^*, \ldots, p_n^*) \in argmax(p_1, \ldots, p_n) \in P_1 \times \cdots \times P_n(U(objective(AM(p_1, \ldots, p_n)))$ subject to $C(AM(p_1, \ldots, p_n))$ and $\forall j = l+1, \ldots, n, p_j^* = p_j$

Claim (follows directly from definitions): Given CFS derived from VP, RS, U and let $(m_1^*, \ldots, m_k^*) = AM(p_1^*, \ldots, p_n^*)$. Then:

- $(p_1^*, \ldots, p_n^*)$ *is Pareto-optimal w.r.t. RS and VP* $\iff$ $(m_1^*, \ldots, m_k^*)$ *are Pareto-optimal w.r.t. RS and CFS*
- $(p_1^*, \ldots, p_n^*)$ *is optimal w.r.t. RS, VP and U* $\iff$ $(m_1^*, \ldots, m_k^*)$ *is optimal w.r.t. RS, CFS and U.*

3.3 Virtual Services

Intuitively, a virtual service is a parameterized transformation function of zero or more virtual things into zero or more virtual things. Virtual service is also associated with an analytic model, which gives metrics of interest and the feasibility Boolean value, for every instantiation of service parameters, which include parameters of input and output virtual things, as well as additional internal parameters.

For example, a supply service transforms zero input virtual things into output things that correspond to a planned purchase (bike supply). Its metrics may include total cost and delivery time, as a function of parameters that capture catalog prices and ordered quantities. Or, a manufacturing service transforms raw materials and parts (virtual things) into products (virtual things), like in the case of assembling bike parts into a bike. Or, a CNC machining service transforms an (virtual) input metal part into a processed (virtual) part, after a series of drilling and milling operations (turning steel material into a bike frame). Or, a transportation service transforms (virtual) items at some locations, into the same (virtual) items at some other locations (shipping a bike from a factory to a customer). Some services are associated with real-world brick-and-mortar services; some may be defined by a service network which involves sub-services. The following definitions formalize these concepts.

Definition 7. *A virtual service or VS is a tuple*

$$VS = (input, output, internalParametersSchema,$$
$$internalMetricSchema, analyticModel, feasMetric)$$

where

1. *input - is a set* $\{VT_i | i \in I\}$ *of input virtual things, where I is an input index set*
2. *output - is a set* $\{VT_i | i \in O\}$ *of output virtual things, where O is an output index set*
3. *internalParametersSchema - is a set*

$$IPS = \{(p_1, D_1), \ldots, (p_n, D_n)\}$$

of pairs, where for $i = 1, \ldots, n$
   (a) p_i *- a parameter name (e.g., quantities of input or output things, prices, coefficients in physics-based equations, control parameters of equipment)*
   (b) D_i *- the domain of* P_i
4. *internalMetricSchema - is a set*

$$IMS = \{(m_1, M_1), \ldots, (m_k, M_k)\}$$

of pairs, where for $i = 1, \ldots, k$
   (a) m_i *- a metric name (e.g., cost, delivery time, profit, carbon emissions)*
   (b) M_i *- the domain of* m_i
   Let the parametersSchema

$$PS = \{(v_1, V_1), \ldots, (v_s, V_s)\}$$

be the union of parametersSchema of input virtual products, output virtual products and internalParametersSchemas.
Let the set
$$MS = \{(mm_1, MM_1), \ldots, (mm_r, MM_r)\}$$

be the union of metricSchemas of input virtual things, output virtual things, and internalMetricSchema.
5. *analyticModel - is a function*

$$AM : V_1 \times \cdots \times V_s \rightarrow MM_1 \times \cdots \times MM_r$$

which gives, for parameter instantiation $(v_1, \ldots, v_s) \in V_1 \times \cdots \times V_s$, *a vector* $(mm_1, \ldots, mm_r)$, *where* $(mm_1, \ldots, mm_r)$ *are metric values in* $MM_1 \times \cdots \times MM_r$. *We will denote resulting metric value* $mm_i, i = 1, \ldots, r$, *using the (.) dot notation as* $AM(v_1, \ldots, v_s).m_i$ *or* $m_i(v_1, \ldots, v_s)$
6. *feasMetric - is a Boolean metric name* $C \in mm_1, \ldots, mm_r$ *such that its corresponding domain* $D = \{T, F\} \in MS$ *(T to indicate that parameter instantiation is feasible, and F otherwise)*

3.4 Specs of Virtual Services

Definition 8. *VS Consumer Spec (projection on metric space of VS) that only reveals partially to the consumers is a tuple*

$$<input, output, internalMetricSchema, MC>$$

where

– *input - is a set of VP (consumer-facing) specs <domain, metricSchema, MFC>*
– *output - is a set of VP (consumer-facing) specs <domain, metricSchema, MFC>*
– *internalMetricSchema is a set*

$$IMS = \{(m_1, M_1), \ldots, (m_k, M_k)\}$$

of pairs, where for $i = 1, \ldots, k$
    1. m_i - a metric name (e.g., cost, delivery time, profit, carbon emissions)
    2. M_i - the domain of m_i
Let metric schema

$$MS = \{(mm_1, MM_1), \ldots, (mm_r, MM_r)\}$$

be the union of metricSchemas of input virtual things, output virtual things, and internalMetricSchema.
– *$MC : MM_1 \times \cdots \times MM_r \to \{T, F\}$ which tells, for every vector of metrics in $MM_1 \times \cdots \times MM_r$, whether it is feasible or not.*

Definition 9. *VS Requirement Feasibility Spec is a tuple*

$$<input, output, internalMetricSchema, objectiveSchema, objectives, MC, OC>$$

where

– *input is a set of (input) VP requirement specs*
– *output is a set of (output) VP requirement specs*
– *internalMetricSchema is a set*

$$IMS = \{(m_1, M_1), \ldots, (m_k, M_k)\}$$

of pairs, where for $i = 1, \ldots, k$
    1. m_i - a metric name (e.g., cost, delivery time, profit, carbon emissions)
    2. M_i - the domain of m_i
Let metric schema

$$MS = \{(mm_1, MM_1), \ldots, (mm_r, MM_r)\}$$

be the union of metricSchemas of input virtual things, output virtual things, and internalMetricSchema.
– *objectiveSchema is a set*

$$OS = (o_1, O_1), \ldots, (o_l, O_l)$$

of pairs, where for $i = 1, \ldots, l$
    1. o_i - a unique objective name (e.g., cost, profit, time, carbon emissions)
    2. O_i - the domain of o_i
– *objectives : $MM_1 \times \cdots \times MM_r \to O_1 \times \cdots \times O_l$ defines objectives as a function of metrics, i.e., objectives$(mm_1, \ldots, mm_r)$ is a vector of objective values $(o_1, \ldots, o_l) \in O_1 \times \cdots \times O_l$ associated with metrics $(mm_1, \ldots, mm_r)$*

- $MC : MM_1 \times \cdots \times MM_r \rightarrow \{T,F\}$
- $OC : O_1 \times \cdots \times O_l \rightarrow \{T,F\}$

Definition 10. *VS Requirements Spec Constraints*

Overall constraints of VS Requirements Spec are the constraints

$$C : MM_1 \times \cdots \times MM_r \rightarrow \{T,F\}$$

defined by: $C(mm_1,\ldots,mm_r) = MC(mm_1,\ldots,mm_r) \wedge OC(objectives(mm_1,\ldots,mm_r))$
Note also that VS Requirements Spec defines a set of all feasible metric vectors R:

$$R = \{(mm_1,\ldots,mm_r) | (mm_1,\ldots,mm_r) \in MM_1 \times \cdots \times MM_r \wedge C(mm_1,\ldots,mm_r)\}$$

Definition 11. *Search for VS*

- *Let*

$$VS = (input, output, internalParametersSchema,$$
$$internalMetricSchema, analyticModel, feasMetric)$$

be a virtual service design spec.
- *Let*
$$PS = \{(v_1,V_1),\ldots,(v_s,V_s)\}$$

be the parameter schema of VS.
- *Let*
$$MS = \{(mm_1,MM_1),\ldots,(mm_r,MM_r)\}$$

be the metric schema of VS. Out of VS parameter vector $(v1,\ldots,v_h,\ldots,v_s)$ *in internalParametersSchema, let* $DV = (v_1,\ldots,v_h)$ *be designated as decision variables (without loss of generality). We use the term fixed parameters for the vector of remaining parameters* $FP = (v_{(l+1)},\ldots,v_s)$.
- *Let*
$$CFS = <input, output, internalMetricSchema, MC>$$

be a virtual service consumer-facing spec.
- *Let*
$$MS = \{(mm_1,MM_1),\ldots,(mm_r,MM_r)\}$$

be the metric schema of CFS.
- *Let*

$$RS = <input, output, internalMetricSchema,$$
$$objectiveSchema, objectives, MC, OC>$$

be a virtual service requirements spec.
- *Let*
$$MS = \{(mm_1,MM_1),\ldots,(mm_r,MM_r)\}$$

be the metric schema of RS.

– Let $U : O_1 \times \cdots \times O_l \to R$ (the set of reals) be a utility function, where $O_1, \ldots, O_n$ are the domains from the objective schema

$$OC = \{(o_1, O_1), \ldots, (o_l, O_l)\}$$

We say that:

– RS and CFS match (or CFS is feasible w.r.t. RS, or CFS and RS are mutually consistent) if
   1. they have the same input, output and metric schemata $\{(mm_1, MM_1), \ldots, (mm_r, MM_r)\}$
   2. the joint constraint $CC(mm_1, \ldots, mm_r)$ defined as the conjunction of $C(mm_1, \ldots, mm_r)$ and $MC(mm_1, \ldots, mm_r)$ is satisfiable, where $C(mm_1, \ldots, mm_r)$ is the overall constraint of the VS Requirements Spec, and $MC(mm_1, \ldots, mm_r)$ is the metric constraint of the VS consumer-facing spec

– A metrics vector $(mm_1^*, \ldots, mm_r^*) \in MM_1 \times \cdots \times MM_r$ is Pareto-optimal w.r.t. to RS and CFS if:
   1. the joint constraint $CC(mm_1^*, \ldots, mm_r^*)$ holds
   2. there does not exist a metrics vector $(mm_1', \ldots, mm_r')$ that
      • satisfies the joint constraint CC
      • for objective vectors $(o_1^*, \ldots, o_l^*) = objective(m_1^*, \ldots, m_k^*)$ and $(o_1', \ldots, o_l') = objective(m_1', \ldots, m_k')$ the following holds: $(\forall i = 1, \ldots, l, o_i' \geq o_i^*)$ and $(\exists j, 1 \leq j \leq l)$ such that $o_j' > o_j^*$
– A metrics vector $(mm_1^*, \ldots, mm_r^*) \in MM_1 \times \cdots \times MM_r$ is optimal w.r.t. RS, CFS and U if: $(mm_1^*, \ldots, mm_r^*) \in argmax(mm_1, \ldots, mm_r) \in MM_1 \times \cdots \times MM_r(U(objective(mm_1, \ldots, mm_r)))$ subject to $CC(mm_1, \ldots, mm_r)$

– Given a vector $FP = (v_{(h+1)}, \ldots, v_s)$ of fixed parameters, parameter vector $(v_1^*, \ldots, v_s^*) \in V_1 \times \cdots \times V_s$ (and the corresponding product $p = template(v_1^*, \ldots, v_s^*)$) are Pareto-optimal w.r.t. VS, RS and FP if:
      • the following constraint is satisfied: $CC(v_1^*, \ldots, v_h^*, \ldots, v_s^*) = MC(AM(v_1^*, \ldots, v_s^*)) \wedge (\forall j = h+1, \ldots, s, v_j^* = v_j) \wedge C(v_1^*, \ldots, v_s^*)$, where MC is the metric constraint from RS, and C is the feasibility metric from VS.
      • there does not exist a parameter vector $(v_1', \ldots, v_s')$ that
         * satisfies the constraint CC, and
         * for objective vectors $(o_1^*, \ldots, o_l^*) = objective(AM(p_1^*, \ldots, p_n^*))$ and $(o_1', \ldots, o_l') = objective(AM(v_1', \ldots, v_s'))$, $(\forall i = 1, \ldots, l, o_i' \geq o_i^*)$ and $(\exists j, 1 \leq j \leq l)$ such that $o_j' > o_j^*$

– Given a vector $FP = (v_{(h+1)}, \ldots, v_s)$ of fixed parameters, parameter vector $(v_1^*, \ldots, v_s^*) \in V_1 \times \cdots \times V_s$ and the corresponding product $p = template(v_1^*, \ldots, v_s^*)$ are optimal w.r.t. VS, RS, U and FP if:
      • $(v_1^*, \ldots, v_h^*, v_{(h+1)}^* \ldots, v_s^*) \in argmax(v_1^*, \ldots, v_s^*) \in V_1 \times \cdots \times Vs$
         $(U(objective(AM(v_1^*, \ldots, v_s^*)))$ subject to $CC(v_1^*, \ldots, v_h^*, \ldots, v_s^*) = MC(AM(v_1^*, \ldots, v_s^*))$ and $(\forall j = h+1, \ldots, s, v_j^* = v_j)$ and $C(v_1^*, \ldots, v_s^*)$, where MC is the metric constraint from RS, and C is the feasibility metric from VS.

Claim (follows directly from definitions): Given CFS derived from VP, RS, U and let $(mm_1^*, \ldots, mm_r^*) = AM(v_1^*, \ldots, v_s^*)$, *then:*

- $(v_1^*, \ldots, v_s^*)$ *is Pareto-optimal w.r.t. VS, RS, and FP if and only if* $(mm_1^*, \ldots, mm_r^*)$ *are Pareto-optimal wr.t. RS and CFS*
- $(v_1^*, \ldots, v_s^*)$ *is optimal w.r.t. VS, RS, FP and U if and only if* $(mm_1^*, \ldots, mm_r^*)$ *is optimal wr.t. RS, CFS and U*

4   *OptiViTh* Decision Guidance System

4.1   Architecture Overview

To enable rapid and convenient creation of V-things, we put the definitions into a more recognizable form of software implementation of the V-things marketplace. The challenge is how to organize the repository so that the artifacts can be stored hierarchically and connected seamlessly to the conceptual schema. In this section we show the organization of the fundamental artifacts (virtual products, virtual services, specs, analytical models, utility functions, etc.) and we demonstrate the composition of a V-thing with the example of a virtual bike product.

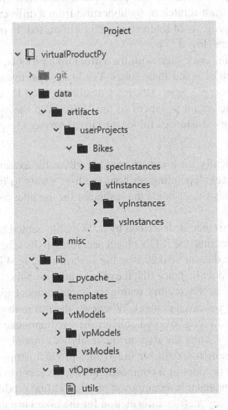

Fig. 2. The Repository Design [7].

We create a software project and separate the abstractions of data objects and of model objects in the repository (shown in Fig. 2 [7]). On the high level, the data folder contains the user-specified artifacts, including the instances of V-things, and the spec instances. These are all organized under the Bikes project associated with the user. In the lib folder we put the more functional elements including the templates, the analytical models and the V-thing operators which can be readily reused.

Each artifact is stored in its own file in the system, and has its unique file identifier in the entire system. The identifier consists of the directory path from the root to the folder containing the file plus the file name. To achieve referential integrity in the file system, it necessities that in the local environment, say, within the same folder, each file and folder must have a unique name, which is ensured by most of the modern operating systems.

4.2   Artifacts in the Repository

We view all the virtual things as the digital instances in the form of parametric data. For broad compatibility, the data is stored as JSON objects in the form of JSON files. And we differentiate the access rights to the data based on the types of the instances or the specific files which the user groups have interests in.

Under the userProjects directory, the users can initiate their virtual thing project ad hoc either by creating from scratch or by importing from a different source. A user can have arbitrarily many projects of interest created or imported. In our example, we have created the Bikes project (Fig. 3 [7]).

For the instance data associated with the virtual bike product, there is a distinction between the V-thing objects and their specs. While the specInstances folder hosts all the specs (CFS, VP Spec, VS Spec, RS, RS Constraints, etc.) for the V-thing, the vtInstances folder hosts the actual instances of the V-things in the vpInstances folder for virtual products and the vsInstances folder for virtual services respectively.

VP Instances. Specifically for a virtual product, vBike for example, it can have zero, one or multiple instances depending on how the user wants to instantiate the virtual product. The vpInstances folder holds the instances for the bike parts which are virtual products as well.

A virtual product instance is described by a JSON file named by the virtual product plus an ID. Each file contains one JSON object according to its schema in the VP Design Spec. A example of the data in vBike0.json file is shown in Fig. 4 [7].

For each virtual product instance file, it can have an optional context header, marked by "@context" annotation. Each entry within the context object specifies a shortcut for a directory path which we call a context. Whenever the program processing data for a virtual product, it would recognize and convert the "@" annotated shortcut to the full path. The "model" entry links the data to the analytical model. And the "bike" sub-object contains the parametric data for the virtual product. Important information as whether it is an atomic product or a composite product is recorded. And if it is a composite product, which means it has composed of other virtual products, the components are also recorded. And key metrics information for the bike virtual product is attached.

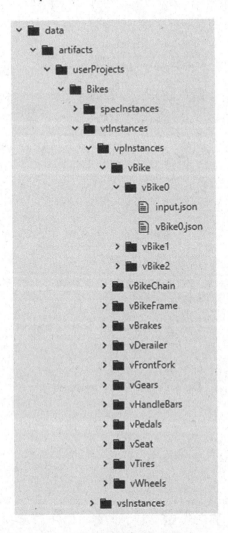

Fig. 3. The Data Artifacts [7].

Another example is shown below for an atomic virtual product in Fig. 5 [7]. The vBikeChain0.json file contains the data of a virtual bike chain product. Instead of being composed by other products, it is an atomic product, which can be used in a stand-alone manner or as a component for other virtual products and services.

VS Instances. For virtual services, the data is organized differently than that of the virtual products. As the virtual services describe the flow of materials and labor, they are more process-based. While the users are interested in the metrics of a service, the essential information for the processes should also be captured by the virtual services. An example of a bike assembly service is shown in file bikeAssembly0.json (Fig. 6 [7]).

```
{
  "@context": {
    "@vsModels": "lib/vtModels/vsModels",
    "@vp": "data/artifacts/userProjects/Bikes...
    "@productRef": "data/artifacts/userProjects...
  },
  "model": "lib/vtModels/vpModels",
  "bike": {
    "type": "assembled",
    "components": {
      "pedals": 2,
      "tires": 2,
      "wheels": 2,
      "bike_frame": 1,
      "handle_bars": 1,
      "seat": 1,
      "breaks": 2,
      "derailer": 1,
      "gears": 1,
      "bike_chain": 1
    },
    "speed": 10,
    "size": 56,
    "weight": 12,
    "material": "composite",
    "name": "RW5000",
    "color": "red",
    "min_weight": 0,
    "max_weight": 12
  }
}
```

Fig. 4. Composite VP Instance [7].

```
{
  "@context": {
    "@vsModels": "lib/vtModels/vsModels",
    "@vp": "data/artifacts/userProjects/Bikes...
    "@productRef": "data/artifacts/userProjects...
  },
  "model": "lib/vtModels/vpModels",
  "bike_chain": {
    "type": "basic",
    "size": 30,
    "material": "iron",
    "weight": 0.254,
    "min_weight": 0,
    "max_weight": 2.1
  }
}
```

Fig. 5. Atomic VP Instance [7].

The information about the flows is reflected in the schematic sections of inflow, outflow, flows and products respectively. The inflows direct what products and materials and how they are passed into the service. The outflows are the virtual products that will be rendered by the virtual service, and there can be none. The flows summarise how each product plays a part role in comprising the final outflow product. And the products section records the relevant parametric data of the inflow products that will be used for processing and analytical purposes in the service.

```
"flows": {
    "bikeChain0": "@productRef/vBikeChain/vBikeChain0",
    "bikeFrame0": "@productRef/vBikeFrame/vBikeFrame0",
    "brakes0": "@vp/vBrakes/vBrakes0",
    "derailer0": "@vp/vDerailer/vDerailer0",
    "frontFork0": "@vp/vFrontFork/vFrontFork0",
    "gears0": "@vp/vGears/vGears0",
    "handleBars0": "@vp/vHandleBars/vHandleBars0",
    "pedals0": "@vp/vPedals/vPedals0",
    "seat0": "@vp/vSeat/vSeat0",
    "tires0": "@vp/vTires/vTires0",
    "wheels0": "@vp/vWheels/vWheels0"
},
"products": {
    "@vp/vBrakes/vBrakes0": {
        "type": "basic",
        "weight": 0.306,
        "material": "steel",
        "min_weight": 0,
        "max_weight": 2
    },
    "@vp/vDerailer/vDerailer0": {=
    "@vp/vFrontFork/vFrontFork0": {=
    "@vp/vGears/vGears0": {=
    "@vp/vHandleBars/vHandleBars0": {=
    "@vp/vPedals/vPedals0": {=
    "@vp/vSeat/vSeat0": {=
    "@vp/vTires/vTires0": {=
    "@vp/vWheels/vWheels0": {=
    }
}
```

Fig. 6. Bike Assembly VS Instance [7].

4.3   V-Thing Creation

While an atomic V-thing can be created from scratch with the parameters schema, a composite V-thing may have parts in it that are also V-things. A composite V-thing may already contain all the data from its sub-components, but if not, that means, some of the data is stored in the V-thing files that may not have been created, and thus instantiation is needed. References are used as placeholders to point each missing part to a "yet-to-exist" V-thing which we call a partial V-thing. In that case, the entrepreneur should solicit in the V-thing Market for the missing parts with a spec. Once a V-thing is provided in accordance to the spec, an instantiation function will be applied that pulls in the data from the component V-thing file, replaces the reference, and renders a full data file for the composite V-thing.

For example, a virtual bike assembly service will create a virtual bike product out of its parts (bike chain, bike frame, brakes, derailers, front fork, gears, handle bars, pedals, seats, tires, and wheels). Among the parts, there is no suitable bike chain or bike frame virtual product that is readily available in the market. Then the entrepreneur lists her demand with the Requirements Specs for the bike chain and bike frame in the market, while she continues building the virtual bike product through the assembly service. The parts that await to be instantiated are marked with "@productRef" annotation indicating the data is not yet plugged in (under flows, bikeChain0 and bikeFrame0 have no correspondent entry under products as shown in Fig. 7 [7]). Once the data is ready for the component V-things, the entrepreneur will go ahead and apply the instantiation function,

```
"flows": {
    "bikeChain0": "@productRef/vBikeChain/vBikeChain0",
    "bikeFrame0": "@productRef/vBikeFrame/vBikeFrame0",
    "brakes0": "@vp/vBrakes/vBrakes0",
    "derailer0": "@vp/vDerailer/vDerailer0",
    "frontFork0": "@vp/vFrontFork/vFrontFork0",
    "gears0": "@vp/vGears/vGears0",
    "handleBars0": "@vp/vHandleBars/vHandleBars0",
    "pedals0": "@vp/vPedals/vPedals0",
    "seat0": "@vp/vSeat/vSeat0",
    "tires0": "@vp/vTires/vTires0",
    "wheels0": "@vp/vWheels/vWheels0"
},
"products": {
    "@vp/vBrakes/vBrakes0": {◼
    "@vp/vDerailer/vDerailer0": {◼
    "@vp/vFrontFork/vFrontFork0": {◼
    "@vp/vGears/vGears0": {◼
    "@vp/vHandleBars/vHandleBars0": {◼
    "@vp/vPedals/vPedals0": {◼
    "@vp/vSeat/vSeat0": {◼
    "@vp/vTires/vTires0": {◼
    "@vp/vWheels/vWheels0": {◼
}
```

Fig. 7. Partial V-Thing [7].

```
"flows": {
    "bikeChain0": "@productRef/vBikeChain/vBikeChain0",
    "bikeFrame0": "@productRef/vBikeFrame/vBikeFrame0",
    "brakes0": "@vp/vBrakes/vBrakes0",
    "derailer0": "@vp/vDerailer/vDerailer0",
    "frontFork0": "@vp/vFrontFork/vFrontFork0",
    "gears0": "@vp/vGears/vGears0",
    "handleBars0": "@vp/vHandleBars/vHandleBars0",
    "pedals0": "@vp/vPedals/vPedals0",
    "seat0": "@vp/vSeat/vSeat0",
    "tires0": "@vp/vTires/vTires0",
    "wheels0": "@vp/vWheels/vWheels0"
},
"products": {
    "@vp/vBrakes/vBrakes0": {◼
    "@vp/vDerailer/vDerailer0": {◼
    "@vp/vFrontFork/vFrontFork0": {◼
    "@vp/vGears/vGears0": {◼
    "@vp/vHandleBars/vHandleBars0": {◼
    "@vp/vPedals/vPedals0": {◼
    "@vp/vSeat/vSeat0": {◼
    "@vp/vTires/vTires0": {◼
    "@vp/vWheels/vWheels0": {◼
    "data/artifacts/userProjects/Bikes/vtInstances/vpInstances/vBikeChain
    "data/artifacts/userProjects/Bikes/vtInstances/vpInstances/vBikeFrame
}
```

Fig. 8. Instantiated V-Thing [7].

and the full data file will be generated for subsequent processing. (V-thing data entries are attached under products for vBikeChain and vBikeFrame as shown in Fig. 8 [7]).

4.4   V-Thing Optimization

Now we demonstrate how the V-thing artifacts would play together through a case study of optimizing a virtual bike production service network (Fig. 9). Generally, a service network models the flow of resources (products, materials, money, information, etc.) throughout a network of services to be generated, processed, transported or consumed. We construct a bike production service network in the context of the V-thing market to model a scenario where virtual things are specified through the requirements specs and optimal instances are created.

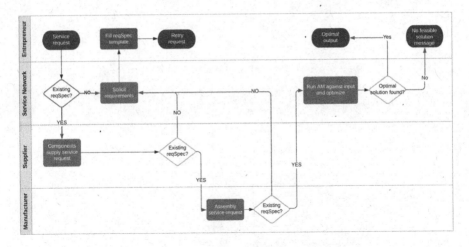

Fig. 9. Bike Production Service Network Flow Chart.

An entrepreneur wants to make a virtual bike product and she has found suppliers of the virtual components of the bike. There are two types of bikes she desires to manufacture and each has different components. She initiates a bike supply service network that has two sub-services: bike components supply service and bike assembly service, by filling out a requirements spec (Fig. 10) to specify what kind of metrics and objectives she expects of the virtual supply service.

```
'@context':
  '@vsModels': lib/vtModels/vsModels/
  '@vp': data/artifacts/userProjects/Bikes/vtInstances/vpInstances/
  '@productRef': data/artifacts/userProjects/Bikes/vtInstances/vpInstances/
  '@reqTemplates': lib/templates/reqTemplates/
input: []
output:
  - bike0
  - bike1
metricSchema: '@reqTemplates/servReqTemplates/bikeSupply0/metricSchemaBikeSupply
objectives:
  schemaAndBounds:
    cost:
      dgalType: floatMetric
      lb: 0
      ub: 5000
    function: >-
      /requirementsTemplates/objFunctionBikeSupply0.py
flows:
  bikeChain0: '@reqTemplates/prodReqTemplates/bikeChain0/reqSpecBikeChain0.json'
  bikeFrame0: '@reqTemplates/prodReqTemplates/bikeFrame0/reqSpecBikeFrame0.json'
  bike0: '@reqTemplates/prodReqTemplates/bike0/reqSpecBike0.json'
```

Fig. 10. Bike Supply Service Requirements Spec.

She is interested in the "cost" metric of the composite service of bikeSupply as well as the costs of the sub-services, so the cost field of the metric schema is annotated with a corresponding dgal type (Fig. 11). Please also note that part of the constraints regarding the metrics can be given at the same time, for example, lower bound and upper bound of cost are provided as optional arguments.

```
cost:
  dgalType: floatMetric
  lb: 0
  ub: 5000
constraints: true
rootService: bikeSupply
services:
  bikeComponentsSupply:
  bikeAssembly:
  bikeSupply:
    cost:
      dgalType: floatMetric
      lb: 0
      ub: 5000
    constraints: true
    inFlow: {}
    outFlow:
      bike0:
        qty:
          dgalType: intMetric
          lb: 0
          ub: 50
      bike1:
        qty:
          dgalType: intMetric
          lb: 0
          ub: 50
    subServices:
      - bikeComponentsSupply
      - bikeAssembly
flows: {}
products: {}
```

Fig. 11. Bike Supply Service Metric Schema.

Then the entrepreneur specifies the objective of the service network, which is to minimize the total cost if the bike products can be assembled and supplied based on the requirements that she provides. She specifies additional constraints (metric constraints, flow constraints, objective constraints, etc.) which are captured in the constraint function (Fig. 12).

```
prod1Constraint = (outFlow["bike0"]["qty"] >= 10)
prod2Constraint = (outFlow["bike1"]["qty"] >= 20)
```

Fig. 12. Bike Supply Service Constraints (part).

Once the requirements spec is fulfilled, an input file is extracted from it as the input for the analytic model of the bike supply service. The entrepreneur also configures which parameters will be used as decision variables. Then the Decision Guidance Management System (Fig. 1 left-hand side) receives the instruction and operates upon the

```
"bikeComponentsSupply": {
  "@context": {
    "@vsModels": "lib/vtModels/vsModels/",
    "@vp": "data/artifacts/userProjects/Bikes/vtInstances
    "@productRef": "data/artifacts/userProjects/Bikes/vtI
  },
  "model": "@vsModels/specialized/bikeComponentsSupply",
  "inFlow": {},
  "outFlow": {
    "bikeChain0": {
      "qty": {"dgalType": "int?"},
      "lb": 0,
      "ppu": 1,
      "item": "bikeChain0"
    },
    "bikeFrame0": {
      "qty": {"dgalType": "int?"},
      "lb": 0,
      "ppu": 2,
      "item": "bikeFrame0"
    },
    "bikeFrame1": {
      "qty": {"dgalType": "int?"},
      "lb": 0,
      "ppu": 5,
      "item": "bikeFrame1"
```

Fig. 13. Bike Supply Service Input with Variables.

V-things by drawing from the extensible, reusable and modular knowledge base of the analytic models. In Fig. 13 the decision variables are annotated by corresponding dgal types with a "?".

The input is fed into the analytic model pipeline. First, the input is instantiated if there are any virtual parts in it, which ensures all the data needed is accessible. Then a general composite analytic model runs against the input to compute the service metrics. During the metrics computation, the system checks whether the service is a composite one (having sub-services) or an atomic one. For a composite service, the pipeline will recursively compute the metrics of the sub-services and aggregate the metrics according to the metric schema in the end. The dgal optimize function formulates the variable input together with the objectives and constraints into a constraint optimization problem, and applies the proper solver to it. The output file is generated with the optimal solution if all constraints are satisfied as shown in Fig. 14. As a result, an optimal virtual service instance is created according to what the entrepreneur has specified.

```
{
  "cost": 1330,
  "constraints": true,
  "rootService": "bikeSupply",
  "services": {
    "bikeComponentsSupply": {
      "cost": 1070,
      "constraints": true,
      "inFlow": {},
      "outFlow": {
        "bikeChain0": {
          "qty": 50,
          "item": "bikeChain0"
        },
        "bikeFrame0": {=},
        "bikeFrame1": {=},
        "brakes0": {=},
        "derailer0": {=},
        "frontFork0": {=},
        "gears0": {=},
        "handleBars0": {=},
        "pedals0": {=},
        "seat0": {=},
        "tires0": {=},
        "wheels0": {=}
      }
    },
    "bikeAssembly": {=},
    "bikeSupply": {=}
  }
}
```

Fig. 14. Bike Supply Service Optimal Output.

5 Conclusions

In this paper we reported on the design and initial development of *OptiViTh* - a decision guidance system to recommend Pareto-optimal instantiation and composition of virtual things - parameterized products and services that can be searched, composed and optimized within a hierarchical service network against customer requirements specs. *OptiViTh* is based on an extensible and reusable repository of virtual things and their artifacts, which adhere to a proposed mathematical framework. We also demonstrated *OptiViTh* on a case study of a bike production service network in the context of cloud manufacturing, including how artifacts are implemented and used for V-thing operations. Specifically, we modeled a scenario where an entrepreneur provides a virtual service requirements spec, and an optimal virtual service instance is created by *OptiViTh*.

We envision powering up a market of virtual things with *OptiViTh* for entrepreneurs to formulate their ideas quickly and easily, and for designers and manufacturers to benefit from fulfilling the specs. We believe that a productivity framework of markets of virtual things is promising in facilitating the idea-to-market process and democratizing innovation.

Many research questions still remain open. They include (1) design and implementation of a more complete set of *OptiViTh* features; (2) studying limitations of existing MILP and NLP solvers for V-things optimization, and developing more efficient specialized algorithms; (3) studying usability of *OptiViTh* for entrepreneurs to formulate their ideas in V-things intuitively; and, (4) studying how effective V-thing markets are in facilitating the idea-to-market process.

References

1. Bottenbley, J., Brodsky, A.: SPOT: toward a decision guidance system for unified product and service network design, pp. 717–728. ICEIS, INSTICC, SciTePress (2021). https://doi.org/10.5220/0010459707170728
2. Brodsky, A., Gingold, Y.I., LaToza, T.D., Yu, L., Han, X.: Catalyzing the agility, accessibility, and predictability of the manufacturing-entrepreneurship ecosystem through design environments and markets for virtual things. In: Proceedings of the 10th International Conference on Operations Research and Enterprise Systems, ICORES 2021, Online Streaming, 4–6 February 2021, pp. 264–272. SCITEPRESS (2021). https://doi.org/10.5220/0010310802640272
3. Brodsky, A., Krishnamoorthy, M., Nachawati, M.O., Bernstein, W.Z., Menascé, D.A.: Manufacturing and contract service networks: composition, optimization and tradeoff analysis based on a reusable repository of performance models. In: 2017 IEEE International Conference on Big Data (IEEE BigData 2017), Boston, MA, USA, 11–14 December 2017, pp. 1716–1725. IEEE Computer Society (2017)
4. Brodsky, A., Luo, J.: Decision guidance analytics language (DGAL) - toward reusable knowledge base centric modeling. In: ICEIS 2015 - Proceedings of the 17th International Conference on Enterprise Information Systems, Barcelona, Spain, 27–30 April 2015, vol. 1, pp. 67–78. SciTePress (2015)
5. Egge, N.E., Brodsky, A., Griva, I.: An efficient preprocessing algorithm to speed-up multistage production decision optimization problems. In: 46th Hawaii International Conference on System Sciences, HICSS 2013, Wailea, HI, USA, 7–10 January 2013, pp. 1124–1133. IEEE Computer Society (2013)
6. Gingold, Y.I., Igarashi, T., Zorin, D.: Structured annotations for 2D-to-3D modeling. ACM Trans. Graph. **28**(5), 148 (2009)
7. Han, X., Brodsky, A.: Toward cloud manufacturing: a decision guidance framework for markets of virtual things. In: Proceedings of the 24th International Conference on Enterprise Information Systems - Volume 2: ICEIS, pp. 406–417. INSTICC, SciTePress (2022). https://doi.org/10.5220/0011114100003179
8. LaToza, T.D., Shabani, E., van der Hoek, A.: A study of architectural decision practices. In: 6th International Workshop on Cooperative and Human Aspects of Software Engineering, CHASE 2013, San Francisco, CA, USA, 25 May 2013, pp. 77–80. IEEE Computer Society (2013)
9. Li, B., et al.: Cloud manufacturing: a new service-oriented networked manufacturing model. Jisuanji Jicheng Zhizao Xitong/Comput. Integr. Manuf. Syst. CIMS **16**, 1–7+16 (2010)
10. Shao, G., Brodsky, A., Miller, R.: Modeling and optimization of manufacturing process performance using Modelica graphical representation and process analytics formalism. J. Intell. Manuf. **29**(6), 1287–1301 (2018)
11. Shin, S., Kim, D.B., Shao, G., Brodsky, A., Lechevalier, D.: Developing a decision support system for improving sustainability performance of manufacturing processes. J. Intell. Manuf. **28**(6), 1421–1440 (2017)
12. Yu, L.F., Yeung, S.K., Tang, C., Terzopoulos, D., Chan, T.F., Osher, S.J.: Make it home: automatic optimization of furniture arrangement. ACM Trans. Graph. **30**(4), 86 (2011)

Efficient Deep Neural Network Training Techniques for Overfitting Avoidance

Bihi Sabiri[1]([⊠]) [iD], Bouchra EL Asri[1] [iD], and Maryem Rhanoui[2] [iD]

[1] IMS Team, ADMIR Laboratory, Rabat IT Center, ENSIAS,
Mohammed V University in Rabat, Rabat, Morocco
{bihi_sabiri,bouchra.elasri}@um5.ac.ma,
bouchra.elasri@ensias.um5.ac.ma
[2] Meridian Team, LYRICA Laboratory, School of Information Sciences, Rabat, Morocco
mrhanoui@esi.ac.ma

Abstract. A deep learning neural network's ultimate goal is to produce a model that does well on both the training data and the incoming data that it will use to make predictions. Overfitting is the term used to describe how successfully the prediction model created by the machine learning algorithm adapts to the training data. When a network is improperly adapted to a limited set of input data, overfitting occurs. In this scenario, the predictive model will be able to offer very strong predictions on the data in the training set, and will also capture the generalizable correlations and the noise produced by the data, but it will predict poorly on the data that it has not yet seen during his learning phase. Two methods to lessen or prevent overfitting are suggested in this publication among many others. Additionally, by examining dynamics during training, we propose a consensus classification approach that prevents overfitting, and we assess how well these two types of algorithms function in convolutional neural networks. Firstly, Early stopping makes it possible to store a model's hyper-parameters when it's appropriate. Additionally, the dropout makes learning the model more challenging. The fundamental concept behind dropout neural networks is to remove nodes to allow the network to focus on other features which reducing the model's loss rate allows for gains of up to more than 50%. This study looked into the connection between node dropout regularization and the quality of randomness in terms of preventing neural networks from becoming overfit.

Keywords: Data overfitting · Early stopping · Dropout regularization · Machine learning · Neural network · Deep learning

1 Introduction

Deep neural networks, which are extremely potent machine learning systems that have a huge number of parameters, suffer from a major issue known as overfitting [1]. Multiple non-linear hidden layers in deep neural networks enable them to learn extremely complex correlations between their inputs and outputs, making them highly expressive

Supported by sabiss.net.

J. Filipe et al. (Eds.): ICEIS 2022, LNBIP 487, pp. 198–221, 2023.
https://doi.org/10.1007/978-3-031-39386-0_10

models. Even though the test data is taken from the same distribution as the training set, many of these complex associations will exist in the training set but not in the real data because of sampling noise when there is insufficient training data. Overfitting results from this, and numerous techniques have been devised to lessen it [2].

A technique known as early stopping aims to halt training before the model overfits or before it learns the model's internal noise [3]. This strategy runs the danger of prematurely stopping the training process, which would cause underfitting, the opposite issue. The ultimate objective is to find the "sweet spot" between underfitting and overfitting. Another approach to tackle overfitting issue is dropout [4]. Dropout is a potent and popular method for regularizing deep neural network training [5]. The basic concept is to randomly remove units and connections from the neural network while it is being trained. Units are prevented from overco-adapting as a result. Dropout samples are drawn from an exponential range of distinct thinning networks during training. By utilizing a single unthinned network with a decreased weight at test time, it is simple to replicate the effect of averaging the predictions of all these thinned networks. This provides significant improvements over conventional regularization techniques and significantly lowers overfitting. But how can we make our machines more efficient by include flaws in their learning process?

In general, making mistakes in a program hinders it from working properly, but oddly, if we construct a neural network, then it is definitely beneficial to purposefully introduce random faults, including the functioning of the network's neurons. We could try to gather more training data to counteract overfitting, but what if we can only access the data that has already been collected for us ? In such case, we could try to produce new data using the data that we already have. In order to gain fresh data that we can still plausibly think of as photographs of cars, animals or faces..., we can slightly rotate the zoom out of the zoom, make modest modifications to the white balance, or create a mirror image of the shot. At this point, we speak of an increase in data or data augmentation. In this post, we will examine two approaches to the overfitting issue in more detail: Dropout and Early Stopping.

The remainder of this paper is structured as follows. In Sect. 2, we briefly assess alternative approaches from the literature and place our answer in relation to them. Section 3 explains the overfitting mechanism for classification. In Sect. 4 we presented the theoretical aspect based on the mathematical approach with illustrative examples. Experiments in Sect. 5 demonstrate the potency of the suggested strategy. To validate the effectiveness, we conduct comprehensive experiments using common datasets and various network designs. While the paper's summary and recommendations for further research are included in Sect. 6.

This article is the extended version of the article that appeared in Sitepress following the ICEIS 2022 conference proceedings [19]. The added content enhances the extended version, such provide brand-new tests (Sect. 5.2, page 18), examples (Sect. 5.2, page 18), additional datasets (Sect. 5.2, page 18), comparisons with the literature (Sect. 2, page 3), algorithm (Sect. 4, page 8), (Sect. 4, pages 9, 10, proofs (Sect. 5, pages 10 to 12), figures (Fig. 1, page 4, Fig. 5, page 10, Fig. 6, page 10, Fig. 7, page 10, Fig. 8, page 12, Fig. 14, page 19, Fig. 15, page 20, Fig. 17, page 21), results (Table 3, page 19) (Table 3, page 19, Table 4, page 22).

2   Related Works

Without making any attempt to be exhaustive, we will highlight a few correlations between ovefitting and prior literature below [6] [4]:

Method of Sochastic Gradient Descent. The earliest description of a dropout algorithm as a useful heuristic for algorithmic regularization can be found in [1]. In order to train neural stall networks on small batches of training examples, the scientists employed the traditional stochastic gradient descent method; however, they changed the penalty term that is typically used to prevent weights from becoming too large. They employed the "average network" for the test, which includes all hidden units but has had its outgoing weights reduced by half to account for the fact that there are now twice as many active units. In actuality, this performs quite similarly to the average over a significant number of dropout networks.

Dropout Regularization for Neural Networks. By minimizing the bidirectional KL-divergence of the output distributions of any two submodels collected via dropout in model training, R-drop is a straightforward yet highly effective regularization technique [5]. To put it more specifically, each data sample undergoes two forward passes during each mini- batch training, and each pass is handled by a different sub model by randomly removing some hidden units. R-Drop reduces the bidirectional Kullback-Leibler (KL) divergence between the two distributions, which compels the two distributions for the same data sample produced by the two sub models to be compatible with one another [5].

Regularization: Implicit and Explicit. In a recent paper, [7] separate the explicit and implicit regularization effects of dropout, i.e., the regularization due to the predicted bias that is created by dropout vs the regularization induced by the noise due to the randomness in dropout. They suggest and demonstrate the viability of an approximate explicit regularizer for deep neural networks. However, their generalization bounds are restricted to linear models and need for norm-bounded weights [8].

Using Dropout Matrix Factorization. The study of dropout was motivated by recent Cavazza papers [9]. Without restricting the rank of the elements or incorporating an explicit regularizer in the objective, Cavazza [9] were the first to investigate dropout for low-rank matrix factorization. They show how dropout in the context of matrices factors out an explicit regularizer whose convex envelope is specified by nuclear norm.

3   Overfitting Machanism

Deep neural networks are models made up of numerous layers of straightforward, non-linear neurons. With a sufficient number of neurons, the model can learn incredibly complex functions that can precisely carry out challenging tasks that are impossible to hard code, such as image classification, translation, speech recognition, etc. Deep neural networks' primary feature is its capacity to autonomously learn the data representation required for feature identification or classification without any prior knowledge.

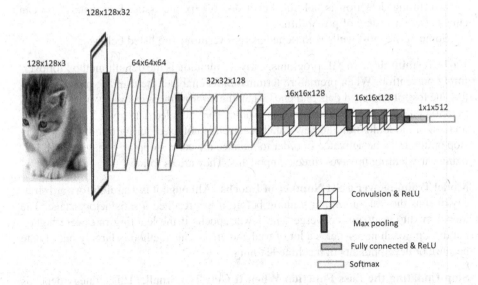

Fig. 1. Convolutional Neural Network: Features Learning.

Convolutional neural networks, (see Fig. 1), were trained using data from the ImageNet Large Scale Visual Recognition Competition. It can accurately classify 1,000 different items and has 16 layers and 138 million parameters.

The data science notion of overfitting describes what happens when a statistical model fits perfectly to its training data but poorly to the test data. Unfortunately, when this occurs, the algorithm's goal is defeated because it is unable to accurately perform against unobserved data [10]. Machine learning algorithms need a sample dataset to train the model when they are created. The model, however, may begin to learn the "noise" or irrelevant information within the dataset if it trains on sample data for an excessively long time or if the model is overly complex. When The model learns the noise and fits it far too well.

The model becomes "overfitted" to the training set and it struggles to generalize effectively to fresh data. If a model cannot adequately generalize to new data, it will be unable to carry out the categorization or forecasting tasks for which it was designed.

High variance and low error rates are reliable indicators of overfitting. A portion of the training dataset is often set aside as the "test set" to look for overfitting in order to stop this kind of behavior. Overfitting is indicated if the test data has a high error rate while the training data has a low error rate. There are insufficiently many feature dimensions, model presumptions, and parameters, as well as an excessive amount of noise. Because of this, the fitting function accurately predicts the training set while producing subpar predictions for the test set of fresh data.

3.1 Popular Solution for Ovefitting

Overfitting and the bias/variance trade-off in neural nets and other machine learning models have been extensively discussed in the literature [11].

Overfitting detection is helpful, but it doesn't fix the issue. Fortunately, we can choose from a variety of possibilities.

Some of the most well-liked remedies for overfitting are listed below:

Early Termination. In ML programs, early termination is the default method for over-fitting prevention. When premature termination is enabled, the training process monitors the loss of excluded data and ends when the improvement in loss over the previous iteration falls below a predetermined threshold. The excluded data gives a good approximation of model loss on the new data because it is not used during training. The early stop option must be activated in order to manage the early stopping behavior [12]. Utilizing early halting involves three components. They are as follows:

Model Training over a Set Number of Epochs. Although it is a straightforward strategy, it runs the risk of ending training before it has reached a satisfactory stage. The model might eventually converge with fewer epochs if the learning rate were higher, but this approach necessitates a lot of trial and error. This method is largely out of date because of developments in machine learning.

Stop Updating the Loss Function When It Gets Too Small. This strategy depends on the fact that the gradient descent weight updates are substantially lower as the model gets nearer to the minima, making it more sophisticated than the first. The drive is typically stopped when the update is as little as 0.001, as doing so minimizes loss and conserves computational resources by avoiding extra epochs. Overfitting is still a possibility, though.

Methodology for Overall Validation. The most frequent early stopping strategy is this clever tactic (see Fig. 2). [20]. Looking at how the training and validation mistakes change with the number of epochs is crucial to comprehending how this works (as in Fig. 2). Until the growing epochs have no longer a significant impact on the error, the learning error diminishes exponentially. With rising epochs, the validation error does, however, initially drop, but beyond a certain point, it starts to grow. Because the model will start to overfit if it is continued past this point, it should be terminated early. Although the validation set technique prevents overfitting the best, it typically takes many epochs before a model overfits, which can use a lot of processing power. Designing a hybrid approach between the commit set strategy and stopping when the loss function update gets modest is a clever method to obtain the best of both worlds. Training might end, for instance, when one of these goals is reached.

Dropout. Dropout is a regularization method for neural networks that, with a predetermined probability, loses a unit along with connections during training. Dropout modified the idea of learning all of the network's weights at once to learning a portion of them in each training iteration [13]. By randomly setting hidden Unit activity to zero with a probability of 0.5 during training, it considerably enhances the performance of deep neural networks on a variety of tasks, including vision issues [14]. Figure 2 shows every neuron is utilized when training a model. Dropout is a straightforward method to avoid overfitting in neural networks, each neuron has a 50/100 probability of being turned off when p = 0.5 (see Fig. 3).

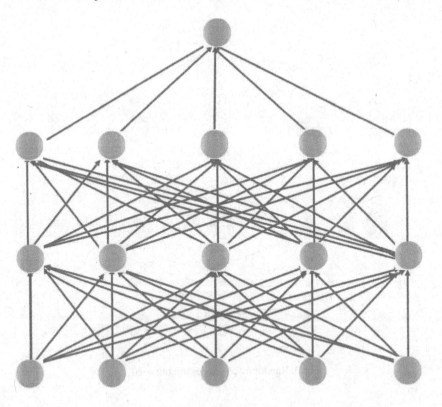

Fig. 2. All nodes are used.

The problem of overfitting in expansive networks was overcome by this issue. Deep Learning architectures that were larger and more accurate suddenly became feasible [15].

Regularization was a significant study topic prior to Dropout. Beginning in the early 2000s, regularization techniques for neural networks, such as L1 and L2 weight penalties, were introduced [16]. However, the overfitting problem was not entirely resolved by these regularizations. Co-adaptation was the root cause.

Neural Network Co-adaptation. Co-adaptation is a significant problem that leads to overfitting in learning big networks. Co-adaptation in a neural network thus refers to the degree to which certain neurons rely on others. If the independent neurons receive "poor" inputs, the dependent neurons may also be impacted, which could lead to a large change in the model's performance (overfitting). If every node in such a network is when learning weights simultaneously, it is typical for some one of the links will be better capable of prediction than the others. In this case, the stronger connections are learned more while the weaker ones are ignored when the network is trained iteratively.

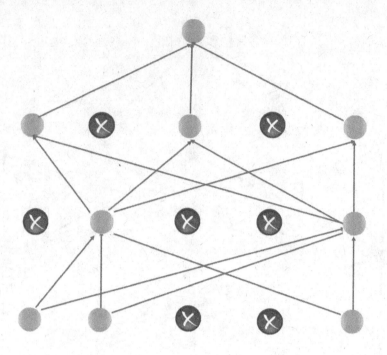

Fig. 3. Randomly p% nodes are not used

Only a portion of the node connections are learned throughout numerous iterations. And the others cease taking part.

The green circles in Fig. 4 represent neurons that are extremely dependent on others, whereas the Yellow circles represent neurons that are independent of others. Overfitting may result in a major change in the model's performance if those independent neurons receive "poor" inputs because the dependent neurons may also be impacted.

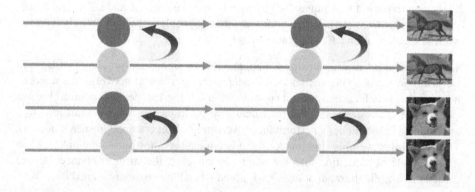

Fig. 4. Co-adaptation features detection-Some neurons depend heavily on others.

Co-adaptation is the term for this occurrence. The classic regularization methods, such as the L1 and L2, were unable to stop this [17]. They also regularize based on the connections' capacity for prediction, which is the cause. They consequently approach determinism in terms of selecting and rejecting weights. Again, the strong become stronger while the weak become weaker. Expanding the size of the neural network would not be helpful, which was a significant result. Consequently, the size and precision of neural networks were constrained. Dropout then followed. a fresh regularization technique. The co-adaptation was resolved by it. We could now construct a deeper foundation (See Fig. 1).

As a result, accuracy and scale of neural networks were constrained. Next was Dropout. a fresh regularization strategy. The co-adaptation was fixed. We might now create wider and deeper networks. And make use of all of it's predictive power. Let's get started on the Mathematics of Dropout with this backdrop. Machine learning techniques always perform better when combined with other models. The obvious idea of averaging the outputs of numerous separately trained nets is prohibitively expensive with big neural networks, though. Combining several models works best when the individual models are distinct from one another, and neural net models can be distinguished from one another by either having unique architectural designs or being trained on unique datasets. Finding the best hyperparameters for each architecture is a difficult operation, and training each huge network requires a lot of compute. This makes training many distinct architectures challenging. Additionally, since big networks typically need a lot of training data, there might not be enough data to train several networks on various subsets of the data. In applications where it is crucial to react fast, even if one were to be able to train a variety of distinct big networks, using them all at test time would be impractical. Each unit is retained with a preset probability p that is independent of other units in the simplest case. p can be set at 0.5 or can be determined using a validation set, which appears to be near to optimal for a variety of networks and applications. However, for the input units, the ideal probability of retention is typically closer to 1 than 0.5. (Sect. 5 below will address the mathematics component of the selection of these values.)

Dropout, which is a remarkably successful method for preventing overinterpretation and consists just of configuring artificial neurons' momentary susceptibility, is a relatively straightforward procedure. In short, neural balls are actually highly intelligent errors since they enable us to combat both the absence of learning data and learning data faults, as well as the learning of a single model rather than an entire forest. Through the concept of "multi start", the initialization process essentially entails starting with numerous initial neural networks whose initialization is random and allowing us to preserve toptimal neural networks for learning, or possibly even better.

4   Maths Behind Dropout

Take into account a neural network with L hidden layers. Let $l \in \{1, 2, \ldots L\}$, let z^l denote the vector of inputs into layer l, W^l and b^l are respectively the weights and biases at layer l [18, 19]. Dropout causes the feed-forward process to change to: (For $l \in \{0, 1, 2, \ldots \ldots L - 1\}$ and any hidden unit i). The following equation can be used to express the conventional network without dropout's training phase mathematically (see Eqs. 1 & 2).

$$Z_i^{l+1} = W_i^{l+1} * Y^l + b_i^{l+1} \tag{1}$$

$$y_i^{l+1} = f(z_i^{l+1}) \tag{2}$$

where f represents any activation function, such as the sigmoid function $f(x) = \frac{1}{1+e^{-x}}$

The result of Eq. 1 is passed through the activation function (f) to incorporate non-linearity to produce the output y^{l+1}, or prediction, demonstrating that all of the neurons are involved in the decision-making process.

Formula for Bernoulli Distribution. In probability, a Bernoulli test with parameter p (real between 0 and 1) is a random experiment (i.e. submitted to chance) with two outcomes: Success or failure The real p represents the probability of a success. The real q = 1 - p represents the probability of failure. The definition of "success" and "failure" is conventional and depends on the conditions of the experiment:

$$f(n,p) = \begin{cases} p & \text{if } n = 1 \\ 1-p & \text{if } n = 0 \end{cases} \tag{3}$$

which can be written

$$f(n,p) = p^n(1-p)^{(1-n)} \tag{4}$$

After dropout, the instruction is revised to become:
(r_i^l indicator variable = 1 if event results in success = 0 otherwise)

$$r_i^l \sim \text{Bernoulli}(p) \tag{5}$$

$$\hat{y}^l = r^l \times y^l \tag{6}$$

$$z_i^{l+1} = w_i^{l+1}\hat{y}^l + b_i^{l+1} \tag{7}$$

$$y_i^{l+1} = f(z_i^{l+1}) \tag{8}$$

Despite the addition of a new term, or neuron, called r, the training is quite identical to the standard network. By assigning a 1 (neuron participates in the training or is turned on) or 0 (neuron is turned off), which maintains the neuron active or shuts it off, the training process proceeds. By doing this, overfitting is lessened, and our model is now capable of making good and precise predictions based on real-world data (data that the model is not aware of). The weights are scaled as W(l)test = pW(l) at test time. Without dropout, the resulting neural network is employed.

A node or unit is a common name for the fundamental computational building block (model neuron). It receives input from various additional devices or even an external source. If you wish to alter synaptic mastering, you can change the weight w that corresponds to each input.

The device calculates a select few features of the weighted sum of its inputs (See Fig. 5).

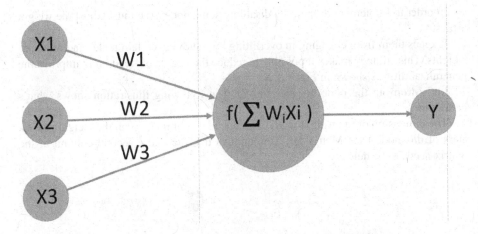

Fig. 5. Full Network without Dropout: All nodes are used for training the model

- • The term "weighted sum" refers to the node i's network input.
- • Wij stands for the weight transferred from node i to unit j.

We will use a network with three layers as inputs, four as hiddens layers, and three output nodes to illuminate this inhibition on the juxtaposed nodes.

Without dropout all the elements of the matrix at each level of the propagation are used during the training of the model and if there are dependencies between the nodes, the ovefitting could make apparition (see Fig. 6) In this case all the nodes participate in the training of the model and the output matrix at the level of the hidden layers for the weights and as follows (see Matrix 9):

$$\text{Input} - \text{Hide}_{4\times 3} = \begin{bmatrix} wi_{11} & wi_{12} & wi_{13} \\ wi_{21} & wi_{22} & wi_{23} \\ wi_{31} & wi_{32} & wi_{33} \\ wi_{41} & wi_{42} & wi_{43} \end{bmatrix} \tag{9}$$

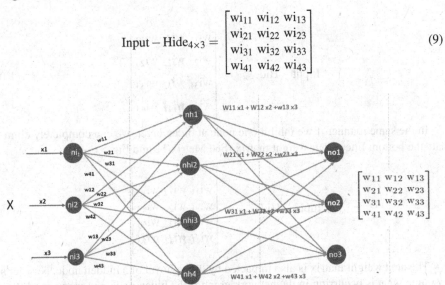

Fig. 6. Full Network without Dropout: All nodes are used for trainining the model.

In order to implement dropout by disabling some nodes, we must adjust the weight arrays.

It keeps them from engaging in overfitting by removing certain nodes and reducing weights. This strategy makes it possible to reduce the degree of freedom to improve the generalizability, as shown in Fig. 7 & 8

Let's drop out the node ni_2 (see Fig. 7). The following illustration shows what's going on:

The dimension of the weight matrix is reduced by elimination all the weight parameters of this node (See Matrix 10) This implies that the entire second column of the matrix needs to be deleted.

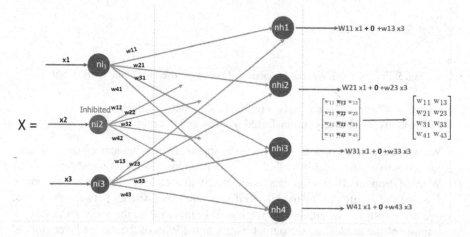

Fig. 7. Dropout: Inhibiting the input node(ni2) decreases the size of the weight matrix(4×2).

$$Input - Hide_{4 \times 2} = \begin{bmatrix} wi_{11} & \cancel{wi_{12}} & wi_{13} \\ wi_{21} & \cancel{wi_{12}} & wi_{23} \\ wi_{31} & \cancel{wi_{32}} & wi_{33} \\ wi_{41} & \cancel{wi_{42}} & wi_{43} \end{bmatrix} \qquad (10)$$

In the same manner if we inhibit one node at hiden layer, we can completely eliminate the bottom line of our weight matrix (See Matrix 12 and Fig. 8).

$$Input - Hide_{3 \times 3} = \begin{bmatrix} wi_{11} & wi_{12} & wi_{13} \\ wi_{21} & wi_{22} & wi_{23} \\ wi_{31} & wi_{32} & wi_{33} \\ \cancel{wi_{41}} & \cancel{wi_{42}} & \cancel{wi_{43}} \end{bmatrix} \qquad (11)$$

The next weight matrix is also impacted by the removal of a hidden node. Now let's examine what is occurring in the network graph if the hiden node nh4 is removed (See Matrix 11):

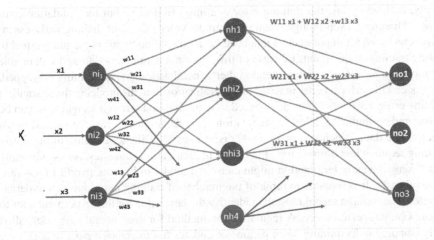

Fig. 8. Dropout: Inhibiting the hiden node(nh4) decreases the size of the weight matrix(3×3).

$$\text{Hide} - \text{output}_{4\times 3} = \begin{bmatrix} wh_{11} & wh_{12} & wh_{13} & \cancel{wh_{14}} \\ wh_{21} & wh_{22} & wh_{23} & \cancel{wh_{24}} \\ wh_{31} & wh_{32} & wh_{33} & \cancel{wh_{34}} \\ wh_{41} & wh_{42} & wh_{43} & \cancel{wh_{44}} \end{bmatrix} \quad (12)$$

In this instance, the entire last matrix column needs to be deleted. As a result, we must eliminate every last product from the summation.

In the dropout method, a predetermined number of nodes from the input and hidden layers are randomly selected to be active while the remaining nodes from these layers are disabled. This means that certain network neurons as well as all of their incoming and outgoing connections are temporarily turned off whenever a training pass occurs. The basic idea behind this technique is to introduce noise into a layer in order to obstruct any non-significant coincidental or interdependent learning patterns that might develop between layer units.

After that, we may use this network to train a portion of our learning set.

5   Experimental Results and Discussions

In this section, we empirically demonstrate that hyper-parameters produce characterizations after a specific number of repetitions. Early termination regularization entails stopping the model's training when its performance on the validation set starts to deteriorate which lowers accuracy or increases loss.

5.1   Simulation Scenario: EarlyStopping

The errors reduce over a number of iterations until the model starts to overfit, which can be seen by showing the error on the training dataset and the validation dataset jointly. Following this, the training error continues to decline but the validation error grows. Therefore, even though training continues after this point, halting early essentially returns the set of parameters that have been utilized up to this point and is equal to ending training at this point. Because of this, the final parameters returned will enable the model to have lower variance and higher generalization. When training is stopped, the model will perform better in terms of generalization than the model with the smallest training error. Early cessation, as opposed to regularization through weight loss, can be considered an implicit kind of regularization. This technique is effective since it needs less training data, which is something that isn't always available. Because of this, early quitting requires less training time than other regularization strategies. As with training data, excessive early termination might cause the model to become overfit to the validation dataset. It is possible to think of the number of training iterations as a hyperparameter. The training model must then identify the best value for this hyperparameter to ensure optimal performance. A regularization method for deep neural networks called early stopping halts training when parameter updates fail to enhance performance over a validation set. As a result, we keep track of the most recent best settings and update them throughout the training process. When parameter modifications no longer result in improvements (after a predetermined number of iterations), we cease training and use the most recent best settings. By restricting the optimization process to a smaller region of parameter space, it acts as a regularizer. The plot below (Fig. 9) shows that the model continues to diverge without Early Stopping [20] after reaching the optimal point where the two errors are equal. If the process does not stop, the model's performance will continue to deteriorate.

The model's performance is optimal at this moment. EarlyStopping requires the validation loss in order to determine whether it has decreased. If so, it will create a checkpoint for the current model.

5.2   Simulation Scenario: Dropout

Some datasets SONAR [21], Diabetes, [22] and MNIST [24] are used in the experiments. Classification tests were carried out using networks of various designs while holding all hyperparameters, including p, unchanged in order to assess the robustness of dropout. The test error rates for these various architectures as training goes on are shown in (Fig. 11). The two distinct groups of trajectories show that the identical designs trained with and without dropout have significantly different test errors. Without employing hyperparameters that were optimized especially for each architecture, dropout provides a significant improvement across all architectures. The algorithm produces the results shown in (Fig. 11) after being tested with some categorization datasets: On the left and right, accuracy and loss are illustrated with and without dropout, respectively): Each neuron in the model has a 50% probability of being deactivated, or p = 0.5, and is made up of 7 hidden layers that alternate with 6 dropout layers. This random deactivation is used at every epoch. In other words, the model will learn with a different configuration of neurons on each pass (forward propagation), with the neurons activating and deactivating at random. With each iteration of this method, somewhat different patterns and neuronal configurations are produced. Disrupting the traits that the model has learned is the goal.

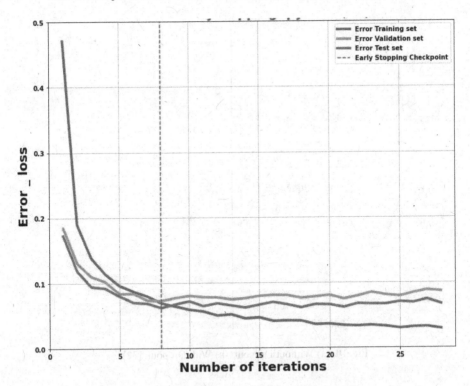

Fig. 9. Dropout : Early Stopping and Loss: checkpoint [32].

The synchronicity of neurons is often the foundation for model learning; however, in the case of dropout, each neuron must be explored separately, and its neighbors may be arbitrarily inhibited at any time. Only during model training is The Dropout active. Each neuron in the test is still alive, and its weight is doubled by probability p. To specify the desired probability of deactivation while using Keras and Tensorflow, just add a dropout layer.

Dropout does, in fact, somewhat impair performance since it mutes neuron activation during training. The neural network cannot rely heavily on any one neuron because each neuron has a proportional chance of being neglected during one forward pass, making dropout a very useful regularization strategy. Since it is more difficult to merely memorize training data with this extra sup-pression, overfitting is prevented, and the neural network is able to generalize to examples it has never seen. This means that even while training results in a decrease in performance, performance improves during inference and validation. But when the neural net is used for prediction, dropout is often eliminated after training is finished.

Figure 10 accuracy and loss curves show that, in the absence of dropout, the predictive model created by the automatic learning algorithm adapts well to the training set, which supports overfitting as the training data has a low error rate and high accuracy. While utilizing dropout lessens the significance of this adaptation.

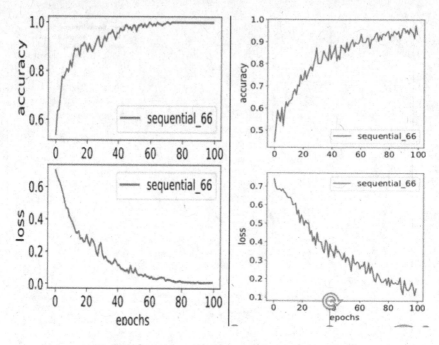

Fig. 10. (a) Without Dropout. (b) With Dropout. [32].

A different dataset was used in this case [22] and the results are shown in (Fig. 10):

In this example, in contrast to the previous one, we show the evolution of precision for the test sample with and without dropout, and we can see that the loss of the test dataset increases continuously when the dropout is not used (81% at iteration 100), whereas it stabilizes at around 45% at the same iteration when the dropout is used. The test sample's accuracy is 75% without dropout and 81% with dropout.

This third example uses MNIST, a set of 28×28 pixel pictures with a single channel, for the experiments. We employ Maxpooling for convolutional and fully-connected layers, softmax activation function for the output layer, and rectified linear function [23] for the dense layer. The CNN (Iraji et al., 2021) architecture for MNIST is $1 \times 28 \times 28$, which represents a CNN with a single input image of size 28×28, a convolutional layer with six feature maps and 5×5 filters, a pooling layer with 2×2 pooling regions, a convolutional layer with sixteen feature maps and 5×5 filters, a fully-connected layer with one thousand hidden units, and an output layer. We show how the precision and loss have changed over time for the training and test samples with and without dropout, and we note that without dropout, the accuracy of the training and test datasets diverge (Fig. 8), and the same is true for the losses of the two datasets, indicating overfitting. However, with dropout, the accuracy and loss converge after ten iterations. Each hidden unit in this case is arbitrarily excluded from the network with a probability of p [0.2, 0.5], so it cannot rely on the presence of other hidden units.

In the second illustration, the dropout strategy entails zeroing each hidden neuron's output with probabilities of 0.25 on the first two levels, 0.5 on the third, and 0.25 on

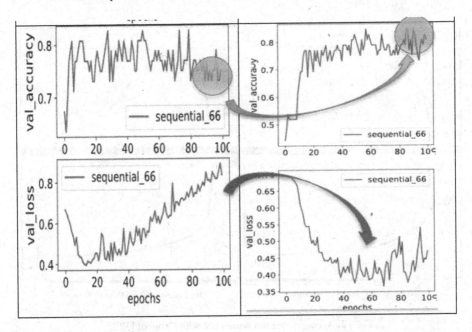

Fig. 11. By using dropout layer test accuracy increased from 0.75 to 0.81 while loss decreased from 81% to 45% [32].

the final layer. The neurons that are "dropped" in this manner do not participate in backpropagation or forward transit. In this manner, a distinct architecture is sampled by the neural network each time an input is given. Mainly because a neuron cannot rely on the existence of certain other neurons. As a result, co-adaptations between neurons are reduced as it is pushed to learn more reliable characteristics that work well with several distinct random subsets of other neurons.

In order to take the geometric mean of the prediction distributions generated by the several dropout networks, we employ all of the neurons by multiplying their outputs by the coefficients mentioned above. We can observe from the upper portion of Fig. 12 above that after a few epochs, the precision of the validation has practically reached a standstill. The accuracy increases linearly at first and then quickly stagnates after that.

Regarding the loss of validation, the Fig. 12 (bottom portion) demonstrates that it first declines linearly before increasing after a few epochs, indicating that overfitting is the cause.

In this situation, it is wise to add a little dropout to our model to see if it might help with reducing over-fitting. Figure 12 shows how the dropout allows us to observe that the validation loss and validation accuracy are synchronized with the training loss and training accuracy. Even if the validation loss and the accuracy line are not linear, the relatively tiny difference between training and validation accuracy demonstrates that the model is not overfitting, and the validation loss lowers when the dropout is activated.

As a result, we can say that the model's generalization capability improved significantly because the loss on the validation set was barely different from the training set

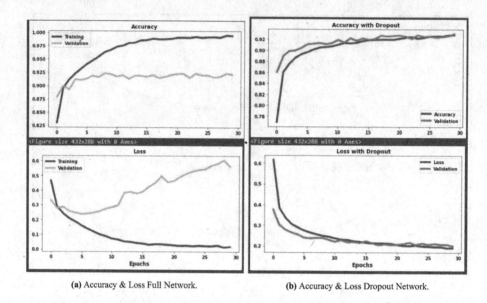

(a) Accuracy & Loss Full Network.   (b) Accuracy & Loss Dropout Network.

Fig. 12. Accuracy & Loss without & with Dropout [32].

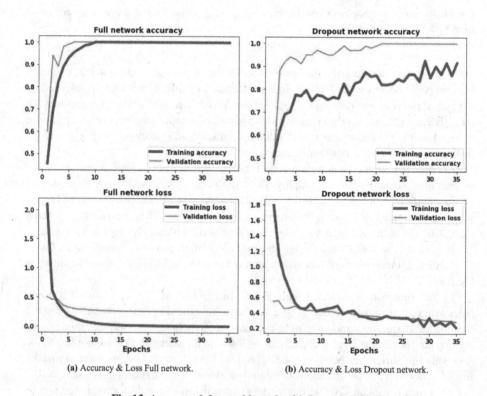

(a) Accuracy & Loss Full network.   (b) Accuracy & Loss Dropout network.

Fig. 13. Accuracy & Loss without & with Dropout [32].

loss. In a second example of classification based on a dataset (Dogs vs. Cats) [25], we will examine the effectiveness of a full network on simulated data and contrast it with conventional dropout. The model is over-fitting since the difference between training accuracy and validation accuracy is very large (See Fig. 12). Training loss approaches 0, whereas training accuracy nears 1. As a result, the model matches the training set exactly, which indicates over-fitting. Even if the model is not entirely stable, we can see that the gap between training and validation loss is less and that the gap between training and accuracy gaps converges slowly using the dropout strategy (See Fig. 12).

Actually, the difference between these two curves is so minor that it can be said that the model is no longer overfitting. The model's lack of overfitting is supported by training and valuation loss. Both loss values are comparable (Fig. 13).

Furthermore, the model no longer accurately predicts the training set. This is another indication that the model does not overfit and that it generalizes more effectively.

These tables demonstrate the change in accuracy [Table 1] and loss [Table 2] of validation datasets with and without dropouts. We see a rather decent performance, which increases the motivation for the potential of Dropout approaches for enhancing the predictive performance [26].

Table 1. Accuracy: validation Dataset [32].

Dataset	Normal Accuracy	Droput Accuracy
SONAR	99%	95%
Diabetes	75%	81%
MNIST	98%	99%
Dogs/Cats	100%	100%

Table 2. Loss: validation Dataset.

Dataset	Normal Loss	Droput Loss
SONAR	99%	45%
Diabetes	50%	30%
Dogs/Cats	55%	23%

Classification with Dropout Regularization. A high variance model overfits the data set and is unable to generalize to new data. By adding regularization or increasing the amount of data and then retraining the model, this issue can be minimized. Since adding extra data is typically too expensive in practice, regularization is frequently a workable solution to the overfitting problem. In this scenario, we covered regularization for logistic regression. Here, regularization is reviewed with an emphasis on neural networks.

The primary actions in creating our neural network are:

1. Clarify the model's structure (such as number of input features)

2. Set up the model's parameters.
3. Loop:
    - • Estimate the current loss (forward propagation).
    - • Find the current gradient (backward propagation).
    - • Revise the parameters (gradient descent)

We'll be putting in place a number of "helper functions" to construct our neural network. To create an L-layer neural network for the following assignment, these auxiliary functions will be utilized. We will be given clear instructions that will guide us through each minor assistance function we implement. For example, we will provide more details on the cost function which is based on cross-entropy. We will now put forward and backward propagation into action. To determine whether our model is genuinely learning, we must determine the cost which neural network seeks to minimize. We'll use the cross-entropy cost, which is provided by the following formula (See Eq. 7):

$$-\frac{1}{m}\sum_{i=1}^{m}(y^{(i)}\log\left(a^{[L](i)}\right)+(1-y^{(i)})\log\left(1-a^{[L](i)}\right))\tag{7}$$

Let's visualize the training data on a dataset for a classification problem in which a class label is predicted for a specific example of input data (see Fig. 14).

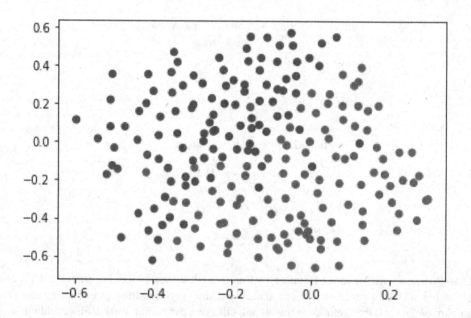

Fig. 14. Model without Dropout.

To distinguish the blue from the red dots, we need a classifier. After the training our model without dropout we obtain the results fo the accuracies shown in Table 3.

Table 3. Accuracy for Training & Test Dataset with no regularization.

Accuracy	Value
Train Accuracy	100%
Test Accuracy	92.5%

Training effectiveness is 100%. This serves as an excellent sanity check to ensure that our model is operational and capable of fitting the training set of data. Testing error is 92.5%, given our tiny dataset and the fact that logistic regression is a linear classifier.

We can also see that the model has obviously overfit the training set of data. Later on in this specialization, we will see how to minimize overfitting, for instance by use dropout.

It is the reference model that will allow to see the impact of the regularization via the dropout method [28]. To draw the decision limit of our model before any regularization see the following figures (See Fig. 15(a) & (b)).

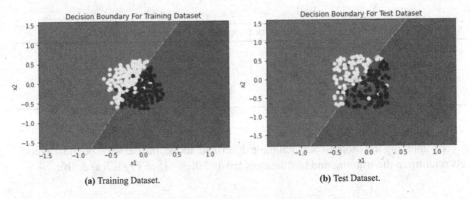

(a) Training Dataset. (b) Test Dataset.

Fig. 15. Decision Boundary Model without Dropout for Training (a) & Test (b) Dataset.

One of the most popular regularization techniques for neural networks is node dropout, which aims to randomly select hidden nodes to become inactive during a training iteration [29]. The overarching goal of this method is to make hidden nodes learn various weights and biases. This imposed diversity between hidden nodes prevents the network from discovering the best match for the training data, which has the effect of improving generalizability to unseen data. The random choice of hidden nodes uses the generation of uniform pseudo-random numbers and is a choice option for the quality of randomness analysis in the context of neural networks.

We will add dropout to the all hidden layers of the neural network we are using. Neither the input layer nor the output layer will experience dropout.

To do so, we create a variable D_1 with the same shape as A_1 using np.random.rand() to generate random numbers between 0 and 1 for D_1. Change A_1 to $A_1 * D_1$. (Some neurons are being turned off.) D_l can be viewed as a mask that disables some of the values when it is multiplied by another matrix (See Algorithm 1).

Algorithm 1. The fundamental algorithm for Dropout: Forward Propagation.

1 **for** $l \leftarrow 1$ to N Hidden layes **do**
2 | $D_l = np.random.rand(A_l.shape[0], A_l.shape[1])$
3 | $D_l = D_l < keep\_rate$
4 | $A_l = A_l Di$
5 | $A_l = A_l / keep\_rate$
6 **end**

By once again applying the same mask D_l to dA_l during backpropagation, the same neurons must be turned off. We had divided A_l by keep_prob during forward propagation. As a result, in order to do backpropagation, we will need to divide dA_l by keep_prob once more) (Fig. 16).

Algorithm 2. The fundamental algorithm for Dropout: Backward Propagation.

1 **for** $l \leftarrow 1$ to N Hidden layes **do**
2 | $dA_l = dA_l D_l$
3 | $dA_l = dA_l / keep\_rate$
4 **end**

By applying the dropout out technique described above, the decision boundary models relating to the training and test datasets are as follows (See Fig. 17(a) & (b):

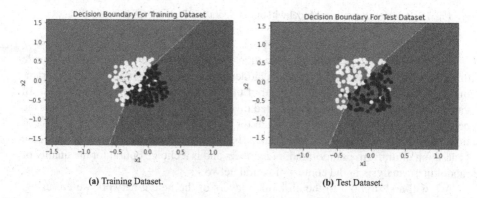

(a) Training Dataset.   (b) Test Dataset.

Fig. 16. Decision Boundary Model with Dropout for Training (a) & Test (b) Dataset.

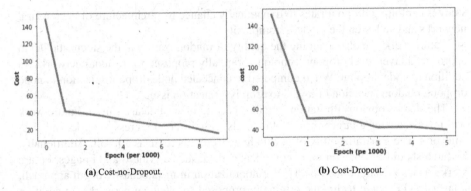

(a) Cost-no-Dropout.   (b) Cost-Dropout.

Fig. 17. Cost Full Network and Dropout.

By comparing the two Tables 3 and 4, we find that the difference between the accuracies of the training and test dataset is smaller with the dropout, which leads us to conclude that the dropout has allowed the model to generalize its manners by having a similar behavior between the data of the training dataset (data already seen) and that of the test dataset (data never seen), which therefore reduces overfitting.

Table 4. Accuracy for Training & Test Dataset with Dropout.

Accuracy	Value
Train Accuracy	94.8%
Test Accuracy	95.0%

6 Conclusion

The primary focus of this research is on the issue of understanding and applying Early halting and Dropout on the entry into the maximum pooling layers of convolutional neural networks. Early stopping is a technique to halt a neural network's training when the validation loss reaches a certain point. Dropout is a method for enhancing the performance of neural networks by minimizing overfitting. By deactivating specific units (neurons) in a layer with a specific probability p, it forces a neural network to learn more resilient properties that are beneficial. For dropout training in layer ReLU networks utilizing the logistic loss, we offer exact iteration complexity results.

In this study, we exclusively take into account neural networks that use the ReLU layer for classification. While we'll keep looking into other neural network topologies, we expect that they'll all use the same fundamental technique.

In order to promote a sparse representation of the data and the neural network structure, we would want to examine the applications of deep network compression [30] and the sparsity property of learning retention rates in the future [31]. Another area of study is exploiting dropout rates to dynamically change the architecture of deep neural networks and so lessen the model's complexity.

Future work can also examine the quality of randomization in the stochastic depth approach, a layer-level dropout technique originally proposed for residual networks, in addition to node dropout. When compared to concealed node dropout selection in node dropout, random selection of layers to drop is a selection issue.

The study's obvious limitations are the sheer number of datasets used. Future work will involve testing a larger variety of datasets with different regression tasks and classification. The generalizability of our findings would increase as a result. Additionally, research should be done on more complicated data, such as videos and images. Future work can also examine the quality of randomization in the stochastic depth approach, a layer-level dropout technique originally proposed for residual networks, in addition to node dropout. When compared to concealed node dropout selection in node dropout, random selection of layers to drop is a selection issue.

Acknowledgements. Thank you for the reviewers of ICEIS 2022 who devoted their precious time to read the initial version of this article. Their comments and suggestions allowed the continuous improvement of the content of this article.

References

1. Hinton, G.E., Srivastava, N., Krizhevsky, A., Sutskever, I., Salakhutdinov, R.R.: Improving neural networks by preventing co-adaptation of feature detectors, vol. 1 (2012)
2. Wu, J.-W., Chang, K.-Y., Fu, L.-C.: Adaptive under-sampling deep neural network for rapid and reliable image recovery in confocal laser scanning microscope measurements. IEEE Trans. Instrum. Meas. **71**, 1–9 (2022). OCLC: 9359636331
3. Yingbin, B., Erkun, Y., Bo, H.: Understanding and Improving Early Stopping for Learning with Noisy Labels. OCLC: 1269561528 (2021)
4. Senen-Cerda, A., Sanders, J.: Almost sure convergence of dropout algorithms for neural networks. OCLC: 1144830913 (2020)
5. Liang, X., Wu, L., Li, J., Wang, Y., Meng, Q.: R-Drop: Regularized Dropout for Neural Networks. OCLC: 1269560920 (2021)
6. Shaeke, S., Xiuwen, L.: Overfitting Mechanism and Avoidance in Deep Neural Networks. OCLC: 1106327112 (2019)
7. Wei, C., Kakade, S., Ma, T.: The Implicit and Explicit Regularization Effects of Dropout. OCLC: 1228392785 (2020)
8. Arora, R., Bartlett, P., Mianjy, P., Srebro, N.: Dropout: Explicit Forms and Capacity Control. OCLC: 1228394951 (2020)
9. Cavazza, J., Morerio, P., Haeffele, B., Lane, C., Murino, V., Vidal, R.: Dropout as a low-rank regularizer for matrix factorization. In: Proceedings of the Twenty-First International Conference on Artificial Intelligence and Statistics, vol. 84, pp. 435–444 (2018). https://proceedings.mlr.press/v84/cavazza18a.html
10. IBM_Cloud_Education. What is Overfitting? (2021)
11. Brownlee, J.: How to Avoid Overfitting in Deep Learning Neural Networks (2018)
12. Maren, M., Lukas, B., Christoph, L., Philipp, H.: Early Stopping without a Validation Set. OCLC: 1106261430 (2017)
13. Moolayil, J.: Learn Keras for deep neural networks: a fast-track approach to modern deep learning with Python. OCLC: 1079007529 (2019)

14. Krizhevsky, A., Hinton, G.E., Sutskever, I.: ImageNet classification with deep convolutional neural networks. Commun. ACM **60**(6), 84–90 (2017)

15. LeCun, Y., Hinton, G., Bengio, Y.: Deep learning. Nature **521**(7553), 436–444 (2015). OCLC: 5831400088

16. Bengio, Y., Lamblin, P., Popovici, D., Larochelle, H.: Greedy layer-wise training of deep networks. In: Advances in Neural Information Processing Systems, vol. 19, pp. 153–160. Morgan Kaufmann Publishers, San Mateo (2007). OCLC: 181070563

17. Ng, A.Y.: Feature selection, L1 vs. L2 regularization, and rotational invariance. OCLC: 8876667046 (2004)

18. Srivastava, N., Hinton, G., Krizhevsky, A., Sutskever, I., Salakhutdinov, R.: Dropout: a simple way to prevent neural networks from overfitting. J. Mach. Learn. Res. **15**, 1929–1958 (2014). OCLC: 5606582392

19. Sabiri, B., El Asri, B., Rhanoui, M.: Mechanism of overfitting avoidance techniques for training deep neural networks. In: Proceedings of the 24th International Conference on Enterprise Information Systems, pp. 418–427 (2022). https://www.scitepress.org/DigitalLibrary/Link. aspx?doi=10.5220/0011114900003179

20. Caruana, R., Lawrence, S., Giles, L.: Overfitting in neural nets. In: 14th Annual Neural Information Processing Systems Conference (2001). OCLC: 5574566588

21. Brownlee, J.: Develop deep learning models on Theano and TensorFlow using Keras. Machine Learning Mastery, Melbourne, Australia, vol. 1 (2017)

22. Larxel. Early Diabetes Classification (2021)

23. Cerisara, C., Caillon, P., Le Berre, G.: Unsupervised post-tuning of deep neural networks. In: IJCNN, Proceedings of the 2021 International Joint Conference on Neural Networks (IJCNN), Virtual Event, United States (2021)

24. Etzold, D.: MNIST—Dataset of Handwritten Digits (2022)

25. Sarvazyan, M.A.: Kaggle: Your Home for Data Science (2022)

26. Artificial intelligence in cancer: diagnostic to tailored treatment. OCLC: 1145585080

27. Iraji, M.S., Feizi-Derakhshi, M.-R., Tanha, J.: COVID-19 detection using deep convolutional neural networks. Complexity **2021**, 1–10 (2021)

28. Lee, G., Park, H., Ryu, S., Lee, H.: Acceleration of DNN training regularization: dropout accelerator. In: 2020 International Conference on Electronics, Information, and Communication (ICEIC), pp. 1–2 (2020). https://ieeexplore.ieee.org/document/9051194/

29. Koivu, A., Kakko, J., Mäntyniemi, S., Sairanen, M.: Quality of randomness and node dropout regularization for fitting neural networks. Expert Syst. Appl. **207**, 117938 (2022). https:// linkinghub.elsevier.com/retrieve/pii/S0957417422011769

30. Li, C., Mao, Y., Zhang, R., Huai, J.: A revisit to MacKay algorithm and its application to deep network compression. Front. Comput. Sci. **14**(4), 1–16 (2020). https://doi.org/10.1007/ s11704-019-8390-z

31. Wang, Z., Fu, Y., Huang, T.S.: Deep learning through sparse and low-rank modeling. OCLC: 1097183504 (2019)

32. Sabiri, B., El Asri, B., Rhanoui, M.: Impact of hyperparameters on the generative adversarial networks behavior. In: Proceedings of the 24th International Conference on Enterprise Information Systems - Volume 1: ICEIS, pp. 428–438 (2022)

Effect of Convulsion Layers and Hyper-parameters on the Behavior of Adversarial Neural Networks

Bihi Sabiri[1]([✉]) [ID], Bouchra EL Asri[1] [ID], and Maryem Rhanoui[2] [ID]

[1] IMS Team, ADMIR Laboratory, Rabat IT Center, ENSIAS, Mohammed V University in Rabat, Rabat, Morocco
{bihi_sabiri,bouchra.elasri}@um5.ac.ma,
bouchra.elasri@ensias.um5.ac.ma
[2] Meridian Team, LYRICA Laboratory, School of Information Sciences, Rabat, Morocco
mrhanoui@esi.ac.ma

Abstract. The most significant models of neuronal networks for unsupervised machine learning now include the generative adversarial networks (GAN) as a separate branch. They have recently shown some degree of success. However, there is still a major obstacle. Many different loss functions have been created to train GAN discriminators, and they all share a similar structure: a total amount of real and fake losses that are solely dependent on the real losses and the generated data respectively. A challenge with a sum equal to two losses is that the training may benefit from one loss while harming the other, as we shall demonstrate causes mode collapse and instability. First, we suggest that the current loss function may prevent learning fine details in the data while trying to contract the discriminator. In this paper, we provide a new class of loss discriminant functions that promotes the training of the discriminator with a higher frequency than that of the generator for make the discriminator more demanding compared to the generator and vice versa. We may choose appropriate weights to train the discriminator in a way that takes advantage of the GAN's stability by using gradients of the real and fake loss parties. Our methodology may be used with any discriminator model where the loss is the total of true and fake pieces. Our approach entails appropriately adjusting the hyper-parameters in order to enhance the training of the two opposing models. The effectiveness of our loss functions on tasks involving the creation of curves and images has been demonstrated through experiments, improving the base results by a statistically significant margin on several datasets.

Keywords: Generative adversarial networks · Machine learning · Recommendation systems · Neural network · Deep learning

1 Introduction

The technology known as Generative Adversarial Networks (GAN) [1] is referred to as an innovative programming approach that enables the construction of generative models, or models that can generate data on their own [2,3]. The fundamental idea behind

Supported by Sabiss.net.

GANs is derived from the two-player zero-sum game, in which the combined gains of the two players are zero and each player's loss or gain in utility is precisely offset by the gain or loss in utility of the other player. One of the most promising recent advancements in machine learning is the rapid development of the GAN [4] and its numerous applications [5,6]. In addition, Yann LeCun, who oversees Facebook's AI research, described it as "The most intriguing discipline-related idea in the last decade of machine learning" [7]. Technically speaking, the GAN are built on the unsupervised learning of the Discriminator and Generator artificial neural networks. With the use of convoluted layers, these two networks mutually induce a contradictory relationship [8,9]: The task of the generator is to produce designs (such as images), and the generator's designs as well as real-world design data are provided to the discriminator.

It is it's responsibility to determine the origin of each design in order to determine whether they are genuine or the product of the Generator.

Following the Discriminator's work, two retroactive loops communicate to two neural networks the information about the designs that need improvement. The designs on which the Discriminator has been deceived by the Generator are returned to the Discriminator.

As a result, the two algorithms continue to progress in a way that benefits both of them: the Generator learns to produce increasingly realistic designs, and the Discriminator learns to distinguish real designs from those produced by the Generator.

In industry, there are numerous applications. Automotive businesses are especially interested in GAN because it may brighten images taken by cameras in autonomous vehicles and increase the effectiveness of artificial vision algorithms. A GAN has been used, for instance, in the building industry to model the behavior of potential building inhabitants and provide input to simulation algorithms that strive to precisely optimize energy use. Finally, the fashion industry has already started using GANs to create materials or shoes that will appeal to consumers. GANs could have significant effects on copyright protection and industrial design because of their capacity to imitate without truly copying.

Regarding the drawbacks of the command systems, specifically: Two research areas—sparsity and noise—have been investigated, and the following notions are shared by both:

1. The key adaptation approach for the issue of data sparsity is data augmentation [10], which is achieved by capturing the distribution of real data under the minimax.
2. Adversarial disturbances and training based on adversarial sampling are frequently employed as solutions to the problem of data noise [11].

In this aticle, we will more closely analyze GAN and the various variations of their loss functions in order to better understand how GAN function while also addressing unexpected performance issues. These two models will be created using the conventional GAN loss function, which is also called the loss min-max function [2]. As opposed to the discriminator, who seeks to maximize this function, the generator attempts to minimize it.

The remainder of this article is structured as follows: In Sect. 2, we briefly assess alternative approaches from the literature and place our answer in relation to them. The solution is described in Sect. 3. In Sect. 4, we outline our suggested technique that may be utilized with any discriminator model, and a loss that is the total of its fake and real components.

This article is the extended version of the article that appeared in sitepress following the ICEIS 2022 conference proceedings [12]. The added content enhances the extended version, such provide brand-new tests (Sect. 4.11, page 13), examples (Sect. 4.11, page 13), additional datasets (Sect. 4.11, page 13), comparisons with the literature (Sect. 2, page 3), algorithm (Sect. 4.2, page 6), proofs (Sect. 4.5, page 6) (Sect. 4.5, page 10) (Fig. 3, page 10), results (Fig. 4, page 14) (Table 2, page 21)

2   Related Works

A series of generative models known as the "Generative Adversarial Networks" aims to identify the distribution that underlies a certain data generation process. It is explained as two competing models that, after being trained, can provide samples that are indistinguishable from samples taken from the normal distribution. An adversarial rivalry between a generator and a discriminator leads to the discovery of this distribution. The discriminator is trained to distinguish between created and actual samples, and the generator is trained to fool the discriminator by creating data that is as convincing and realistic as possible. Two networks of neurons, D and G, which will be referred to as the discriminator and the generator, respectively, are put in competition by this generative model. We give a succinct overview of the generative adversarial network literature in this section: The usefulness of GANs as generative models is illustrated in the Goodfellow article [1, 13]. Based on the effective application of deep convolution layers, GANs for picture synthesis gained popularity [13, 14].

Generic GANs: GANs in general are the subject of the majority of survey publications [15, 16]. Beginning at the end of 2017 [17], one of the initial revisions to the generic GANs was made. An additional broad overview of GANs that uses an analogy between GAN principles and signal processing concepts to make GANs easier to grasp from a signal processing perspective was released in the beginning of 2018 [18]. In the same period, Hitawala et al. [19]discussed the key advancements made to GAN frameworks. There have been other articles [15, 20] that look at GANs in general in 2019 and 2020

Classical Algorithms: Classical image processing techniques are unsupervised algorithms that enhance low-light photos using solid mathematical models. They have straightforward calculations and are effective. However, they are not reliable enough and must be manually calibrated to be used in some circumstances, as noted in [21]

Model for Generation Implied. Other than the descriptive models, a well-liked subset of deep Black-box models known as "generative models" map a top-down CNN to connect latent variables to signals, the Generative Adversarial Network, for example (GAN) and variations [1]. These models have had outstanding success in producing realistic images and learning the generator network with a helper discriminator network..

Adversarial Networks: As demonstrated by Generative Adversarial Network [1], function well enough for several supervised and unsupervised issues with guided learning. In [22] The authors suggest a model under which the requirement has increased for paired images, and image trans- One method of relating two domains is lack of cycle consistency. These strategies have used to a variety of different uses, such as dehaz- super-resolution, ing, etc. It has recently been used. To EnlightenGAN's low-light image enhancing [23] has produced encouraging findings, and this influenced our GAN model. Generic advertising sarial networks have also been used [1]. Convolutional decoder networks were advantageous for the network module for a generator. Denton [24] utilized a Laplacian pyramid of adversaries to Using an image synthesizer and discriminators, at different resolutions. This piece attracted comments, pelling photos of high resolution and could also condition discussion of class labels for manageable generation. Alec Radford et al. used a common convolutional neural network decoder [25], but created a very powerful and reliable architecture that uses a batch. -

Totally Connected GANs: Fully connected neural networks were utilized in the initial GAN architectures for both the discriminator and generator [1]. The use of this style of architecture in real estate very straightforward picture data sets: Kaggle MNIST (handwritten numbers), the CIFAR-10 (natural pictures), and the Set of Toronto Face Data (TFD).

3   How Are Generative Models Implemented?

In this article, we'll conduct a comparison between two ways to figure out the loss: the first entails estimating this loss on the generator and the discriminator globally and contrasting it with the hyper-parameterized discriminator (see Fig. 1). The second involves adding up the losses of the two competing models are attempting to identify the best hyper-parameters ((se Fig. 1). Convoluted networks make up the discriminator network. thematic layers. We are able to do this for each network layer. performing a convolution, after which we are batch normalization to make the network more reliable. faster, more precisely, and lastly, we will carry out a Leaky ReLu.

The generator will produce an image with the same geometry as the image that is good with deconvolution processing to make a processing equivalent to the processing of the discriminator, and we will give it a noise vector, which will be numbers generated as a vector of z numbers between −1 and 1 drawn randomly according to a normal distribution. Here, we'll start with a small vector, vector 100 that will serve as a noise vector, and we'll gradually generate an image with the same shape as the one used as the discriminator's input. How best to teach this generator is now the question. We have a neural network that gives us an output, but we would like it if the output was typically something different even if, in the long run after learning, it gave us the output we expected. But there is always a discrepancy between what we would like from the neural network and what we really have. In order to fix the weights of our neural network, we first generate a loss function from this difference and then compute its gradient. The assembly of the generator and discriminator will be used to calculate the loss. After determining the gradient of the entire system in this manner, we will then be able to adjust the generator and discriminator variables. Higher computational power is needed for GANs, but the infrastructure is the same. GANs take a lot of computational power and memory during training because these models will be quite huge and contain a lot of parameters. Therefore, we need to have more data traffic while utilizing GANs.

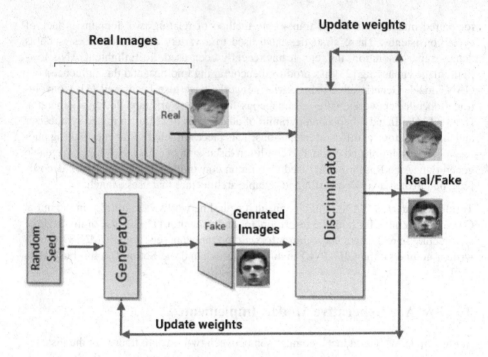

Fig. 1. GAN Forward & Backpropagation [12].

4   Theoretical and Mathematical Basis for GAN Networks

4.1   Gradient Descent

The distribution of the samples G(z) obtained when z $\sim$ p_z is defined implicitly by the generator G as a probability distribution p_g. Therefore, if given sufficient capacity and training time, we want method 1 to converge to a good estimator of p_{data}. Results from this section were obtained in a nonparametric environment, such as investigating convergence in the space of probability density functions allows us to represent a model with limitless capacity. The global optimum of this minimax game for $p_g = p_{data}$ will be demonstrated in Subsect. 4.6 (The probability distributions of data samples are expressed by $p_{data}(x)$).

Algorithm: Generative adversarial nets are trained using a minibatch stochastic gradient descent method. kk is a hyperparameter that specifies the number of steps to be applied to the discriminator. In our studies, we chose the least expensive alternative, kk = 1 (see Algorithm 1) [27, 28].

In order to increase D's ability to determine if a sample is from G or alternatively real data, we need to set an aim for training G and D that will encourage G to produce better samples.

Algorithm 1. The fundamental algorithm and Policy updating for GAN.

```
 1  for i ← 1 to N iterations do
 2      for t ← 1 to kk steps do
 3          Sample minibatch of t noise examples {z⁽¹⁾ , . . . , z⁽ᵗ⁾} from noise prior pₘ(z).
 4          Sample minibatch of t real samples {x⁽¹⁾ , . . . , x⁽ᵗ⁾} from data generating
              distribution pᵈᵃᵗᵃ(x).
 5          Ascending the stochastic gradient of the discriminator will update it :
```

$$\vec{\nabla}_\theta d \frac{1}{t} \sum_{i=1}^{t} [\log\Big(D(x^{(i)})\Big) + \log\Big(1 - D(G(z^{(i)}))\Big)]$$

```
 6      end
 7      for t ← 1 to kk steps do
 8          Sample minibatch of t noise examples {z⁽¹⁾ , . . . , z⁽ⁿ⁾} from noise prior pₘ(z).
 9          Descend the stochastic gradient to update the generator :
10
```

$$\vec{\nabla}_\theta g \frac{1}{t} \sum_{i=1}^{t} [\log\Big(1 - D(G(z^{(i)}))\Big)]$$

```
11      end
12  end
```

Note 1. An analogy can be made to the falsification of works of art. D and G are referred to as the "forger" and the "critic", respectively. The goal of G is to transform a random noise into a sample that is similar to the real observations. The goal of D is to learn to recognize false samples from true observations. It will be demonstrated mathematically how the GAN loss function [29] works and how the error is calculated and propagated to update the parameters to meet various network goals. (This explanation is significantly based on [1] and [30])

4.2   Generator

The generator attempts to confuse the discriminator as much as possible in order to have it mistake produced images for genuine ones (see Table 1).

Table 1. Objective of each model [12].

Objective of Discriminator	Objective of Generator
To maximize $D(x)$	To maximize $D(G(z))$
To minimize $D(G(z))$	

The important thing to keep in mind in this situation is that we want to minimize the loss function. The generator should make an effort to reduce noise as much as possible between the label for true data and the one for false data as judged by the discriminator.(see Eq. 1)

$$L_G = Error(D(G(z)), 1) \tag{1}$$

In binary classification problems, binary cross-entropy is a commonly used loss function. The cross entropy equation is as follows [31] (see Eq. 2):

$$H(p, q) = \mathbb{E}_{x \sim p_{data}(x)}[-\log q(x)] \tag{2}$$

The random variable in classification problems is discrete. A summation can therefore be used to indicate the expectation (see Eq. 3):

$$H(p, q) = -\sum_{x=1}^{M} p(x) \log(q(x)) \tag{3}$$

Since there are only two possibilities for binary cross entropy, there are two labels: 0 and 1. As a result, the Eq. 3 can be formulated as (Eq. 4):

$$H(y, \hat{y}) = -\sum y \log(\hat{y}) + (1-y) \log(1-\hat{y}) \tag{4}$$

Binary cross-entropy satisfies our goal in that it measures the difference between two distributions in the context of a binary distribution, which is to determine whether a values is true or false. By applying it to the loss functions in Eq. 4 we obtain (see Eq. 5):

$$L_D = -\sum_{x \in \chi, z \in \zeta} \log(D(x)) + \log(1 - D(G(z))) \tag{5}$$

The same procedure applies to Eq. 3 (see Eq. 6):

$$L_G = -\sum_{x \in \chi, z \in \zeta} \log(D(G(z))) \tag{6}$$

The 2 equations above of the loss functions for the generator and the discriminator are represented on the figure below (see Fig. 2).

4.3 Discriminator

The discriminator's objective is to accurately classify the actual pictures as truthful, fake pictures as false (see Table 1) Consequently, we could think about the discriminator's loss function will be as follows (see Eq. 7):

$$L_D = Error(D(x), 1) + Error(D(G(z)), 0) \tag{7}$$

Here, we refer to a function that tells us the separation or difference between the two functional parameters by utilizing a very general, unspecific notation for Error.

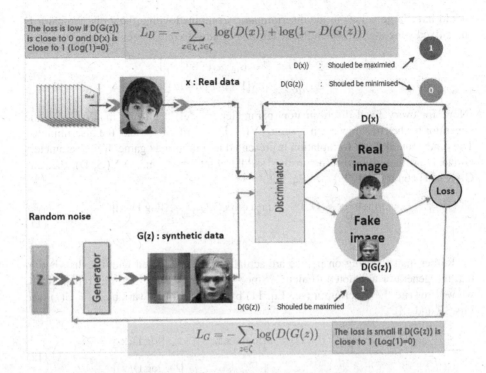

Fig. 2. Updated weight for Discriminator & Generator for optimal loss

4.4 Optimizing the Model

Now that the loss functions for the generator and discriminator have been established, it is time to use mathematics to address the optimization issue, i.e., determine the parameters for the generator and discriminator such that as optimization of loss functions. This is equivalent to the practical development of the model.

4.5 Cost of the GAN

Conceptually, the objective of training is to increase the expectation that D, the discriminator, correctly classifies the data as either authentic or fraudulent and for the generator, G, it has the objective of deceiving the discriminator. One model is typically trained at a time when training a GAN (see Fig. 2).

Consider the classifier D, which returns 1 for a real image and 0 for a fake one. The discriminator then looks for big values of:

$$\mathbb{E}_{x \sim P_{\text{data}}(x)}[-\log D(x)] \tag{8}$$

Additionally, we have a fictitious sample G(Z) for every Z, and D(G(Z)) is the discriminator classification, which should ideally be 0. Consequently,

$$\mathbb{E}_{z \sim P_{\text{generated}}(z)}[1 - \log D(G(z)) \tag{9}$$

should have large values in the discriminator. Combined with a fixed discriminator D and a fixed generator G, which the discriminator aims to maximize:

$$V(G, D) = \mathbb{E}_{x \sim P_{data}(x)}[\log D(x)]$$
$$+ \mathbb{E}_{z \sim P_{generated}(z)}[1 - \log D(G(z))] \tag{10}$$

Now, for every fixed discriminator (parameters/model) achieving maxV(G, D), the generator wishes to achieve $\min_G \max_D V(G, D)$ in order to "trick" the discriminator. The fundamental GAN formulation is presented as a mini-max game. If the parameters G and D are determined to attain the "saddle point" minG maxD V(G, D), then the GAN has been trained [26] (see Eq. 11).

$$\min_G \max_D V(G, D) = \min_G \max_D \mathbb{E}_{x \sim P_{data}(x)}[\log D(x)]$$
$$+ \mathbb{E}_{z \sim P_{generated}(z)}[1 - \log D(G(z))] \tag{11}$$

Rather than focusing on p_z, we are actually more concerned with the distribution that the generator attempted to model. Since this is the case, Using this replacement, we will rewrite the function in (see Eq. 11) by creating a new variable, y = G(z) (see Eqs. 12 and 13).

$$\min_G \max_D V(G, D) = \min_G \max_D \mathbb{E}_{x \sim P_{data}(x)}[\log D(x)]$$
$$+ \mathbb{E}_{z \sim P_{generated}(z)}[1 - \log D(y)] \tag{12}$$

$$= \int_{x \in \chi} P_{data}(x)\log(D(x)) + p_g(x)\log(1 - D(x))dx \tag{13}$$

The function $y \rightarrow a * \log(y) + b * \log(1 - y)$ reaches its maximum in [0, 1] at $\frac{a}{a+b}$ for any $(a, b)- \in (\mathbb{R}^*, \mathbb{R}^*)$

Proof. $z = a * \log(y) + b * \log(1 - y) \Leftrightarrow z' = d(z)/d(y) = a/y - b/(1 - y)$. Find the optimum y* by putting z '= 0 see curve of the green function below (see Fig. 3):

$$z' = 0 \Leftrightarrow \frac{a}{y*} = \frac{b}{1 - y*} \Leftrightarrow \frac{1 - y*}{y*} = \frac{b}{a} \Leftrightarrow \frac{1}{y*} = \frac{a+b}{a} \Leftrightarrow y* = \frac{a}{a+b}$$

The discriminator's goal is to increase the value of the value Eq. 12 and 13). The optimal discriminator, denoted D*(x), occurs when the derivative with respect to D(x) is zero (see Eq. 14), so,

$$\frac{P_{data}(x)}{D(x)} - \frac{P_g(x)}{(1 - D(x))} = 0 \tag{14}$$

The discriminator's performance is at its best when it cannot tell the difference between the real input and the synthesized data. By condensing Eq. 14, the ideal discriminator D for a fixed Generator is as follows (see Eq. 15):

$$D^*(x) = \frac{P_{data}(x)}{(P_{data}(x) + P_g(x))} \tag{15}$$

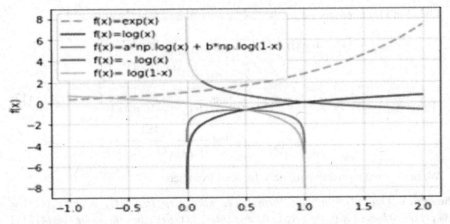

Fig. 3. The function f(y) = a * log(y) + b * log(1 − y) reaches its maximum at $y* = \dfrac{a}{a+b}$ $(a = b \Rightarrow y* = \dfrac{1}{2})$.

The optimal discriminator must meet this requirement for G to be fixed ! Observe that the formula makes intuitive sense: If a sample x is highly authentic, we would anticipate that $p_{data}(x)$ will be near to 1 and $p_g(x)$ will converge to zero; in this scenario, the optimal discriminator would give that sample a value of 1. The label of the real photographs corresponds to (D(x) = 1), which corresponds to the label of the authentic images. On the other hand, since $p_{data}(G(z))$ must be near to zero for a generated sample x = G(z), we would anticipate that the best discriminator would award a label of zero.

4.6 Cost of the Generator

Assuming that the discriminator is fixed, we examine the value function to train the generator. Start by entering the outcome we discovered above, namely (Eq. 15), into the value function to observe the outcome.

It should be noted that the training goal for D can be understood as maximising the log-likelihood for calculating the conditional probability $P(Y = y\|x)$, where Y denotes whether x originates from p_{data} (with y = 1) or from p_g (with y = 0). According to Eq. 15 we can deduce the Eq. 16:

$$1 - D^*(x) = \frac{p_g(x)}{(p_{data}(x) + p_g(x))} \tag{16}$$

Equation 11's minimax game can now be rewritten as (see Equation 17):

$$C(G) = \max_D V(G, D^*) = \mathbb{E}_{x \sim p_{data}(x)}[\log D^*(x)]$$
$$+ \mathbb{E}_{z \sim p_{generated}(z)} \log[1 - D^*(y)]$$
$$= \mathbb{E}_{x \sim p_{data}(x)}[\log \frac{p_{data}(x)}{(p_{data}(x) + p_g(x))}]$$
$$+ \mathbb{E}_{x \sim p_{generated}(x)}[\log \frac{p_g(x)}{(p_{data}(x) + p_g(x))}] \tag{17}$$

We need some motivation to move forward from here.

Theorem 1. *Overall Minimum Virtual Learning Criteria C(G) = maxV(G,D)= V(G,D*) is achieved if* $p_g = p_{data}$. *At this stage, C(G) reaches the value -log(4)* [41] *(see Eq. 18)*

$$C(G) = -\log(4) + D_{KL}(p_{data}||\frac{p_{data}(x) + p_g(x)}{2})$$
$$+ D_{KL}(p_g||\frac{p_{data}(x) + p_g(x)}{2}) \tag{18}$$

where KL represents the Kullback-Leibler divergence.

We can see the Jensen-Shannon divergence between the model's distribution and the method used to generate the data in the preceding expression (see Eq. 19):

$$C(G) = -\log(4) + 2JSD(p_{data}||p_g) \tag{19}$$

C(G) is always positive and zero if two parts are equal. We showed that $C^* = -\log(4)$ is the global minimum of C(G) and the only solution is $p_g = p_{data}$. We have demonstrated that $C* = -\log(4)$ is the global minimum of C(G) and that the only solution is $p_g = p_{data}$, i.e., the generative model fully reproducing the data distribution. In essence, we are extracting a -log4 that was previously undiscovered by taking advantage of the peculiarities of logarithms. The parameters of expectation must be changed in order to extract this value, and one such adjustment is to divide the denominator by two. Why was it required, exactly ? What's remarkable about this situation is that we can now characterize the expectations as a Kullback-Leibler divergence.

The analysis's straightforward conclusion is that we want the JS divergence between the distribution of the data and the distribution of the generated samples to be as little as is practical given that the purpose of learning the generator is to reduce the value function V(G, D). The generator should be able to learn the underlying distribution of the data from sampled training instances, and this result certainly matches our understanding. In other words, p_g and p_{data} ought to be as close as feasible to one another. Therefore, the best generator G is the one that can mimic p_{data} to model a plausible model distribution p_g.

4.7 Loss Function

The binary cross-entropy loss formula can be used to generate the loss function that Ian Goodfellow et al. described in their original publication (Eq. 2). The binary cross-entropy loss is denoted by the formula (see Eq. 20):

$$L(y, \hat{y}) = -\sum_{i=1}^{N_c} y_i \log(\hat{y}_i) \tag{20}$$

Cross-entropy in binary classification, where $N_c = 2$, can be calculated as follows (see Eq. 21):

$$L(y, \hat{y}) = -(y \log(\hat{y}) + (1 - y) \log(1 - \hat{y})) \tag{21}$$

4.8 Generator Loss

Discriminator is in competition with the generator. As a result, it will attempt to minimize Eq. 5, and its loss function is as follows (see Eq. 22):

$$L^{(G)} = \min[\log(D(x)) + \log(1 - D(G(z)))] \tag{22}$$

4.9 Discriminator Loss

The label data from $P_{data}(x)$ during discriminator training are $y = 1$ (real data) and $\hat{y} = D(x)$. By substituting this in the loss function previously mentioned (Eq. 21), we obtain (see Eq. 23):

$$L(D(x), 1) = \log(D(x)) \tag{23}$$

On the other hand, the label for data coming from the generator is $y = 0$ (false data) and $\hat{y} = D(G(z))$. The loss in this instance starting from the Eq. 21 is therefore given by Eq. 24

$$L(D(G(z)), 0) = \log(1 - D(G(z))) \tag{24}$$

The discriminator's current goal is to correctly distinguish between the real and fake datasets. In order to do this, Eqs. 2 and 3 should be maximized, and the discriminator's final loss function can be expressed as (see Eq. 25):

$$L^{(D)} = \max[\log(D(x)) + \log(1 - D(G(z)))] \tag{25}$$

4.10 Combined Loss Functions

Combining Eqs. 22 and 25, we can write (see Eq. 26):

$$L = \min_G \max_D [\log(D(x)) + \log(1 - D(G(z)))] \tag{26}$$

Keep in mind that the above loss function is only applicable for a single data point; therefore, to take the full dataset into consideration, we must use the following expectation for the above Eq. 26 (see Eq. 27):

$$\min_{G} \max_{D} V(G, D) = \min_{G} \max_{D} \mathbb{E}_{\mathbf{x} \sim p_{data}(\mathbf{x})}[\log D(\mathbf{x})]$$
$$+ \mathbb{E}_{\mathbf{z} \sim p_{generated}(\mathbf{z})}[1 - \log D(G(\mathbf{z}))] \tag{27}$$

which is the objective function in formal sense as shown previously (see Eq. 11)

4.11 Experimental Setup

Either fully connected (FC) GANs or deep convolutional GANs are the two major types of networks to build. Depending on the training data we are giving the network, we will use one over the other. An FC network is more appropriate when working with single data points, and a DC- GAN is more ideal when working with images [32]. The first type is used in the first 2 examples and the second type will be used in the rest of the use cases.

Typical Use Case 1. In the first example, the random samples are generated using the randomization function, and for added simplicity, the second coordinate is generated using the quadratic function. This code can be updated to construct datasets with more dimensions and/or more complex connections between their characteristics, such as sine, cosine, or higher degree polynomials.

We create a function for the generator that accepts two arguments: a reuse variable for reusing the same layers, and a hsize array for the number of units in the hidden layers in the random sample placeholder (Z). With the help of these inputs, it builds a fully connected neural network with a specified number of nodes and x hidden layers. This function produces a two-dimensional vector whose dimensions correspond to those of the actual dataset from which we are trying to learn. The aforementioned function is easily altered to incorporate more hidden layers, distinct layer types, distinct activations, and distinct output mappings in an effort to converge the two antagonists G and D. This function accepts a placeholder as input to hold samples from the vector space of the actual dataset. Real samples and samples produced by the Generator network are both acceptable types of samples. In terms of the generator, we experiment with the number of hidden layers to identify the GAN network's ideal convergence point.

After training the GAN network and as illustrated in (Fig. 4), we conclude that the number of hidden layers has a impact on the convergence of the GAN network; the more abundant the hidden layers, the more convergence is established (see Fig. 4 (c)).

Typical Use Case 2. The discriminator net utilized sigmoid activations whereas the generator nets included rectifier linear activations and tanh activations [34]. We only used noise as the input to the lowest layer of the generator network, where dropout and additional noise were applied at intermediate layers of the generator. The images lack labels since generative modeling is an unsupervised learning problem. A vector or a matrix of random numbers, sometimes known as a latent tensor, serves as the generator's usual input and serves as the image's seed. A latent tensor of the kind (100, 1, 1)

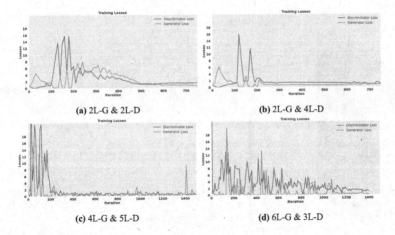

Fig. 4. Impact of changing number of hidden layers on the behavior of the GAN (a): 2 hidden layers in G and 2 in D (b): 2 hidden layers in G and 4 in D (c): 4 hidden layers in G and 5 in D (d): 6 hidden layers in G and 3 in D.

will be transformed by the generator into an image tensor of the form $(4 \times 4 \times 128)$. Rectified Linear Unit (ReLu) activation is used to map the latent tensor (z) and a matrix for actual pictures to hidden layers, with layer sizes of 16 and 64 on the generator and discriminator, respectively. The discriminator output layer then has a final sigmoid unit. An image is fed into the discriminator, which attempts to categorize it as "genuine" or "manufactured". It is comparable to other neural networks in this regard. We'll employ convolutional neural networks (CNNs), which produce a single number for each individual [35]. Stride of 2 will be used to gradually shrink the output feature map's size (see Fig. 5). The stride S indicates how many pixels the window advances after each operation for a convolutional or a pooling operation.

Since the discriminator is a binary classification model, we can measure how effectively it can distinguish between actual and produced images using the binary cross entropy loss function. The RMSprop optimizer, which is a variant of the gradient descent technique with momentum, is used in this case (Rome, 2017). The vertical oscillations are limited by the this optimizer. As a result, we may speed up our learning rate, allowing our algorithm to take more horizontal steps while convergent faster and forcing us to train the GAN as follows:

1. A batch of photos is created using the generator and then run through the discriminator.
2. To "fool" the discriminator, calculate the loss by setting the target labels to 1.
3. Using the loss to do a gradient descent, or altering the generator's weights, to make it better at producing realistic images to "trick" the discriminator.

Now that the models have been trained, we have the following outcomes (see Fig. 6):

Debugging the training process can be greatly aided by visually representing losses. We anticipate that with GANs, the discriminator's loss won't get too high while the generator's loss gradually decreases. As shown in Fig. 6, the loss at the discriminator level

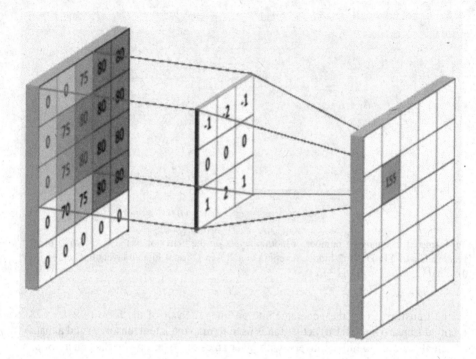

Fig. 5. Convulsion with a slight step: Filter size = 3 × 3 Output = 5 × 5.

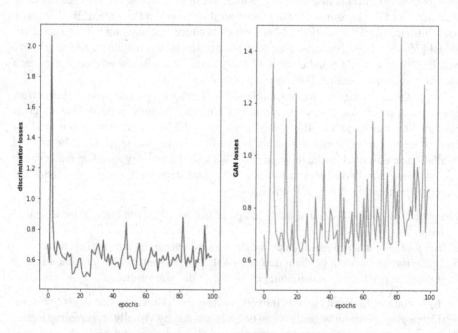

Fig. 6. Left: Discriminator Loss. Right: Gan Loss [12].

soon stabilizes at 0.6, whereas the GAN (generator and discriminator) level oscillates at 0.8. So starting from noise at the beginning, this is how learning looks:

Therefore, there Fig. 6 is something that we can't see but is slightly rudimentary, and after a given number of learning cycles, we can see images that appear to be deformed rather quickly. However, even if they are poorly drawn, there will still be something that genuinely resembles faces (see Fig. 7).

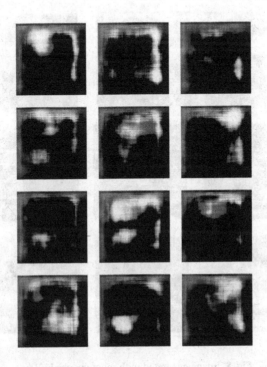

Fig. 7. Images generated at the beginning of Gan's training [12].

Where we can see that the faces generated are becoming increasingly exact with more epochs (see Fig. 8).

When the refiner network starts producing blurry versions of the synthetic input images, we introduce the adversarial loss. The refiner network is first trained with only self-regularization loss. The output of the refiner network at different stages of training is shown in Fig. 6 and 8. It produces a hazy image at first, which gets more lifelike as training goes on. The discriminator and generator losses during various training iterations are shown in Fig. 6. It is important to note that the discriminator loss is initially low, indicating that it can distinguish between genuine and refined objects with ease. As training progresses, the discriminator loss gradually rises and the generator loss falls, producing more authentic images.

Typical Use Case 3. To ensure that the input and output dimensions are the same, we include the encoder block section in this instance and use the same padding in

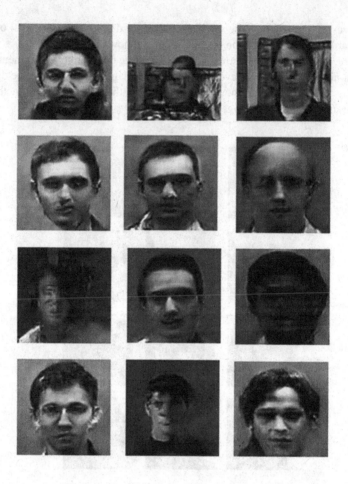

Fig. 8. Image generated with more epochs [12].

addition to batch normalization, and leaky Rectified Linear Units(Leaky ReLU). Then, to address the issue raised in the previous scenario, we recycled the encoder block segment and gradually increased the size of the filter in the discriminator itself. We are completing the reverse of convolutional layers. The steps, and for convenience of implementation, padding ans the sides are the same, and batch normalization and leaky ReLU are employed. On the We employ decoder blocks and gradually reduce the filter size on the generator side. A deep learning model's optimization algorithm selection can make the difference between successful outcomes in a matter of minutes, hours, or days. Recently, the Adam optimization approach has gained more popularity for deep learning applications in computer vision and natural language processing. It is an extension to stochastic gradient descent. In this second instance, it is therefore utilized to create the disciminator. We can see that consensus optimization produces significantly better results on this architecture (see Fig. 9). The two losses also converge to a figure that is generally acceptable (0.2). The quality of the photos generated by the generator is noticeably higher than it was in the first instance (see Fig. 10).

Typical Use Case 4. Based on the Minifaces dataset, this illustration. In addition, we will scale, crop, and normalize the images to $3 \times 64 \times 64$ and apply a mean and standard deviation of 0.5 for each channel to the pixel values. In order to make it easier to train the discriminator, this will guarantee that the pixel values are in the range $(-1, 1)$. A vector of random numbers is often utilized as the generator's input and serves as the image's seed. A latent shape tensor $(128, 1, 1)$ will be transformed into a $3 \times 28 \times 28$ shape image tensor by the generator. The Adam optimizer is used in the adjustment function to train the discriminator and the generator jointly for each batch of training

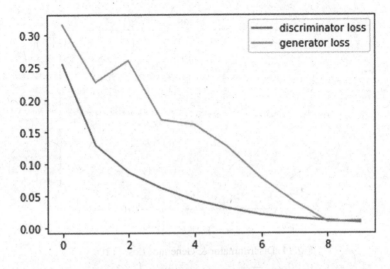

Fig. 9. Discriminator & Generator Loss [12].

Fig. 10. Images generated after many epochs of traininig [12].

data. Additionally, we will retain sample created photos for inspection at regular intervals. The variations in image quality between epochs are explained by the oscillations in the generator's scores and errors, which are shown in Fig. 11 and 12.

Typical Use Case 5. In this example, we show how a GAN can provide samples that are indistinguishable from those taken from samples of the normal distribution after being trained (see Fig. 14). A unique type of normal distribution with a mean of 0 and a standard deviation of 1 is known as the standard normal distribution, also known as the z-distribution (see Fig. 14). The process of learning to sample from the normal distribution is shown in Fig. 15. One hidden layer of 16 ReLU units on the generator and 64 on the discriminator makes up the network.

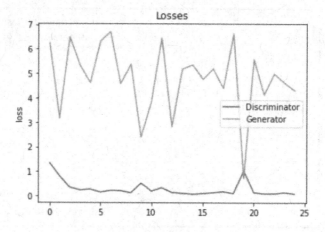

Fig. 11. Discriminator & Generator Loss [12].

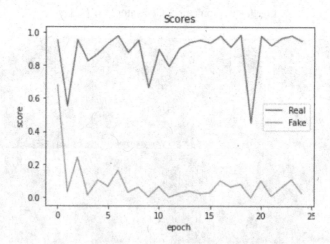

Fig. 12. Discriminator & Generator Score [12].

We received the results after training the gan for around 600 epochs (Fig. 15. In the following three cases, Table 1's optimization of the hyperparameters and losses. It is essential to label the true images as "true" and the false images as "false" when learning the discriminator network. After some training, however, the system will be activated by the employment of misleading strategies that "surprise" the discriminator. And impose restrictions on it that force it to revise prior knowledge. These tactics involve naming the created photos as though they were "genuine". By using this technique, the generator is enabled to make the appropriate adjustments, resulting in faces that are more similar to those provided as a model.

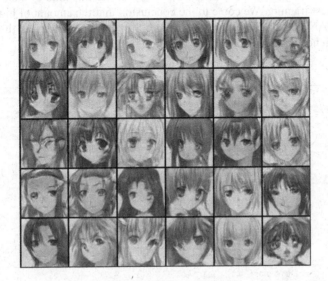

Fig. 13. Images generated with long time of training [12].

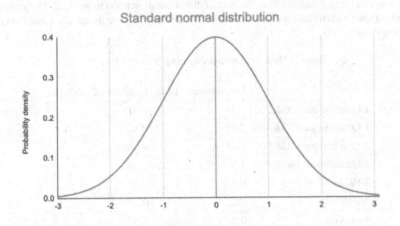

Fig. 14. z-distribution, is a special normal distribution with a mean of 0 and a standard deviation of 1 [12].

The final example uses the same misleading technique, forcing the discriminator to generate curves that resemble the model that was supplied as input (see Fig. 14). The created curve eventually leans toward a distribution curve with certain deformations after 600 epochs (Fig. 15 Right) The amount of data handled affects the GAN's performance. The models demand a significant amount of connectivity and memory as well as significant computational power due to their size and complexity (Fig. 13).

Table 2 shows an excessively high average error from the generator for the dataset Minifaces. Some photographs consequently appear rather distorted. The processing time required for the convergence of the gan takes several hours for datasets having images. We have adjusted the optimizer's parameters for this kind of data in order to provide the best training. We come to the conclusion that the amount of hidden layers affects the GAN network's ability to converge; the more hidden layers there are, the more convergence has been obtained (see Table 2).

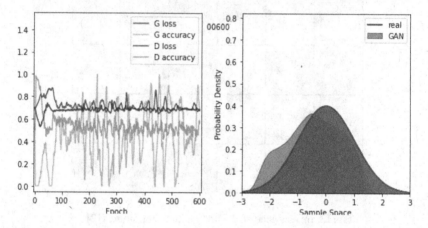

Fig. 15. GAN learning to sample from the normal distribution over 600 times. Left: The generator and discriminator's accuracies and losses. Right: The true N(0, 1) density and the observed GAN probability density [12].

Table 2. Hyper-parameters optimization and losses.

Scenario	Discriminator Loss	Generator/GAN Loss
1-Quadratique (2G&2D)	1.24	0.71
1-Quadratique (2G&4D)	1.37	0.67
1-Quadratique (4G&5D)	1.15	0.39
1-Quadratique (6G&3D)	1.37	0.54
2-Faces	0.6	0.8
3-Celeb	0.2	0.2
4-Minifaces	0.2	4.5
5 z-distribution	0.7	0.7

5 Conclusion

GAN networks are not fully autonomous creative systems because they are unable to express an aim or evaluate their own performance. The dataset that the GAN is trained on holds the secret to its inventiveness. Although GAN is a strong framework that encourages feature extraction, it is unstable because to training-time convergence uncertainty. The model's contradictory insight serves as the primary impetus for improvement, yet because of the model's very simplistic structure and the large number of unknowable variables influencing its output, it is unstable.

We suggest a set of designs that are largely stable for training generative advertisement networks, and we show that advertisement networks can acquire useful picture representations for both supervised learning and generative modeling. There are still some types of model instability, though. The current state of the art for image GANs is plagued by poor frame quality, a lack of frames, or both. Images provide high dimensional data, necessitating networks with several parameters, which could be one cause for the increased processing power requirement. Such data require handling by a complicated architecture that takes into account both spatial and temporal data.

More work is needed to address this instability problem. The expanded functional coverage, which includes speech and other areas like video (pre-trained features for speech synthesis), should be extremely intriguing, in our opinion. Additionally, more research needs to be done on how the actions conducted on enormous amounts of data. It would also be interesting to conduct more research on the characteristics of learned latent space.

Acknowledgements. Thank you for the reviewers of ICEIS 2022 who devoted their precious time to read the initial version of this article. Their comments and suggestions allowed the continuous improvement of the content of this article.

References

1. Goodfellow, I.J., et al.: Generative adversarial networks. Commun. ACM **63**(11), 139–144 (2020). OCLC: 8694362134
2. Brownlee, J.: How to code the GAN training algorithm and loss functions (2020)
3. Parthasarathy, D., Backstrom, K., Henriksson, J., Einarsdottir, S.: Controlled time series generation for automotive software-in-the-loop testing using GANs. In: 2020 IEEE International Conference On Artificial Intelligence Testing (AITest), pp. 39–46 (2020). OCLC: 8658758958
4. Abdollahpouri, H., et al.: Multistakeholder recommendation: survey and research directions (2020). OCLC: 1196494457
5. Aldausari, N., Sowmya, A., Marcus, N., Mohammadi, G.: Video generative adversarial networks: a review. ACM Comput. Surv. **55** (2023). https://www.scopus.com/inward/record.uri?eid=2-s2.0-85128166282, https://doi.org/10.1145/3487891
6. Nguyen, R., Singh, S., Rai, R.: Physics-infused fuzzy generative adversarial network for robust failure prognosis. Mech. Syst. Signal Process. **184** (2023). https://www.scopus.com/inward/record.uri?eid=2-s2.0-85136116185, https://doi.org/10.1016/j.ymssp.2022.109611
7. Rocca, J.: Understanding generative adversarial networks (GANs) (2021)

8. Barua, S, Erfani, S.M., Bailey, J.: FCC-GAN: a fully connected and convolutional net architecture for GANs. arXiv (2019). OCLC: 8660853988

9. Sun, H., Deng, Z., Parkes, D.C., Chen, H.: Decision-aware conditional GANs for time series data. arXiv (2020). OCLC: 8694375343

10. Sandy, E., Ilkay, O., Dajiang, Z., Yixuan, Y., Anirban, M.: Deep generative models, and data augmentation, labelling, and imperfections. In: Proceedings of First Workshop, DGM4MICCAI 2021, and First Workshop, DALI 2021, Held in Conjunction with MICCAI 2021, Strasbourg, France, 1 October 2021, Livre numérique (2021). [WorldCat.org]

11. Mayer, C., Timofte, R.: Adversarial sampling for active learning. In: 2020 IEEE Winter Conference on Applications of Computer Vision (WACV), Snowmass Village, CO, USA, pp. 3060–3068. IEEE (2020)

12. Sabiri, B., El Asri, B., Rhanoui, M.: Impact of hyperparameters on the generative adversarial networks behavior. In: Proceedings of the 24th International Conference on Enterprise Information Systems - Volume 1: ICEIS, pp. 428–438 (2022)

13. Mao, X., Li, Q.: Generative adversarial networks for image generation (2021). OCLC: 1253561305

14. [noa, 2015] (2015). Unsupervised Representation Learning with Deep Convolutional Generative Adversarial Networks. OCLC: 1106228480

15. Hong, Y., Hwang, U., Yoo, J., Yoon, S.: How generative adversarial networks and their variants work. ACM Comput. Surv. **52**, 1–43 (2020). https://doi.org/10.1145/3301282

16. Gui, J., Sun, Z., Wen, Y., Tao, D., Ye, J.: A review on generative adversarial networks: algorithms, theory, and applications. arXiv (2020). https://arxiv.org/abs/2001.06937

17. Wang, K., Gou, C., Duan, Y., Lin, Y., Zheng, X., Wang, F.: Generative adversarial networks: introduction and outlook. IEEE/CAA J. Autom. Sinica **4**, 588–598 (2017)

18. Creswell, A., White, T., Dumoulin, V., Arulkumaran, K., Sengupta, B., Bharath, A.: Generative adversarial networks: an overview. IEEE Signal Process. Mag. **35**, 53–65 (2018). https://doi.org/10.1109/MSP.2017.2765202

19. Hitawala, S.: Comparative study on generative adversarial networks. arXiv (2018). https://arxiv.org/abs/1801.04271

20. Saxena, D., Cao, J.: Generative adversarial networks (GANs): challenges, solutions, and future directions (2021)

21. Tanaka, M., Shibata, T., Okutomi, M.: Gradient-based low-light image enhancement. In: 2019 IEEE International Conference on Consumer Electronics (ICCE), pp. 1–2 (2019). OCLC: 8019257222

22. Zhu, J.-Y., Park, T., Isola, P., Efros A.A.: Unpaired image-to-image translation using cycle-consistent adversarial networks. arXiv (2017). OCLC: 8632227009

23. Jiang, Y., Gong, X., Ding, L., Yu, C.:EnlightenGAN: deep light enhancement without paired supervision (2019). OCLC: 1106348980

24. Denton, E., Chintala, S., Szlam, A., Fergus, R.: Deep generative image models using a laplacian pyramid of adversarial networks (2015). OCLC: 1106220075

25. Alec, R., Metz, L., Chintala, S.: Unsupervised representation learning with deep convolutional generative adversarial networks (2015). OCLC: 1106228480

26. Zhou, K., Diehl, E., Tang, J.: Deep convolutional generative adversarial network with semi-supervised learning enabled physics elucidation for extended gear fault diagnosis under data limitations. Mech. Syst. Signal Process. **185**, 109772 (2023). https://www.sciencedirect.com/science/article/pii/S0888327022008408

27. Wang, X., Li, J., Liu, Q., Zhao, W., Li, Z., Wang, W.: Generative adversarial training for supervised and semi-supervised learning. Front. Neurorobot. **16**, 859610 (2022). https://www.frontiersin.org/articles/10.3389/fnbot.2022.859610/full

28. Li, Y., et al.: Curricular robust reinforcement learning via GAN-based perturbation through continuously scheduled task sequence. Tsinghua Sci. Technol. **28**, 27–38 (2023). https://www.scopus.com/inward/record.uri?eid=2-s2.0-85135316788&doi=10.26599

29. Brophy, E., De Vos, M., Boylan, G., Ward, T.: Multivariate generative adversarial networks and their loss functions for synthesis of multichannel ECGs. IEEE Access **9**, 158936–158945 (2021). OCLC: 9343652742

30. Rome, S.: An annotated proof of generative adversarial networks with implementation notes (2017)

31. Tae, J.: The math behind GANs (2020)

32. Stewart, M.: Introduction to turing learning and GANs. Ph.D. Researcher, Towards Data Science

33. Jessica, L.: CelebFaces attributes (CelebA) dataset

34. Jarrett, K., Kavukcuoglu, K., Ranzato, M., LeCun, Y.: What is the best multi-stage architecture for object recognition. In: 2009 IEEE 12th International Conference on Computer Vision (ICCV), pp. 2146–2153 (2009). OCLC: 8558012250

35. Zhang, X.-J., Lu, Y.-F., Zhang, S.-H.: Multi-task learning for food identification and analysis with deep convolutional neural networks (2016) (2016). OCLC: 1185947516

36. Author, F.: Article title. Journal **2**(5), 99–110 (2016)

37. Author, F., Author, S.: Title of a proceedings paper. In: Editor, F., Editor, S. (eds.) CONFERENCE 2016, LNCS, vol. 9999, pp. 1–13. Springer, Heidelberg (2016). https://doi.org/10.10007/1234567890

38. Author, F., Author, S., Author, T.: Book title, 2nd edn. Publisher, Location (1999)

39. Author, A.-B.: Contribution title. In: 9th International Proceedings on Proceedings, pp. 1–2. Publisher, Location (2010)

40. LNCS Homepage. http://www.springer.com/lncs. Accessed 4 Oct 2017

41. Goodfellow, I.: NIPS 2016 tutorial: generative adversarial networks (2016). OCLC: 1106254327

Information Systems Analysis
and Specification

Modelling Software Tasks for Supporting Resource-Driven Adaptation

Paul A. Akiki[1]([⊠]), Andrea Zisman[2], and Amel Bennaceur[2]

[1] Department of Computer Science, Notre Dame University-Louaize, Zouk Mosbeh, Lebanon
paul.akiki@ndu.edu.lb

[2] School of Computing and Communications, The Open University, Milton Keynes, UK
{andrea.zisman,amel.bennaceur}@open.ac.uk

Abstract. Software systems execute tasks that depend on different types of resources. The variability of resources hinders the ability of software systems to execute important tasks. For example, in automated warehouses, malfunctioning robots could delay product deliveries and cause financial losses due to customer dissatisfaction. Resource-driven adaptation addresses the negative implications of resource variability. Hence, this paper presents a task modelling notation called SERIES, which is used for representing task models that support resource-driven adaptation in software systems. SERIES is complemented by a tool that enables software practitioners to create and modify task models. SERIES was evaluated through a study with software practitioners. The participants of this study were asked to explain and create task models and then provide their feedback on the usability of SERIES and the clarity of its semantic constructs. The results showed a very good user performance in explaining and creating task models using SERIES. These results were reflected in the feedback of the participants and the activities that they performed using SERIES.

Keywords: Task modelling notation · Resource-driven adaptation

1 Introduction

Software systems execute tasks that rely on different types of resources. The variability of resources, for reasons such as unexpected hardware failure, excess workloads, or lack of materials, prevents software systems from executing important tasks. This can cause enterprises and individuals to suffer from financial and personal losses. For example, in automated warehouses, malfunctioning robots could delay product deliveries causing customer dissatisfaction and, therefore, reducing an enterprise's sales. The unavailability of medical materials hinders the ability of hospitals to perform medically-critical operations causing loss of life.

Resource-driven adaptation, which is a type of self-adaptation [1, 2], addresses the negative implications of resource variability. In the case of resource variability, software systems should adapt in different ways to manage the lack of resources. The adaptation can include (i) the execution of a similar task that requires fewer resources, (ii) the

J. Filipe et al. (Eds.): ICEIS 2022, LNBIP 487, pp. 249–272, 2023.
https://doi.org/10.1007/978-3-031-39386-0_12

substitution of resources with alternative ones, (iii) the execution of tasks in a different order, or (iv) even the cancellation of non-critical tasks.

The consideration of tasks in resource-driven adaptation provides granularity in the adaptation decision-making because there could be differences in (a) the priorities that specify the level of importance of each task, (b) the applicability of an adaptation type to a task, (c) the types of resource that are used by a task, and (d) the resource consumption of task variants that represent different versions of a task. Software system tasks can be represented as task models that comprise tasks and their properties and relationships. Hence, task models can be used as input for adaptation decision-making. Task models are represented using a task modelling notation. To support resource-driven adaptation, a task modelling notation should support the modelling of characteristics (a)–(d) above.

Existing approaches that focus on tasks (e.g., [3, 4]) do not consider characteristics that are useful for resource-driven adaptation and do not offer a task modelling notation. Furthermore, existing task modelling notations such as ConcurTaskTrees (CTT) [5] and HAMSTERS [6], also have missing characteristics (a)–(d) above, which are useful for resource-driven adaptation. Furthermore, many existing resource-driven adaptation approaches do not consider tasks and work by either disabling optional components [7, 8], reducing data returned by a query [9, 10], changing system configurations through policies [11, 12], or reducing source-code that consumes a lot of computational resources [13, 14]. However, as previously explained, considering tasks provides more granularity in comparison to making adaptation decisions for a whole software system and applying these decisions to all tasks in the same way.

Given the abovementioned limitations, this paper presents a task modelling notation called SERIES (taSk modElling notation for Resource-driven adaptatIon of softwarE Systems). SERIES is based on CTT, which is a notation for representing task models hierarchically using a graphical syntax. The graphical syntax of SERIES is also inspired by UML class diagrams [15]. SERIES has a supporting tool that enables software practitioners to create and modify task models.

We presented the SERIES notation in previous work [16], which included the meta-model and supporting tool. Additionally, that work included an assessment of the usability of SERIES based on the cognitive dimensions framework [17], an evaluation of the overhead of identifying tasks and task variants from a SERIES task model when a task is invoked in a software system, and an evaluation of the intrusiveness of integrating resource-driven adaptation into software systems. The overhead and intrusiveness were evaluated with adaptation components that are part of a resource-driven adaptation framework, which we proposed in another previous work [18].

The work in this paper complements our previous work [16] by presenting a study with software practitioners to evaluate the usability of SERIES and the clarity of its semantic constructs. The participants were asked to explain and create SERIES task models and then provide feedback. The results of this study showed very good user (software practitioner) performance in terms of interpreting and creating task models using SERIES. These results were reflected in the activities that the participants performed using SERIES and the feedback that they provided.

The remainder of this paper is structured as follows. Section 2 discusses existing task modelling notations. Section 3 presents the meta-model of SERIES as a class diagram.

Section 4 illustrates the semantic constructs of SERIES using an example from an automated warehouse system. Section 5 describes the supporting tool of SERIES. Section 6 presents the study that was conducted to evaluate SERIES with software practitioners. Finally, Sect. 7 provides a conclusion for this paper and an overview of future work.

2 Related Work

Task modelling notations have useful characteristics such as tasks and relationships that are used to represent task models. Existing task modelling notations have been used by various model-based development approaches that target user interfaces [19], serious games [20], and collaborative learning systems [21]. Moreover, several surveys [22–24] have discussed and compared existing task modelling notations.

Existing task modelling notations are represented in two ways: (i) *textual* such as UAN [25] and GOMS [26], and (ii) *graphical* such as CTT [5] and HAMSTERS [6]. Moreover, both representations decompose task models in a hierarchical structure. A graphical representation can be used to visualise a hierarchy of tasks and relationships, which makes it easier to understand. Existing task modelling notations support different modelling operators such as *interruption, optionality*, and *concurrency*. The *interruption* operator either suspends a task until another task finishes execution or disables a task by another one. The *optionality* operator either sets a task as optional or provides a choice between multiple tasks to start one of them and disable the rest. The *concurrency* operator allows two tasks to execute simultaneously. Some task modelling notations such as AMBOSS [27], HTA [28], GTA [29], and Diane+ [30] support more operators than others such as TKS [31], GOMS [26], and UAN [25]. CTT [5] and UsiXML [32] support most operators.

Existing task modelling notations provide useful characteristics to represent tasks and relationships [24]. However, they lack the characteristics to support resource-driven adaptation. For example, existing task modelling notations do not support the *association of resource types and priorities to tasks*, which are significant to identify adaptable tasks due to variations in certain resource types. Additionally, task modelling notations do not support *task variants*, which represent different versions of a task that vary according to resource consumption, priorities, user roles, and parameters. Moreover, there is a lack of *stereotypes* to indicate the adaptation types that can be applied to a task. For example, if a task requires a specific resource type, then resource substitution as an adaptation type cannot be applied. The software system should consider another adaptation type instead; for example, delaying the task execution until the resource type becomes available. Furthermore, *feedback properties*, which can be used to receive and display feedback from and to end-users respectively, are not supported by existing task modelling notations. The feedback properties can be useful to provide feedback when an adaptation type is applied to a task and to display the reason for applying an adaptation type to a task.

In conclusion, existing task modelling notations lack characteristics, which are useful for representing task models that serve as input for resource-driven adaptation decision-making. These characteristics include task priorities, task variants, task execution types, resource types, and user feedback. Given these limitations, SERIES is proposed to support resource-driven adaptation in software systems.

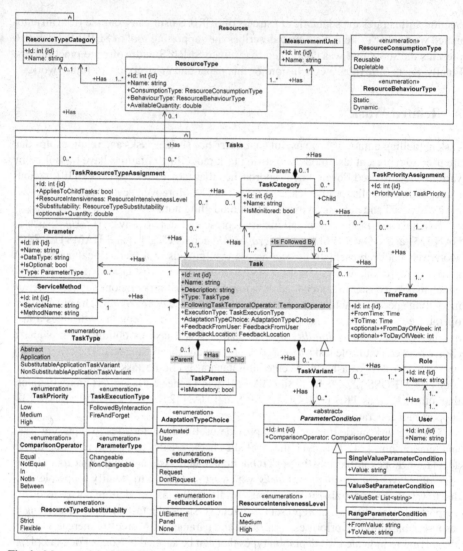

Fig. 1. Meta-model of SERIES represented as a class diagram (the grey parts are from the CTT notation).

3   Meta-model of SERIES

The meta-model of SERIES represented as a class diagram is shown in Fig. 1. SERIES is based on CTT given CTT's useful characteristics like application tasks that SERIES extends with task variants and a hierarchical graphical representation that SERIES extends with additional task properties and relationships. Furthermore, CTT tasks can be associated with User Interface (UI) elements [19]. This is useful for requesting and displaying adaptation-related feedback to end-users on the UI. SERIES' meta-model includes various constructs that are needed for supporting resource-driven adaptation in software systems. As shown in Fig. 1, the part of the meta-model that is highlighted in grey is incorporated from CTT while the rest are new concepts proposed in SERIES to support resource-driven adaptation. In the following subsections, we explain SERIES meta-model and use *italic* to emphasise the names of its main constructs. We start by describing the constructs that have been incorporated from CTT and follow by describing the different types of tasks, namely: abstract task, application task, and application task variant.

3.1 Constructs that SERIES Incorporates from CTT

SERIES incorporates tasks and relationships with temporal operators from CTT. These constructs are represented in the meta-model by the *Task* class and its self-association "*is followed by*" and property *FollowingTaskTemporalOperator*. Tasks are connected using relationships that are annotated with temporal operators, which express how the tasks relate to each other. These temporal operators are important when performing adaptation decisions on a task that has dependents (e.g., cancelling a task could affect other tasks that depend on it).

SERIES also incorporates from CTT two task types, namely *abstract task* and *application task*, which are represented by the enumeration *TaskType*. An *abstract task* involves complex actions and is broken down into a sequence of child (sub) tasks, which are represented in the class *Task* by the self-composition parent-child and the association class *TaskParent*. The property "*IsMandatory*" on this association class indicates whether a subtask is mandatory or optional. An *application task* is executed by the software system without user interaction. The child (sub) tasks of an *abstract task* are represented as *application tasks*.

3.2 Abstract Task

In SERIES, we extend the notion of an *abstract task* to include a *description, execution type, feedback properties* (i.e., *adaptation type choice, feedback from user*, and *feedback location*), *resource types*, and *parameters*.

In addition to the *name*, tasks have a *description*, which is a longer text that provides more explanation. The *task name* is mandatory while the *description* is optional.

The *execution type* specifies whether or not end-users require an immediate result from the task to perform additional interaction with the software system. The possible values for the *execution type* are represented by the *TaskExecutionType* enumeration and include "*followed-by-interaction*" and "*fire-and-forget*". If an immediate result from a task is not required to execute another task after it, then the task's *execution type* will be specified as "*fire-and-forget*", otherwise it will be specified as "*followed-by-interaction*". The *execution type* of tasks affects adaptation decisions during resource variability. In this regard, a task that is "*followed-by-interaction*" cannot be delayed because the user is expecting an immediate result from it. On the other hand, a "*fire-and-forget*" task can be delayed to be executed later when there is less strain on the resource types that it requires and are facing variability.

The *adaptation type choice* indicates whether the type of adaptation is selected manually by the end-user or automatically by the software system. The possible values for adaptation type choice are represented by the *AdaptationTypeChoice* enumeration and include "*automated*" and "*user*". The "*automated*" selection of the type of adaptation relieves end-users from having to frequently make manual choices. Additionally, if multiple types of adaptation share the lowest cost and the adaptation type choice property was set to "*user*", then the software system prompts end-users to select one of the least costly types of adaptation. Otherwise, the software system automatically selects one of the least costly types of adaptation.

The property *feedback-from-user* specifies whether the software system shall ask end-users to provide their feedback on the outcome of the task when adaptation is performed. This feedback enables the software system to improve its adaptation choices. The possible values are represented by the enumeration *FeedbackFromUser* and include "*request*" and "*don't request*". If *feedback-from-user* was set to "*request*", then the software system would request feedback from end-users on how the adaptation affected their work and use this feedback to adjust the costs of applying various types of adaptation. In the case where feedback from the user is not required, then the property *feedback-from-user* will be set to "*don't request*".

The *feedback location* specifies where the feedback shall be requested and given from and to end-users respectively. The possible values for *feedback location* are represented by the enumeration *FeedbackLocation* and include "*UI element*", "*panel*", and "*none*". The "*UI element*" is a suitable option when end-users need to provide immediate feedback as they work. Hence, the feedback will be given on the part of the UI that corresponds to a task (e.g., as a popup window next to the button that the end-user presses to initiate the task). On the other hand, the "*panel*" option groups multiple feedback messages to be checked later. In case no feedback is required then the *feedback location* will be set to "*none*".

Resource types represent a variety of entities that are required to execute a task. A *resource type* is represented by the class *ResourceType*, which has several properties. The property *consumption type* specifies whether the resource type is "*reusable*" or "*depletable*" as indicated by the enumeration *ResourceConsumptionType*. A "*reusable*" *resource type* is available to another task after the task that is using it is done, whereas a "*depletable*" *resource type* is used once. A *resource type* also has a *behaviour type* that is either "*static*" or "*dynamic*" as indicated by the enumeration *ResourceBehaviourType*. A "*static*" *resource type* does not have a behaviour, whereas a "*dynamic*" *resource type* has a behaviour. A *resource type* has an available quantity specified in terms of a measurement unit that is represented by the class *MeasurementUnit*. The available quantity indicates how many resources of a certain type are available on hand to identify whether certain resource types are facing shortages.

A *task resource type assignment* represents an association between a *task* and the *resource types* that it requires. The property *AppliesToChildTasks* is set to either "true" or "false" to indicate whether or not the *resource type* that is assigned to a *task* also applies to its subtasks. By setting this property to "true", there would be no need to duplicate the effort and re-associate the same resource type will all the sub-tasks. Additionally, the *ResourceIntensiveness* property specifies whether the concerned task has a *low, medium*, or *high* consumption for the *resource type* that is assigned to it. This is useful to identify which tasks place more strain on certain types of resources to make adaptation decisions accordingly.

The *task resource type assignment* also specifies the resource type's substitutability for the task. The substitutability is either *"strict"* or *"flexible"* as indicated by the *ResourceTypeSubstitutability* enumeration. A *"strict"* resource type cannot be substituted with alternatives. Hence, during resource variability situations, the software system should seek a type of adaptation that does not involve resource substitution. On the other hand, a *"flexible"* resource type is substitutable with alternative resource types.

Resource types are categorisable under *resource type categories*, which are instances of the *ResourceTypeCategory* class. Hence, if a task requires all the resource types in a category then it would be associated with the category rather than with each resource type individually, to speed up the work.

Similar tasks are categorisable under *task categories*. Like *resource type categories*, *task categories* speed up the work by facilitating the association of *resource types* with tasks. Hence, if all the tasks in a *task category* use a *resource type* then the *task category* is associated with this *resource type*. Additionally, if all the tasks in a *task category* use all the *resource types* in a *resource type category* then the *task category* is associated with the *resource type category* rather than performing individual associations among the *tasks* and *resource types*.

Parameters represent a task's expected input data. A parameter has a *name*, *data type*, and *parameter type*. The *data type* specifies what kind of data the parameter holds (e.g., decimal and string). The *parameter type* specifies whether a *parameter's value* is *"changeable"* or *"non-changeable"* as indicated by the enumeration *ParameterType*. A parameter's value is changed for performing adaptation to execute the task differently. However, in certain cases, parameter values should remain as provided to the software system (e.g., as user input). Hence, both parameter types are needed.

3.3 Application Task

An *application task* has the same characteristics as an *abstract task* in addition to priorities and services.

The *priority* that is assigned to a task represents the task's importance in a domain. Hence, a priority is either *"low"*, *"medium"*, or *"high"* as indicated by the *TaskPriority* enumeration. Priorities are useful for adaptation during resource variability to keep scarce resources available for the most important tasks. The priorities that are specified in the task model are based on domain knowledge.

Priorities could differ among timeframes, which represent intervals of time that are meaningful for a certain domain. For example, a task could have a "low" *priority* in the morning from 8:00 AM to 12:00 PM and a "high" *priority* in the afternoon from 12:01 PM to 5:00 PM. The classes *TimeFrame* and *TaskPriorityAssignment* represent timeframes and the assignment of priorities to tasks respectively.

Priority assignments can be applied to *task categories* as well. This helps in speeding up the work when tasks that belong to the same *task category* have the same *priority assignment*. Hence, when a priority is assigned to a category it would automatically apply to all the tasks within it.

The *service method* represents the function in the software system's source code that is called when the task is executed. A *service method* is represented by the *ServiceMethod* class, which has a *method name* and *service name*. A *method name* represents the name of the function that is called to execute a task, whereas a *service name* represents the name of the class where the function is implemented.

Moreover, when the software system receives a request to initiate a task it checks whether adaptation is needed before executing the task. To make adaptation decisions, the software system requires information from the task model (e.g., which type of adaptation applies to the initiated task). Hence, upon receiving a task initiation request the software system identifies the corresponding task from the task model by comparing the names of the class and function that are invoked from the source code to the service methods in the task model.

3.4 Application Task Variant

An *application task variant* is a special case of an application task and is needed to (1) avoid treating all executions of an application task in the same way when adapting and (2) identify how to execute an application task with fewer resources. An *application task variant* has the same characteristics as an *application task*, with the addition of parameter conditions, resource intensiveness, roles, and substitutability.

Parameter conditions specify the parameter values that distinguish application task variants from each other. A *parameter condition* is represented by the class *ParameterCondition* and its subclasses "*single value*", "*value set*", and "*range*". *Single value*, *value set*, and *range parameter conditions* compare the value of a parameter to a single value (e.g., 10), set of values (e.g., 10, 15, and 20), and range of values (e.g., 10 to 20) respectively. Moreover, the comparisons performed in the parameter conditions use one of five comparison operators as indicated by the enumeration *ComparisonOperator*. The operators "*equal*" and "*not equal*" are used for *single-value parameter conditions*, the operators "*in*" and "*not in*" are used for *value-set parameter conditions*, and the operator "*between*" is used for *range parameter conditions*.

Resource intensiveness indicates the level of resource consumption of an *application task variant* for a certain *resource type* (i.e., the strain that an application task variant places on a *resource type*). The value of *resource intensiveness* is either *low*, *medium*, or *high* as indicated by the enumeration *ResourceIntensivenessLevel*. A high *resource intensiveness* means that more resources are required to execute a task, and vice-versa. The *resource intensiveness* is represented in the class *TaskResourceTypeAssignment* since its value is specified per combination of task variant and resource type. *Resource intensiveness* helps in adaptation decision-making when a software system needs to execute a similar task variant that consumes fewer resources to reduce the strain on the resources that are facing variability.

Roles represent the different groups of end-users who are eligible to initiate the *application task variant*. Although an *application task variant* is performed by the software system, it could be initiated after a request from an end-user. A typical example of a role in an enterprise is a job title like manager or clerk. However, roles are not necessarily related to job titles. For example, a role could be used to indicate different age groups of end-users. Roles and end-users are represented by the classes *Roles* and

Users respectively. The association between these two classes denotes the assignment of roles to end-users.

Roles are important for adaptation because they can be used by a software system to identify whether privileged users invoke a certain task variant thereby affecting its priority. For example, a task could be considered more important if it was invoked by a manager in an enterprise software system or by a senior citizen in a public service software system (e.g., transportation).

An *application task variant* can be either *substitutable* by another *application task variant* or *non-substitutable*. This depends on the type of parameter that is used in the parameter condition. If the parameter is "*changeable*", then its value can be changed to invoke an alternative application task variant that expects a different value. On the other hand, if the parameter type is "*non-changeable*", then its value cannot be changed. Hence, it is not possible to invoke an alternative *application task variant* that expects a different parameter value.

A substitutable task variant could be executed instead of another one that has higher *resource intensiveness*. However, non-substitutable task variants are also important for addressing resource variability. Although non-substitutable task variants cannot be interchanged like their substitutable counterparts, they have different priorities and are associated with parameter conditions that specify in which cases they are executed. This helps software systems in making adaptation decisions that reduce the strain on limited resources to keep them available to the most important task variants.

In order to illustrate, in the following section we present an example of an automated warehouse system represented in SERIES task model.

4 Automated Warehouse System Example

Consider an example of a warehouse for a retail store that receives customer orders throughout the day. In this example, the warehouse is automated by robots that perform order preparation tasks and pack items into boxes. To prepare an order, robots locate the respective items in the warehouse, pack the items in boxes, and decorate the boxes to prepare them for delivery (e.g., seal boxes, and attach labels with addresses).

Consider that robots should place items of the same type together in a pile inside the boxes. Robots and boxes are essential resources for the retail store's operations. Robots can temporarily go out of service due to unexpected errors or due to the need for recharging, thereby, delaying order fulfilment and causing financial losses. Similarly, due to a high number of orders, the warehouse may run out of boxes.

Figure 2 illustrates an excerpt of a task model, which is represented using SERIES, for the automated warehouse system example. Additionally, Fig. 2 includes a legend that shows the different constructs used in the task model. These constructs include task types and a relationship that SERIES uses from CTT, as well as task types, task execution types, relationships, and task properties that SERIES adds to CTT.

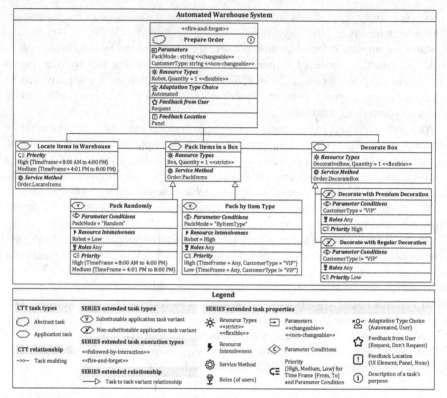

Fig. 2. Excerpt of a task model example from an automated warehouse system.

4.1 Abstract Task: Prepare Order

The example task model has an abstract task called "Prepare Order", which represents the activity of preparing customer orders for delivery. This task has a *description*, which is accessed via the information icon displayed in the top right corner of the box that represents the task. This description explains further the purpose of the task.

The task "Prepare Order" has a "*fire-and-forget*" execution type since end-users do not require a result to be returned from the task to interact with the system and execute another task. For example, a batch of order preparation tasks could be executed, and when it is done the outcome would be checked by a warehouse control employee.

Two *parameters* are given as input for "Prepare Order". The first one is a *changeable parameter* called "PackMode", which specifies the mode for packing items in a box. The second one is a *non-changeable parameter* called "CustomerType", which specifies the type of customer for whom the order is being prepared. The "PackMode" is set to be *changeable* because it is possible to change its value to specify that the packing should be done differently (e.g., items are sorted by their type or randomly). The "CustomerType" is set to be *non-changeable* because, in this case, it is not possible to downgrade a customer from VIP to regular when performing adaptation.

A single *flexible resource type* called "Robot" is used to process a customer order. The robot is set as a *flexible resource type* to indicate that it is substitutable by another robot during resource variability (e.g., due to unexpected robot malfunctions).

The *adaptation type choice* is set to "*automated*" for the task "Prepare Order". This means the software system will automatically select the least costly adaptation type to apply to this task. The property *feedback-from-user* is set to "*request*" for the task "Prepare Order". This means when an adaptation type is applied to this task, feedback will be requested from the end-user to provide input on the outcome of the adaptation. The *feedback location* for "Prepare Order" is set to "*panel*" to indicate that end-users can provide feedback on a separate side panel that lists messages underneath each other to inform the end-users about the decisions of the system.

4.2 Application Tasks

The abstract task "Prepare Order" is divided into three application tasks "Locate Items in Warehouse", "Pack Items in a Box", and "Decorate Box". These application tasks represent the sequence of actions that are required for preparing an order.

Locate-Items-in-Warehouse has a "high" *priority* between 8:00 AM and 4:00 PM and a "medium" *priority* between 4:01 PM and 8:00 PM. This example shows how a task can have different priorities for different timeframes.

Pack-Items-in-a-Box has a single *resource type* called "Box", which represents a container used to pack the items of a customer order. The "Box" is set to *strict* to indicate that it is not substitutable. For example, in this case, it is not possible to substitute a box with another container such as a bag because the box provides better protection. However, the choice depends on the requirements for a particular domain.

Decorate-Box has a single *flexible resource type* called "DecorativeBow", which is used to decorate the boxes after packing. The "DecorativeBow" is set to *flexible* since it can be substituted with other types of decorations during resource variability.

4.3 Application Task Variants for Pack Items in a Box

The *application task* "Pack Items in a Box" has two *application task variants*, which are "Pack Randomly" and "Pack by Item Type". These *application task variants* are substitutable because it is possible to change the value of the "PackMode" parameter to execute one variant instead of another.

Furthermore, the variants of "Pack Items in a Box" are linked to it via a "task to task variant" relationship to indicate that each variant performs the packing differently. The "task to task variant" relationship is similar to a generalisation relationship in UML. Hence, the *application task variants* use (inherit) the information found in their parent application task (e.g., resource type and service method). In this example, both "Pack Randomly" and "Pack by Item Type" use the *resource type* "Box". Furthermore, both of these variants call the same service method (Order.PackItems) but with a different value for the parameter "PackMode".

Pack-Randomly is executed when the "PackMode" parameter is equal to "Random". This variant has a "low" *resource intensiveness* on the robot *resource type* since the robot will perform the packing randomly without taking additional time to sort the items before placing them in the box. Additionally, end-users with any role (e.g., any warehouse employee) can initiate "Pack Randomly". Hence, its importance is not affected by who initiated it. Furthermore, "Pack Randomly" has a "high" *priority* between 8:00 AM and 4:00 PM and a "medium" *priority* value between 4:01 PM and 8:00 PM. These priorities and timeframes were chosen because customer orders are mostly fulfilled between 8:00 AM and 4:00 PM and random packing of items reduces the strain on resources during this busy time.

Pack-by-Item-Type is executed when the "PackMode" parameter is equal to "ByItem-Type". This variant has a "high" *resource intensiveness* on the robot *resource type* since it would take the robot more time to identify the items and pack each type in a separate pile in comparison to packing items randomly. Furthermore, "Pack by Item Type" has a "high" *priority* for VIP customers and a "low" *priority* for non-VIP customers. Pack-by-Item-Type places more strain on the robots in comparison to Pack-Randomly. Hence, it was given a low *priority* for non-VIP customers.

4.4 Application Task Variants for Decorate Box

The *application task* "Decorate Box" has two *non-substitutable application task variants*, which are "Decorate with Premium Decoration" and "Decorate with Regular Decoration". These variants are non-substitutable because the value of the "CustomerType" parameter cannot be changed. This means that it is not possible to change an order that is designated for a VIP customer to an order for a regular customer.

Decorate-with-Premium-Decoration is executed when the parameter "Customer-Type" is equal to "VIP". This means that VIP customers will receive a premium decoration for their order. Moreover, this task variant has a "high" *priority* value at any time-frame, which means it is important for VIP customers to receive a premium decoration on their order regardless of when the order is prepared.

Decorate-with-Regular-Decoration is executed when the parameter "Customer-Type" is not equal to "VIP". Hence, non-VIP customers will receive regular decoration for their orders. Furthermore, this variant has a "low" *priority* at any timeframe, which means it is not important for non-VIP customers to receive a decoration for their order. For example, if there is a shortage in decorative bows, then the task would be cancelled, and non-VIP customers would receive an undecorated box.

5 Supporting Tool of SERIES

To support the creation and modification of task models represented in SERIES, we developed a tool as shown in Fig. 3. The tool is composed of three panels, namely: (a) Task Model Explorer, (b) Visual Task Model, and (c) Properties panels. The "Task Model Explorer" panel (Fig. 3a) offers a hierarchical view of a task model. This panel shows how the tasks are ordered under each other using a compact tree structure that enables users to navigate to the different parts of the model. The "Visual Task Model" panel (Fig. 3b) displays the task model graphically using the SERIES notation. Moreover, it is possible to zoom in and out of the task model using the "Zoom" slider.

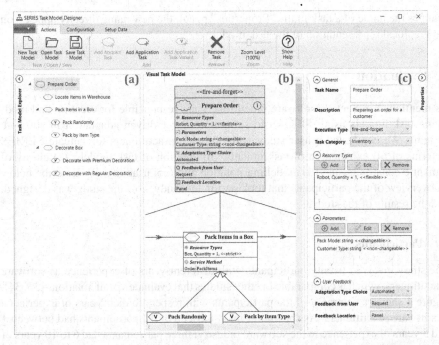

Fig. 3. Supporting tool for creating and modifying task models using SERIES.

The "Properties" panel (Fig. 3c) displays the properties of the selected task. Users access this panel to edit the values of the properties (e.g., change the name of a task). A property can either have a single value like task name and description or a set of values like resource types and parameters. Properties with lists of values are shown in the "Properties" panel with "Add", "Edit", and "Remove" buttons. The buttons "Add" and "Edit" open windows for inputting and editing the data of the respective property. On the other hand, the button "Remove" deletes the selected value.

The automated-warehouse-system example shown in Fig. 2 is displayed in the supporting tool in Fig. 3. The "Task Model Explorer" panel shows the task model as a hierarchical view, which starts from the abstract task "Prepare Order", followed by the application tasks and their corresponding application task variants. The "Visual Task Model" panel displays the task model, which is partially shown in Fig. 3b to keep the text on the figure readable in the limited space on the page. The "Properties" panel shows the properties and their values for the abstract task "Prepare Order" that is selected in the task model.

The name and description are editable as text, execution type and user feedback are selected from combo boxes, while resource types and parameters can be added, edited, or removed. Different properties and values are displayed depending on which part of the task model is selected. For example, if a task variant was selected, the "Properties" panel would also display parameter conditions, resource intensiveness, and roles. Furthermore, when a user edits the values of these properties, the new values are directly reflected in the "Task Model Explorer" and the "Visual Task Model". For example, when a user changes the name of a task in the "Properties" panel, the new name directly appears on the graphical representation of the task model.

6 Evaluation

As mentioned in Sect. 1, software practitioners are responsible for creating and modifying task models using SERIES to support resource-driven adaptation in software systems. We conducted a user study with software practitioners to evaluate SERIES in terms of (a) explanation of task models, (b) creation of task models, (c) clarity of semantic constructs, and (d) modelling notation and tool usability. We describe below an overview of the participants that took part in the study, how the study was designed, and the results of the study.

6.1 Participants

The study involved twenty participants with different years of experience as software practitioners, which is comparable to other studies that evaluate visual notations [33, 34]. Figure 4 shows a summary of the participants with respect to their years of experience as software practitioners. As shown in Fig. 4, thirteen of the participants had between 1 and 5 years of experience in the software industry; three participants had 6 to 10 years of experience; three other participants had less than one year of experience; and only one participant never had any experience in the software industry. However, this participant had some personal experience in developing non-commercial software applications. The participants' collective experience includes the development of software systems for multiple domains such as business, education, electronics, games, government, and multimedia. These software systems cover several software paradigms including web, mobile, desktop, and virtual reality.

Figure 5 shows the demographics of the participants in which sixteen participants are currently working as software practitioners at software companies in different countries including Lebanon, the United States, Canada, Denmark, Egypt, Germany, and the Netherlands; and four participants are currently working as researchers at The Open University in the United Kingdom, in which three of them had previous experience in the software industry.

All the participants had previous experience in using visual modelling notations. As shown in Fig. 6, their collective experience includes using UML diagrams, flow charts, ER models, relational models, logic circuits, and architecture diagrams. The participants had also used various modelling tools including StarUML, Draw.io, Cadence, Dia, Jira, Lucid Chart, PgAdmin, and Umlet. As shown in Fig. 7, thirteen participants 63%) had used visual modelling notations both at work and in a course, and seven participants (37%) had only used visual modelling notations in a course.

The diversity in the participants' backgrounds including their levels and types of experience in the software industry and with visual modelling notations and the places where they have worked enriches the study's outcome. This is due to feedback obtained from people with different capabilities and perspectives.

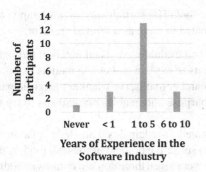

Fig. 4. Experience of the participants in the software industry.

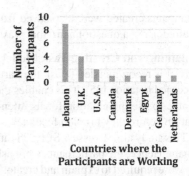

Fig. 5. Countries where the participants are working.

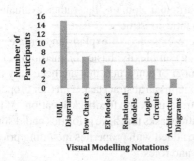

Fig. 6. Experience of the participants with visual modelling notations (each participant could list multiple notations).

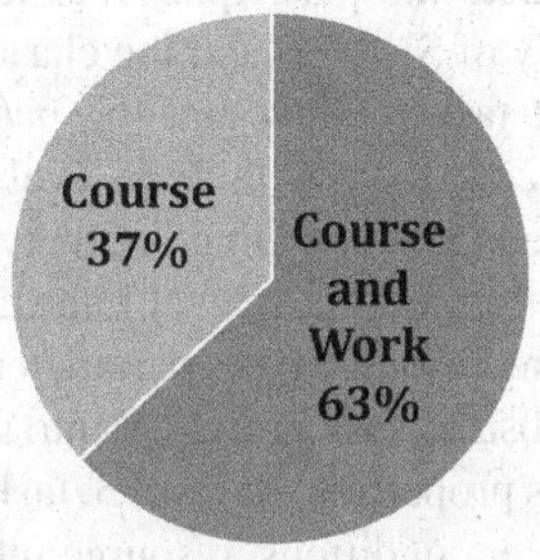

Fig. 7. Capacity in which the participants used visual modelling notations.

6.2 Design of the Study

The study involved a brief video tutorial about SERIES and its supporting tool, the explanation and creation of task models using SERIES and its tool, and the completion of a questionnaire about the usability of SERIES and the clarity of its semantic constructs. Each participant took on average 55 min to complete the study and all the activities were done in English.

Environment and Data Collection. The study was conducted online due to the Covid-19 pandemic. However, this did not obstruct any of the planned activities. The participants were given remote control of the researcher's computer via Zoom (videoconferencing software program). They were able to use the supporting tool of SERIES (refer to Sect. 5). Zoom was chosen because it provided the best performance for remote access and videoconferencing with minimal to no lagging over an internet connection. Moreover, the participants' verbal input (feedback) and their work on the supporting tool of SERIES were captured using audio and screen recordings respectively. The task models that the participants created were also saved from the tool. The participants' written input was captured using a questionnaire that was presented to them as a Word document.

Explaining and Creating Task Models. When evaluating a visual notation through a study, it is recommended to ask the participants to explain and create models using the concerned notation [35]. This enables the researcher to observe whether the participants are able to use the notation for its intended purpose and enables participants to try the notation before they provide feedback on it.

This study aims to assess SERIES rather than the participants' knowledge of a certain domain. Hence, the participants were asked to select the domain of the task models that they were required to explain and create. They were given three domain options including hospital, manufacturing, and surveillance. Although three domain options were given, there was no discrepancy among them in the level of difficulty of the task models that the participants were asked to explain and create. Furthermore, the study does not aim to evaluate the participant's expertise in a particular domain, but rather evaluate the usability of SERIES and the clarity of its semantic constructs.

The task models that the participants were expected to explain and create were presented to them with three levels of increasing complexity: basic (1), medium (2), and advanced (3), depending on the types and number of constructs that are necessary to be used in the models. At level 1, the task model contains an abstract task with four properties including resource type, parameter, feedback-from-user, and feedback-location. At level 2, the abstract task is divided into three application tasks that have priorities and resource types as properties. At level 3, task variants are added with properties including priority, parameter conditions, resource intensiveness, and roles.

Questionnaire. After the participants explained and created SERIES task models, they were asked to provide their feedback by completing a questionnaire (available at https:// bit.ly/series-questionnaire). The questionnaire contains questions about how well the semantic constructs of SERIES convey a clear meaning that enables a software practitioner to explain and create task models using this notation (Q1.1 and Q1.2); questions about the usability of SERIES (Q2.1), including how easy it is to use SERIES to develop the requested tasks at its various levels of complexity. Following, the participants were asked to select three Product Reaction Cards (PRCs) [36] that they thought were most suitable for describing SERIES (Q2.2). The PRCs enable researchers to understand the desirability resulting from a user's experience with a product. We used the PRCs because they allow participants to qualitatively express their opinion about SERIES by selecting descriptive terms, which complement the quantitative ratings that they were asked to give in other parts of the evaluation. The participants were given a set of 16 PRCs (eight positive and eight negative ones). The selection was not restricted to either positive or negative. Hence, the participants were able to select any three PRCs, all positive, all negative, or a mixture of both.

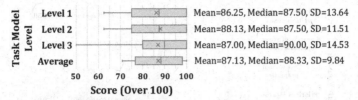

Fig. 8. Scores on the participants' explanation of task models (an "x" on the boxplot represents a mean and a circle represents an outlier)

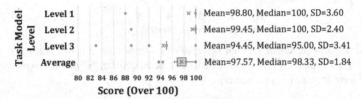

Fig. 9. Scores on the participants' creation of task models (an "x" on the boxplot represents a mean and a circle represents an outlier).

6.3 Results

In this section, we present the results of the various analysis of the study.

Task Model Explanation Results. The participants took on average 1.86 min to explain the task models for each one of the three levels presented. The participants averaged close scores on the explanation of all three task model levels. As Fig. 8 shows, their scores (over 100) on the explanation were on average 86.25, 88.13, and 87, for levels 1, 2, and 3 respectively, and their average score across all three levels was 87.13. These results show that the participants exhibited very good performance when explaining task models that are represented using SERIES. The close scores on the three task model levels indicate that the participant did not face difficulty when the complexity of the task model hierarchy increased. The average score on level 1 was slightly lower than the scores on the other two levels which are more complex. A reason for this could be that the purpose of the abstract task from level 1 became clearer once sub-tasks were shown at level 2 (several participants mentioned this).

Considering the overall explanation scores and that the mistakes were not focused on a particular semantic construct, it is possible to say that there is no major ambiguity in the meaning of a particular semantic construct in SERIES. Examples of the participants' explanation mistakes include explaining a task without discussing its execution type, explaining a parameter property without discussing its parameter kind, and explaining a priority property without discussing its corresponding timeframe. Although there is room for improving the explanations to reach perfect scores (100/100), it is important to mind that the participants were only given a brief tutorial about SERIES and that they never had any previous experience with this notation.

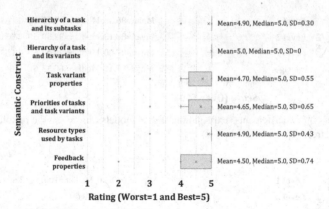

Fig. 10. Participants' feedback on their perception of whether the semantic constructs convey a clear meaning for explaining and creating SERIES task models (an "x" on the boxplot represents a mean and a circle represents an outlier).

Task Model Creation Result. The participants took on average 3.47 min to create the task models that correspond to the requirements for each one of the three levels. The participants averaged close scores on the creation of task models for all levels. As Fig. 9 shows, their scores (over 100) were on average 98.80, 99.45, and 94.45, for levels 1, 2, and 3 respectively, and the average score across all three levels was 97.57.

These results show that the participants exhibited excellent performance when creating task models using SERIES via its supporting tool. The mean score on level 3 was slightly lower than the scores of the other two levels. Nonetheless, it is still very high (94.45). There were some outlier scores as shown in Fig. 9. However, these scores were mostly ≥88. The mistakes were overall minor. For example, some participants did not specify the "non-substitutable" property on a pair of the task variants or specified the "role" property on one of the task variants but forgot to specify it on the other variant. Another example of a mistake is giving a parameter the wrong name (typo). These mistakes are mostly due to lapses that do not indicate the existence of any major difficulties in using SERIES to create task models.

Comparison between Task Model Explanation and Creation. As explained above, both the explanation and the creation of task models yielded high scores in terms of performance of these activities. However, the performance mean scores for the creation of task models (Fig. 9) were higher than the performance mean scores for the explanation of task models (Fig. 8). This could be due to the participants performing the explanation of the task models first. Hence, this gave them some additional exposure to examples of SERIES task models before they used this notation.

As mentioned in Sect. 6.1, the level of experience of this study's participants is diverse. The scores of the participants were compared to see whether the level of experience affected their ability to explain and create task models. Concerning the explanation of task models, the four participants with little or no experience (<1 year - never) averaged a score of 85 across the three task model levels. The thirteen participants with medium experience (1 to 5 years) averaged a close score of 86.02. The three participants

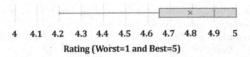

4   4.1   4.2   4.3   4.4   4.5   4.6   4.7   4.8   4.9   5

Rating (Worst=1 and Best=5)

Fig. 11. Feedback given by the participants on the ease of use of SERIES (the "x" on the boxplot represents the mean).

with high experience (6 to 10 years) averaged a score of 94.72 across the three task model levels. Although the average score of the participants with the higher experience level is higher, the other participants still averaged high scores ($\geq$85) considering it is the first time they work with SERIES.

Concerning the creation of task models, there was little to no difference among the scores for participants for all levels of experience. This is expected considering that the overall scores are very high. The four participants with little or no experience (<1 year - never) averaged a score of 97.33 across the three task model levels. The thirteen participants with medium experience (1 to 5 years) averaged the same score (97.33). On the other hand, the participants with a high level of experience (6 to 10 years) averaged a score of 98.88 across the three task model levels. Since the scores are overall very close, the level of experience did not affect the ability of the participants to create task models using SERIES.

Clarity of Semantic Constructs Results. The participants rated the clarity of the semantic constructs of SERIES on a scale between 1 and 5, where 1 is the worst and 5 is the best. Figure 10 shows these ratings. We can see that the mean values for the various semantic constructs ranged between 4.5 and 5.0. These results indicate that the participants considered the semantic constructs to convey a clear meaning that enables software practitioners to explain and create task models. As Fig. 10 shows, four of the semantic constructs, i.e., task variant properties, priorities of tasks and task variants, resource types used by tasks, and feedback properties, had one outlier with a rating of three or two over five. In this case, the participants felt that it would be useful to have some more clarification about the respective constructs. However, these values do not affect the participants' overall positive perception of the clarity of the semantic constructs as shown by the mean ratings that is $\geq$4.5.

Usability Results. The usability of SERIES was measured in terms of the participants' ratings of its ease of use and selected PRCs, as well as qualitative feedback. More specifically, the participants were given five questions on ease of use that covered difficulty, clarity, understandability, frustration, and mental effort. The result reported in Fig. 11 is computed from the means of the answers that the participants gave to all five questions. The mean rating given by the participants on the ease-of-use questions was 4.79 over 5 (1 is the worst and 5 is the best).

Each participant was asked to select three PRCs that best described SERIES. As Fig. 12 shows, the PRCs that the participants selected were mostly positive (58 out of 60). The selected PRCs complement the results related to the ease of use, which indicates that the participants have a very positive perception of the usability of SERIES.

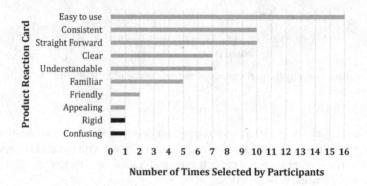

Fig. 12. Product Reaction Cards (PRCs) selected by the participants to describe SERIES – each participant was asked to select three PRCs (positive PRCs are shown grey while the negative ones are shown in black).

The qualitative feedback was given in the form of comments on the clarity of the semantic constructs and the usability of SERIES and its supporting tool. These comments were categorised under six themes, namely: "Notation is Usable" (31%), "Notation is Learnable" (29%), "Tool is Usable" (14%), "Tool is Learnable" (8%), "Minor Change Suggestions" (10%), and "Minor Clarification Suggestions" (8%). These comments were overall positive. A sample of the comments are presented next as italicised text and grouped under the abovementioned themes. For anonymity, the participants are quoted using a reference number (e.g., P1).

The comments from the first theme are related to the usability of the notation. In this regard, P12 said, "*This is actually one of the things that I liked about SERIES. It is the ability to create the subtasks very clearly and to define how they follow each one after the other*". P14 said "*The three-level hierarchy is very clear and helpful in analysing and formalising a complex task. The visual representation of the task, subtasks, and variants of the sub-tasks is very well organised and lets the user have every information at hand*". This indicates that the visual presentation of the hierarchy helped the participants to understand the meaning of the relationship between a task and its subtasks and among subtasks and relates to the principles of *semantic transparency* and *perceptual discriminability* from the Physics of Notations [37].

The comments from the second theme are related to the learnability of the notation. In this regard, P10 said, "*The constructs are clear once we see the tutorial*". Additionally, P8 noted the same thing and elaborated by saying "*Notation is easy to use after a quick tutorial. Terms mean the same throughout (e.g., resources for tasks and subtasks)*". This indicates that software practitioners could learn SERIES without significant time and effort. The idea presented by P8 is due to SERIES adhering to the dimension of *consistency* from the Cognitive Dimensions Framework [17] and the principle of *semiotic clarity* from the Physics of Notations. P20 said "*Certain elements of the task model's visual representation resemble UML class diagrams. Tasks are represented as boxes like classes and task variants are related to tasks using an arrow that resembles generalisation relationships*". Although the purpose of SERIES is different than that of UML class diagrams, the visual resemblance in parts of the notation created familiarity that helps software practitioners in learning SERIES.

The comments from the third theme are related to the usability of the tool. In this regard, P7 said *"The UI is very user-friendly. All functionality is easily visible"*. Additionally, P14 said *"The user interface lets the user reach every window in an easy and organised manner – i.e., the user does not need to scroll menus to find and fill in the necessary information"*. These comments reflect a positive perception of the supporting tool's usability, which complements the usability of the notation.

The comments from the fourth theme are related to the learnability of the tool. In this regard, P2 said, *"The tool resembles other tools making it very easy to learn"*. P13 said, *"The UI of the tool is familiar because it resembles existing IDEs"*. This was a design choice because the panels of the tool were designed to resemble panels that are common in existing integrated development environments (IDEs). For example, the "Task Model Explorer" and the "Properties" panels in the tool of SERIES resemble the solution explorer and properties box respectively in the Visual Studio IDE.

The fifth and sixth themes are related to a few minor change suggestions and minor clarification suggestions respectively. These minor suggestions do not hinder the ability of software practitioners to use SERIES and its supporting tool.

Regarding the changes, P17 suggested having *"colour coding for priorities"*, and P8 suggested adding a *"small UI"* under the selection boxes of resources, priorities, and parameters so these properties can be added quickly without having to open a new (popup) window. The implementation of P17's suggestion is simple and could be useful to attract attention to the high-priority tasks and task variants. As for P8's suggestion, the addition of "small UIs" could over-clutter the properties box if these UIs are always visible. Hence, one possibility could be to use dropdown UIs that the user opens via the selection boxes or buttons without navigating to a new window.

As for the clarifications, P12 said *"Everything was mostly clear, maybe a bit more information on the priority times"*, and P14 said *"At the beginning, the timeframe and its relation with the priority was a bit confusing"*. P12 and P14 mentioned this point because the brief tutorial that was given to the participants only demonstrated the association of priorities with a specific time frame. However, during the study, the participants were asked to set the priorities for any time frame. Nonetheless, even with the brief coverage of this concept, the participants' explanation, use, and feedback demonstrated that they know how to work with priorities and timeframes. Hence, this point did not cause a major issue overall. For example, both P12 and P14 rated the corresponding semantic constructs (priorities of tasks and task variants) with a 4 over 5 (i.e., "clear") because the concepts got clarified after some reflection.

Conclusions. The results of the study demonstrated a good performance from the participants to explain and create task models of different levels of complexity using SERIES. Overall, the participants agree that the semantic constructs of SERIES are clear, and that SERIES is usable. The participants suggested a few minor changes to SERIES, which do not hinder the ability of software practitioners to use this notation.

7   Conclusion and Future Work

This paper presented a task modelling notation called SERIES, which offers semantic constructs for representing the tasks of software systems that perform resource-driven adaptation. SERIES represents task models graphically. An example of an automated warehouse system was used to illustrate the work. SERIES is complemented by a tool that supports software practitioners in working with task models.

SERIES was evaluated through a study with twenty software practitioners from different backgrounds and geographic distribution. The study was divided into four parts, namely: (i) video-based tutorial about SERIES and its supporting tool; (ii) explanation of task models represented using SERIES; (iii) creation of task models using SERIES via its supporting tool; and (iv) usability analysis based on questionnaire. The results of the study showed a very good performance of its participants in explaining and creating SERIES task models. Moreover, the participants gave positive feedback about the usability of SERIES and the clarity of its semantic constructs.

Currently, we are considering the feedback given by the participants to amend the notation's tool to use drop-down windows instead of popup windows for editing task properties. In the future, we plan to extend SERIES to be used as a more general task modelling notation and not only to support resource-driven adaptation situations. We are also investigating ways to motivate a broader use of SERIES in organisations.

Acknowledgements. This work was supported by the Engineering and Physical Sciences Research Council [grant numbers EP/V026747/1, EP/R013144/1].

References

1. Cheng, B.H.C., et al.: Software engineering for self-adaptive systems: a research roadmap. In: Cheng, B.H.C., de Lemos, R., Giese, H., Inverardi, P., Magee, J. (eds.) Software Engineering for Self-Adaptive Systems. LNCS, vol. 5525, pp. 1–26. Springer, Heidelberg (2009). https://doi.org/10.1007/978-3-642-02161-9_1

2. de Lemos, R., et al.: Software engineering for self-adaptive systems: a second research roadmap. In: de Lemos, R., Giese, H., Müller, H.A., Shaw, M. (eds.) Software Engineering for Self-Adaptive Systems II. LNCS, vol. 7475, pp. 1–32. Springer, Heidelberg (2013). https://doi.org/10.1007/978-3-642-35813-5_1

3. Sousa, J.P., Poladian, V., Garlan, D., Schmerl, B., Shaw, M.: Task-based adaptation for ubiquitous computing. IEEE Trans. Syst. Man Cybern. Part C (Appl. Rev.) **36**(3), 328–340 (2006). https://doi.org/10.1109/TSMCC.2006.871588

4. Perttunen, M., Jurmu, M., Riekki, J.: A QoS model for task-based service composition. In: Proceedings of the 4th International Workshop on Managing Ubiquitous Communications and Services, p. 11 (2007)

5. Paterno, F., Mancini, C., Meniconi, S.: ConcurTaskTrees: a diagrammatic notation for specifying task models. In: Howard, S., Hammond, J., Lindgaard, G. (eds.) INTERACT 1997. ITIFIP, pp. 362–369. Springer, Boston (1997). https://doi.org/10.1007/978-0-387-35175-9_58

6. Martinie, C., Palanque, P., Winckler, M.: Structuring and composition mechanisms to address scalability issues in task models. In: Campos, P., Graham, N., Jorge, J., Nunes, N., Palanque, P., Winckler, M. (eds.) INTERACT 2011. LNCS, vol. 6948, pp. 589–609. Springer, Heidelberg (2011). https://doi.org/10.1007/978-3-642-23765-2_40

7. Xu, M., Buyya, R.: Brownout approach for adaptive management of resources and applications in cloud computing systems: a taxonomy and future directions. ACM Comput. Surv. (CSUR). **52**, 1–27 (2019)
8. Klein, C., Maggio, M., Årzén, K.-E., Hernández-Rodriguez, F.: Brownout: building more robust cloud applications. In: Proceedings of the 36th International Conference on Software Engineering - ICSE 2014, Hyderabad, India, pp. 700–711. ACM Press (2014)
9. Gotz, S., Gerostathopoulos, I., Krikava, F., Shahzada, A., Spalazzese, R.: Adaptive exchange of distributed partial Models@run.time for highly dynamic systems. In: 2015 IEEE/ACM 10th International Symposium on Software Engineering for Adaptive and Self-managing Systems, Florence, Italy, pp. 64–70. IEEE (2015)
10. Viswanathan, L., Jindal, A., Karanasos, K.: Query and resource optimization: bridging the gap. In: 2018 IEEE 34th International Conference on Data Engineering (ICDE), pp. 1384–1387. IEEE (2018)
11. Keeney, J., Cahill, V.: Chisel: a policy-driven, context-aware, dynamic adaptation framework. In: Proceedings POLICY 2003. IEEE 4th International Workshop on Policies for Distributed Systems and Networks, pp. 3–14. IEEE (2003)
12. Efstratiou, C., Friday, A., Davies, N., Cheverst, K.: A platform supporting coordinated adaptation in mobile systems. In: Proceedings Fourth IEEE Workshop on Mobile Computing Systems and Applications, pp. 128–137. IEEE (2002)
13. Christi, A., Groce, A., Gopinath, R.: Resource adaptation via test-based software minimization. In: 2017 IEEE 11th International Conference on Self-adaptive and Self-organizing Systems (SASO), pp. 61–70. IEEE (2017)
14. Christi, A., Groce, A.: Target selection for test-based resource adaptation. In: International Conference on Software Quality, Reliability and Security, pp. 458–469. IEEE (2018)
15. Fowler, M.: UML Distilled: A Brief Guide to the Standard Object Modeling Language. Addison-Wesley, Boston (2003)
16. Akiki, P., Zisman, A., Bennaceur, A.: SERIES: a task modelling notation for resource-driven adaptation. In: Proceedings of the 24th International Conference on Enterprise Information Systems, pp. 29–39. SCITEPRESS - Science and Technology Publications, Online Streaming (2022). https://doi.org/10.5220/0011001800003179
17. Green, T.R.G., Petre, M.: Usability analysis of visual programming environments: a 'cognitive dimensions' framework. J. Vis. Lang. Comput. **7**, 131–174 (1996)
18. Akiki, P.A., Zisman, A., Bennaceur, A.: Work with what you've got: an approach for resource-driven adaptation. In: 2021 IEEE International Conference on Autonomic Computing and Self-organizing Systems Companion (ACSOS-C), DC, USA, pp. 105–110. IEEE (2021). https://doi.org/10.1109/ACSOS-C52956.2021.00030
19. Calvary, G., Coutaz, J., Thevenin, D., Limbourg, Q., Bouillon, L., Vanderdonckt, J.: A unifying reference framework for multi-target user interfaces. Interact. Comput. **15**, 289–308 (2003)
20. Vidani, A.C., Chittaro, L.: Using a task modeling formalism in the design of serious games for emergency medical procedures. In: 2009 Conference in Games and Virtual Worlds for Serious Applications, pp. 95–102. IEEE (2009)
21. Molina, A.I., Redondo, M.A., Ortega, M., Lacave, C.: Evaluating a graphical notation for modeling collaborative learning activities: a family of experiments. Sci. Comput. Program. **88**, 54–81 (2014)
22. Guerrero-García, J., González-Calleros, J., Vanderdonckt, J.: A comparative analysis of task modeling notations. Acta Universitaria **22**, 90–97 (2012)
23. Limbourg, Q., Vanderdonckt, J.: Comparing task models for user interface design. Handb. Task Anal. Hum.-Comput. Interact. **6**, 135–154 (2004)
24. Martinie, C., Palanque, P., Bouzekri, E., Cockburn, A., Canny, A., Barboni, E.: Analysing and demonstrating tool-supported customizable task notations. Proc. ACM Hum.-Comput. Interact. **3**, 1–26 (2019)

25. Hartson, H.R., Gray, P.D.: Temporal aspects of tasks in the user action notation. Hum.-Comput. Interact. **7**, 1–45 (1992)
26. Kieras, D.: GOMS models for task analysis. In: Diaper, D., Stanton, N.A. (eds.) The Handbook of Task Analysis for Human-Computer Interaction (2004)
27. Giese, M., Mistrzyk, T., Pfau, A., Szwillus, G., Detten, M.: AMBOSS: a task modeling approach for safety-critical systems. In: Forbrig, P., Paternò, F. (eds.) Engineering Interactive Systems 2008, pp. 98–109. Springer, Heidelberg (2008). https://doi.org/10.1007/978-3-540-85992-5_8
28. Annett, J.: Hierarchical task analysis. Handb. Cogn. Task Des. **2**, 17–35 (2003)
29. Van Der Veer, G.C., Lenting, B.F., Bergevoet, B.A.: GTA: groupware task analysis—modeling complexity. Acta Physiol. (Oxf) **91**, 297–322 (1996)
30. Tarby, J.-C., Barthet, M.-F.: The DIANE+ method. In: CADUI, pp. 95–119 (1996)
31. Johnson, H., Hyde, J.: Towards modeling individual and collaborative construction of jigsaws using task knowledge structures (TKS). ACM Trans. Comput.-Hum. Interact. (TOCHI) **10**, 339–387 (2003)
32. Limbourg, Q., Vanderdonckt, J., Michotte, B., Bouillon, L., López-Jaquero, V.: USIXML: a language supporting multi-path development of user interfaces. In: Bastide, R., Palanque, P., Roth, J. (eds.) DSV-IS 2004. LNCS, vol. 3425, pp. 200–220. Springer, Heidelberg (2005). https://doi.org/10.1007/11431879_12
33. Batra, D., Hoffler, J.A., Bostrom, R.P.: Comparing representations with relational and EER models. Commun. ACM **33**, 126–139 (1990). https://doi.org/10.1145/75577.75579
34. Shoval, P., Shiran, S.: Entity-relationship and object-oriented data modeling — an experimental comparison of design quality. Data Knowl. Eng. **21**, 297–315 (1997). https://doi.org/10.1016/S0169-023X(97)88935-5
35. Bork, D., Roelens, B.: A technique for evaluating and improving the semantic transparency of modeling language notations. Softw. Syst. Model. **20**(4), 939–963 (2021). https://doi.org/10.1007/s10270-021-00895-w
36. Benedek, J., Miner, T.: Measuring desirability: new methods for evaluating desirability in a usability lab setting. Proc. Usability Professionals Assoc. **2003**, 57 (2002)
37. Moody, D.: The "physics" of notations: toward a scientific basis for constructing visual notations in software engineering. IEEE Trans. Softw. Eng. **35**, 756–779 (2009)

A Critical View on the OQuaRE Ontology Quality Framework

Achim Reiz[(✉)] ⓘ and Kurt Sandkuhl ⓘ

Rostock University, 18051 Rostock, Germany
{achim.reiz,kurt.sandkuhl}@uni-rostock.de

Abstract. Creating and maintaining high-quality ontologies is crucial as they encapsulate the logic for rule-based artificial intelligence. One way to objectively assess ontologies is ontology metrics, and the research community has already proposed measurement frameworks for assessing ontology quality. Arguably, the most holistic framework available is OQuaRE. Not only does it suggest metrics, but it also links these metrics to quality characteristics and interprets them using a school-like grading system.

However, in an effort to implement the framework for the metric calculation software NEOntometrics, serious drawbacks came to light: (1) Important parts of the framework are only available at online resources (thus not part of the scientific record and without ensured permanent availability). (2) The proposing authors have been involved in all applications of the framework and corresponding research, an outside perspective is missing. (3) At times, the meaning of metrics changed throughout the papers without notice, leading to conflicting definitions. (4) A thorough evaluation of the quality grades is missing.

This research aims to fill this gap: We first scrutinize the various metric proposals with their conflicting versions and resolve these conflicts. Afterward, we present OQuaRE to its full extent to create a permanent, lasting record. At last, we use a large body of evaluated ontologies to assess whether OQuaREs quality ratings sufficiently cover the ontology models used in the wild.

Keywords: OQuaRE · NEOntometrics · Ontology Quality · Quality Framework

1 Introduction

Ontologies are the kit to glue together various enterprise data sources in large knowledge graphs. They allow us to explain a domain of our world to a computer using description logic, which is then able to augment the information by inferring tacit knowledge. Thus, they are a central technology to get the most value out of data.

Developing ontologies requires a deep understanding of the formalization technology used. The complexity puts quality control activities at the forefront to ensure the development or selection of high-quality artifacts. Ontology metrics offer a reproducible and objective assessment of key aspects of the ontology and are a great way to guide quality control activities. However, we can put countless aspects of ontologies into numbers, like the depth of the graph, the number of classes, annotations, object properties,

J. Filipe et al. (Eds.): ICEIS 2022, LNBIP 487, pp. 273–291, 2023.
https://doi.org/10.1007/978-3-031-39386-0_13

and more. And the selected metrics then need interpretation to understand which value ranges are desirable.

Quality frameworks guide the metric selection and interpretation. To name a few, Gangemi et al. presented metrics that assess graph-related attributes [1], Yao et al. focused on the cohesion of an ontology [2]. OntoQA by Tartir et al. comprises several metrics for the schema and individuals and gives textual interpretation guidelines for their metrics [3].

OQuaRE is probably the most holistic ontology quality framework. It was first introduced as an adaption of the SQuaRE software evaluation framework in 2011 [4] and has since been part of several publications [5–10], always involving the proposing authors Duque-Ramos and Fernandez-Breis.

OQuaRE proposes measurable metrics and links them to abstract quality characteristics like *precision* or *interoperability*. The authors further give recommended value ranges based on a school-like grading system and provide a public endpoint for ontology evaluation. In that sense, one could argue that OQuaRE offers all that a developer needs to assess whether a developed ontology suffices.

These advantages made the framework an ideal candidate for integration into our metric calculation software NEOntometrics [11], which aims to be a one-stop resource for informing on and calculating ontology metrics. During this effort, though, we identified four challenges that make the applicability of OQuaRE questionable.

At first, the metric implementations differ throughout the various versions. Even though the same two authors are part of all the previous research on OQuaRE, metrics with the same name are defined differently throughout the various papers. Section two presents these heterogeneities and resolves them.

Secondly, to its full extent, the framework is only available online. Here, the first papers build on a webpage, and the others on a wiki. Both do not ensure long-term availability, and during the time of publishing this article, both those resources have gone offline. Thus, this paper collected all the framework aspects, creating a lasting record for OQuaRE in section three.

Third, an outside perspective is missing. Actual evaluations and work on the framework was always carried out by the same group of authors, and an outside perspective is missing.

At last, even though parts have been evaluated, the studies so far have not examined whether the school-like grading system fits the ontologies that are being developed and used by the knowledge engineering community. This analysis of the framework is carried out in section four.

This paper extends the contribution presented at the International Conference on Enterprise Information Systems (ICEIS 2021) [12]. The changes to the paper are as follows: First, we identified the need to clarify some argumentations in section two and to correct some of the formulas in Table 2. Secondly, the OQuaRE framework is now part of the NEOntometrics [11] suite. The tool allowed us to gather data on 4094 ontologies and fuels the newly performed evaluation in section four.

2 Inconsistent Metric Definitions of OQuaRE

The papers referencing OQuaRE propose 19 ontology metrics, even though not all are used in every paper. For this section, we checked whether the quality metrics proposed in the papers [4–9], in the online documentation and wiki [13, 14], and in the tool [15] are consistent with each other.

Twelve of the OQuaRE metrics are well-defined. However, the newer papers proposed six metrics differently, even though they sometimes recalled the previous ones as their foundation. One metric was defined homogeneously, but its naming is ambiguous.

To homogenize the papers, we selected the metrics with the most acceptance in the community measured by citations. This approach emphasizes the definitions by [4], cited 85 times (at the time of writing this paper), then [5] and its associated documentation [14] with 38 citations, and [6] with 19 citations[1].

2.1 Number of Children (NOCOnto)

The first metric that is inconsistent in its definitions is *NOCOnto*. It is defined by [7, 8] as the *"Mean number of direct subclasses per class minus the subclasses of thing"*. [4], as well as the documentation web page [14] proposes the same metric but uses the name "relationship" for the direct sub-class *relations*: *"Mean number of direct subclasses. It is the number of relationships divided by the number of classes minus the relationships of Thing"*. However, as this paper consistently uses the word *"relationship"* where other papers declare *subclasses* (cf. *INROnto*), we assume they mean the same.

[5, 6] use the same metric, not for subclasses but for superclasses. They describe the metric as the *"Average number of the direct **superclasses** per class minus the subclasses of Thing"* The recent papers [9, 10], the wiki [13], as well as the tool calculation [15] use the first published definition, but subtract the leaf classes: *"Number of the direct **subclasses** divided by the number of classes minus the number of **leaf classes**"*.

We propose to use the metric by [4, 7, 8, 14]. At first, because it represents the most commonly cited definition. Secondly, the name alone suggests the use of subclass relationships.

2.2 Response for a Class (RFCOnto)

This metric is defined by [4, 14] as the *"Number of properties that can be directly accessed from the Class"*. Even though this definition would include annotation properties, later versions state annotation properties explicitly. Thus, we do not consider annotations for this calculation. The definition also emphasizes the connection of classes and properties *"directly accessed..."*. Its calculation formula divides the sum of properties and superclasses by the number of classes minus the relationships of the root class.

[5, 6] describe *NOCOnto* as the *"Number of **Datatype** Properties and **Object** Properties that can be directly accessed from the class"*. We assume that the intention from the second definition does not differ from the previous and that both do not target the definition of object properties but their assertion into the classes.

---

[1] Retrieved from Scopus on November 3rd 2022.

[7, 8] describe RFCOnto as the *"Number of **usages** of object and data properties and superclasses divided by the number of classes minus the subclasses of Thing"*. We assume that the *"number of usages"* is equivalent to the *"accessed by"*. Thus, the definition in these papers does not differ from each other. The recent papers [9, 10] dropped the subtraction of the subclasses of thing, otherwise stating the same: *"Number of usages of **object** and **data properties** and **superclasses** divided by the **number** of classes"*. The wiki and the tool both implemented the latest calculation methodology [13, 15].

For this metric, we chose the widely used definition [4–8, 14] that includes the subtraction of the subclasses of the root class.

2.3 Relationship Richness (RROnto)

The first heterogeneity is in the naming: [4, 14] define *RROnto* as *Property Richness*. [5, 6, 13] describe the metric as *PROnto*, *Property Richness*. Afterward, the naming is reverted to the abbreviation *RROnto* and the meaning *Relationship Richness*. While there exists another metric called *PROnto*, it was first introduced metric-wise in 2017 by [9] and is further utilized by [10] (see below). The wiki and the tool [13, 15] also describe *PROnto* but swap the meaning of *RROnto* and *PROnto*. To untangle the confusion for the different naming conventions, we use the newest term *RROnto* and *Relationship Richness* for this metric.

Regarding metric definitions, the first paper and documentation [4, 14] describe *RROnto* as the *"Number of **properties** defined in the ontology divided by the number of relationships and properties"*. [5, 6, 8, 9] describes the metric differently as *"Number of usages of **object** and **data properties** divided by the number of **subclassof** relationships and properties"*. The formula presented in the online documentation [14] specifies that the definition uses the number of properties per class (object property assertions), not the object properties defined in the ontology generally. The terminology of *relationship* in the paper [4] is the same as in *NOCOnto* and other metrics and is seen as equivalent to subclass relationships. Thus, we can assume that the two definitions describe the same thing.

The newest publication [10] exchanges the subclass to a superclass relationship and puts it into the dividend: *"Number of usages of object and data properties and **superclasses divided** by the number of **classes**"*. The tool does not follow the paper definitions and calculates it as the number of used properties divided by the sum of the used properties and the direct ancestor classes.

Again, we select the most cited metric, defined by [4–9, 14].

2.4 Properties Richness (PROnto)

At first, the definitions for *PROnto* seemed very diverse. However, as shown for the *RROnto* metric, the name *PROnto* was sometimes assigned for the metric *RROnto*. *PROnto*, in its distinctive form, was first introduced by the papers [9, 10] as the: *"Number of **subclassof** relationships **divided** by the number of **subclassof** relationships and properties"*. The wiki and the tool implement this definition but swapped the names of

PROnto and *RROnto* [13, 15]. *PROnto* is, thus, not inconsistent in the definitions but in the naming. We suggest using the *PROnto* for the definition named above.

2.5 Tangledness (TMOnto)

The definitions of tangledness are not as widespread as some of the other metrics. The first paper [4] and the documentation [14] define it as the *"Mean number of **parents** per **class**"*. The papers [7–10] define it as the *"Mean number of **classes** with **more** than 1 direct **ancestor**"*. We choose the older definition, as it is more broadly accepted, having more than six times the citations compared to the newer papers.

2.6 Weighted Method Count (WMCOnto)

The metric stays relatively consistent throughout the first five papers and the documentation. The paper published in 2011, 2015, and 2016 [4, 7, 8, 14] define it as *"Mean number of properties and relationships per class"*, the ones published in 2013 and 2014 [5, 6] declare it as the *"mean (2013 paper: average) number of Datatype Properties, Object Properties and subclasses per class"*. As in *NOCOnto* and *INROnto*, we assume that "relationships" are equivalent to subclass declarations.

The latest papers, the wiki, and application [9, 10, 13, 15] shift heavily in their meaning of *WMCOnto* and define it as the *"Mean **length** of the **path** from thing to a (!sic) leaf classes.* However, we suggest using the older, often-cited version.

2.7 Attribute Richness (AROnto)

The challenge with *AROnto* is not the heterogeneity of its definition; it is defined by [4, 6–10] as the *"number of **restrictions** of the ontology per classes"*, [5, 14] define it as the *"mean number of **attributes** per class"*, we assume that they both mean the same. However, the meaning of restriction or attribute in the context of ontologies is not fully clear if it is not concerned with properties (as the analysis of properties is already covered in other metrics). An implementation effort [16], thus, interpreted the metric as the number of property restrictions (*owl:someValuesFrom, owl:hasvalue, ...*) divided by the number of classes.

[9] give more insights into the meaning of *AROnto* and describes it as *"the number of elements that can be related by properties"*. This metric is implemented by the tool as well [15]. The tool considers the *domain* axioms of *data* and *object properties* and counts how many classes (including subclasses) can be linked with the given properties. We adopted this calculation method.

3 Harmonized OQuaRE Framework

As shown in the previous section, the available OQuaRE papers are inconsistent in some of their proposed metrics. An imprecise terminology further contributes to the fuzziness in the metric definitions. The following section targets to translate the proposed metrics to a precise notation, clear heterogeneities, and build a joint base for future implementations of the OQuaRE metrics.

First, we present the homogenized metric·calculation. Subsection two is concerned with the collected quality characteristics. Subsection three recapitulates the metric interpretations that are part of OQuaRE.

3.1 Metric Definitions

The metrics are at the core of the OQuaRE framework. The following two tables collect the harmonized metrics. First, Table 1 introduces the fundamental ontology attributes on which the OQuaRE metrics build. Every measurement is connected to a distinct symbol and comes with an example. Table 2 then presents the harmonized OQuaRE metrics using the symbols previously introduced. While this table is similar to the one presented

Table 1. Ontology Attributes Needed for Calculating the OQuaRE Metrics (Adapted from [12]).

Symbol	Meaning
c_i	The ith class of the ontology *E.g.: Class "Mother"*
$a_i(O)$	Annotation i on ontology O, does not include $a_i(c)$ *E.g.: This ontology is about family relations*
$a_i(c)$	Annotation i on class c *E.g.: Mother, Description: A Mother is a female who has at least one child*
$ind_i(c)$	Individual i of a class c *E.g.: Karen is instanceOf Mother*
$sub_i(c)$	Subclass i of the class c *E.g.: Mother is subClassOf Parent*
$dp_i(c)$	Data property assertion i on class c *E.g.: Person subClassOf (Age exactly 1 sxd:integer)*
$op_i(c)$	Object Property assertion i on class c *E.g.: Daughter isRelativeOf some Mother*
$dom_i(DP)$	Classes in the domain i of data property DP (incl all subclasses) *E.g.: Age Domain Person*
$dom_i(OP)$	Classes in the domain i of object property OP (incl all subclasses) *E.g.:isRelativeOf Domain Person*
DP_i	Data property i declared in ontology *E.g.: Age*
OP_i	Object property i declared in ontology *E.g.: isRelativeOf*
$sup_i(c)$	Superclass i of the class c *Parent superClassOf Mother*
root	Root class of ontology *E.g., owl:thing*
$leaf_i$	Leaf class i, a leaf class does not have a subclass *E.g.: Mother*
$\overrightarrow{path_i(c)}$	Path i from *root* to c *E.g.: root → Parent → Mother*
$\left\|\overrightarrow{path_i(c)}\right\|$	Length of a path i from *root* to class c *E.g.: |root → Parent → Mother | = 3*

Table 2. OQuaRE Metrics (Adapted and corrected from [12]).

Metric	Description	Formula
ANOnto	Annotation richness	$\frac{\sum_{i,j} a_i(c_j) + \sum_i a_i(O)}{\sum_i c_i}$
AROnto	Attribute richness	$\frac{\sum_{i,j} dom_i(DP_j) + \sum_{i,j} dom_i(OP_j)}{\sum_i c_i}$
CBOnto	Coupling between objects	$\frac{\sum_i \sup(c_i)}{\sum_i c_i - sub(root)}$
CROnto	Class richness	$\frac{\sum_{j,i} ind_j(c_i)}{\sum_i c_i}$
DITOnto	Depth of subsumption hierarchy	$\max_i \left| \overrightarrow{path} \right| (c_i)$
INROnto	Relationships per class	$\frac{\sum_{i,j} sub_i(c_j)}{\sum_i c_i}$
NACOnto	Number of ancestor classes	$\frac{\sum_{i,j} sup_i(leaf_j)}{\sum_i leaf_i}$
NOCOnto	Number of children	$\frac{\sum_{i,j} sub_i(c_j)}{\sum_i c_i - \sum_j sub_j(root)}$
NOMOnto	Number of properties	$\frac{\sum_{i,j} dp_i(c_j) + \sum_{i,j} op_i(c_j)}{\sum_i c_i}$
LCOMOnto	Lack of cohesion in methods[a]	$\frac{\sum_{i,j} \left| \overrightarrow{path_i} \right| (leaf_j)}{\sum_{i,j} \overrightarrow{path_i}(leaf_j)}$
RFCOnto	Response for a class[b]	$\left(\sum_{i,j} dp_i(c_j) + \sum_{i,j} op_i(ci_j) \right) \frac{\sum_{i,j} sub_i(c_j)}{\sum_i c_i - \sum_i sub_i(root)}$
RROnto	Relationship richness	$\frac{\sum_{i,j} dp_i(c_j) + \sum_{i,j} op_i(c_j)}{\sum_{i,j} sub_i(c_j) + \sum_{i,j} dp_i(c_j) + \sum_{i,j} op_i(c_j)}$
TMOnto	Tangledness[a]	$\frac{\sum_{i,j} sup_i(c_j)}{\sum_i c_i}$
TMOnto2	Tangledness 2	$\frac{\sum_{k,i,j} sup_k(sup_i(c_j))}{\sum_{i,j} sup_i(c_j)}; \{ \sum_i sup_i(c) > 1 \}$
WMCOnto	Weighted method count	$\frac{\sum_{i,j} dp_i(c_j) + \sum_{i,j} op_i(c_j) + \sum_{i,j} sub_i(ci_j)}{\sum_i c_i}$
WMCOnto2	Weighted method count 2	$\frac{\sum_{i,j} \overrightarrow{path_i}(leaf_j)}{\sum_i leaf_i}$
PROnto	Property richness	$\frac{\sum_{i,j} sub_i(c_j)}{\sum_{i,j} dp_i(c_j) + \sum_{i,j} op_i(c_j) + \sum_{i,j} sub_i(c_j)}$
POnto	Ancestors per class	$\frac{\sum_{i,j} sup_i(c_j)}{\sum_i c_i}$

[a] Retrieved from Scopus on November 3rd 2022.
[b] This metric differs in its calculation to the corresponding paper in [12].

in [11], the old publication had some errors in translating the findings of Sect. 2 into the tables below. These errors are corrected in this publication.

The metrics presented below are either homogeneously described in the OQuaRE publications or previously discussed in section two.

3.2 OQuaRE Quality Characteristics

OQuaRE defines a quality model on top of the quality metrics, which informs on desirable ontology features and is based on the software evaluation framework SQuaRE. The quality model comprises high-level quality characteristics, each having further sub-characteristics. Parts of the quality characteristics are described in [4, 5, 7, 8]. The data is available to a full extent on the OQuaRE webpage and wiki [13, 14].

The first published quality model, extensively described in [14] and used by [4, 5], comprises 51 sub-characteristics. The wiki, referenced by the rest of the papers, dropped two characteristics *C8-Reliability* (3 sub-characteristics) and *C9-Performance Efficiency* (2 sub-characteristics), and split two up (C2.6, 2.7 & C.2.12, C2.13), resulting in a total of 48 quality characteristics.

The descriptions of the characteristics presented in the tables below are shortened and refined. The elements presented below are the union out of [13, 14] and the papers [4, 5, 7, 8], resulting in a total of 53 sub-characteristics. The ordering of the item encapsulates no further meaning.

The "**structural**" characteristic evaluates the connections within an ontology and the attributes of the graph (Table 3).

Table 3. Sub-Characteristics for Characteristic "Structural" [12].

#	Sub-Characteristic	Description
C1.1	Formalization	The ontology is built on top of a formal model (*e.g., OWL, OBO*) to support reasoning
C1.2	Formal Relations Support	The ontology supports formal relations beyond taxonomy
C1.3	Redundancy	All knowledge items are informative
C1.4	Structural Accuracy	The terms are correct
C1.5	Consistency	No set of items is contradictory or conflicting
C1.6	Tangledness	The fewer multi-inheritance relationships are declared, the better
C1.7	Cycles	The existence of cycles is usually bad design and shall be avoided
C1.8	Cohesion	The classes are strongly related
C1.9	Domain Coverage	The ontology covers the specified domain (*requires an expert evaluation*)

Functional Adequacy describes the capability to provide concrete functions. Regarding this characteristic, the two documentations have minor differences. In [14], the elements *Clustering* and *Similarity* are fused, as well as *Guidance* and *Decision Trees* (Table 4).

Table 4. Sub-Characteristics for Characteristic "Functional Adequacy" [12].

#	Sub-Characteristic	Description
C2.1	Reference Ontology	The ontology can be used as a reference source
C2.2	Controlled Vocabulary	Heterogeneity is avoided. The ontology provides terminology management (*e.g., through the use of labels*)
C2.3	Schema and Value Reconciliation	The ontology provides a common data model and integrations to achieve semantic interoperability
C2.4	Consistent search and query	The formal model and structure guide the search process for data by providing concepts, machine-computable properties, and axioms
C2.5	Knowledge Acquisition	The capability of the ontology to represent the knowledge acquired in the form of instances
C2.6	Clustering	The annotations of terms enable clustering
C2.7	Similarity	The components can be compared for (*e.g., taxonomy, relation, attribute, or semantic*) similarity
C2.8	Indexing and Linking	The classes can act as indexes for fast information retrieval
C2.9	Result Representation	The ontologies capability to analyze complex results
C2.10	Classifying Instances	Instances can be recognized as class members with defined properties
C2.11	Text Analysis	Structure supports association detection between words and concepts to classify word types
C2.12	Guidance	Capability to guide the specification of domain theories through capturing knowledge and constraints about a domain
C2.13	Decision Trees	Capability to build decision trees
C2.14	Knowledge Reuse	The knowledge base can be used to build other ontologies
C2.15	Inference	The capability to use reasoners to make implicit knowledge explicit
C2.16	Precision	The ontology provides the right results with the needed accuracy

Compatibility describes the ability of at least two software components to exchange information and perform their required functions while sharing a hardware or software environment (Table 5).

Table 5. Sub-Characteristics for Characteristic "Compatibility" [12].

#	Sub-Characteristic	Description
C3.1	Replaceability	The ontology can replace another ontology with the same purpose in the same environment
C3.2	Interoperability	The ontology can cooperatively combine its knowledge with other ontologies

Transferability describes the degree to which software can be transferred from one environment to another (Table 6).

Table 6. Sub-Characteristics for Characteristic "Transferability" [12].

#	Sub-Characteristic	Description
C4.1	Portability	The ontology or parts of it can be transferred between environments
C4.2	Adaptability	The ontology can be adapted to different specified environments (*e.g., languages, expressivity levels*)

Operability is concerned with the effort that is needed to use the ontology by stated or implied users (Table 7).

Table 7. Sub-Characteristics for Characteristic "Operability" [12].

#	Sub-Characteristic	Description
C5.1	Appropriateness Recognizability	The ontology enables the users to detect faults
C5.2	Learnability	The ontology enables users to learn its applications
C5.3	Ease of Use	It is easy for the users to operate and control the ontology
C5.4	Helpfulness	The application assists the users

Maintainability describes the capability of ontologies to be modified for changing environments, requirements, or functional specifications (Table 8).

Table 8. Sub-Characteristics for Characteristic "Maintainability" [12].

#	Sub-Characteristic	Description
C6.1	Modularity	The ontology is composed of discrete components. Changing one has minimal effect on the others
C6.2	Reusability	A part of the ontology can be used in other ontologies
C6.3	Analyzability	The ontology can be diagnosed regarding deficiencies, inconsistencies
C6.4	Changeability	The ontology can be easily modified
C6.5	Modification Stability	Unexpected effects from modifications are avoided
C6.6	Testability	The ontology can be validated

The characteristic **Quality in Use** measures how a product used by specific users meets their needs to achieve their goals. It is the quality in a particular context of use. Unlike the other quality criteria, this characteristic does describe additional sub-characteristics (Table 9).

Table 9. Sub-Characteristics for Characteristic "Quality in Use" [12].

#	Sub-Characteristic	2nd Sub Characteristic	3rd Sub Characteristic	Description
C7.1.1	Usability in Use	Effectiveness in Use		A specified user can achieve their goals with accuracy and completeness in their context of use
C7.1.2		Efficiency in Use		The used resources match the ontologies effectiveness
C7.1.3		Satisfaction in Use		The user is satisfied in their specified context of use
C7.1.3.1			Likability	Cognitive satisfaction
C7.1.3.2			Pleasure	Emotional satisfaction
C7.1.3.3			Comfort	Physical satisfaction
C7.1.3.4			Trust	*Not further described*
C7.2.1	Flexibility in Use	Context Conformity in Use		Usability in use meets requirements of the intended context of use
C7.2.2		Context Extendibility in Use		Usability in use in a context beyond initially intended

The following two quality characteristics are part of the first version of the quality framework published in [14]. They were later dropped in the wiki [13].

Performance Efficiency describes the relationship between software performance and resource consumption under stated conditions (Table 10).

Table 10. Sub-Characteristics for Characteristic "Performance Efficiency" [12].

#	Sub-Characteristic	Description
C8.1	Response Time	The ontology provides appropriate response times and throughput rates
C8.2	Resource Utilization	The application uses the appropriate amount and types of resources when using the ontology

Reliability is concerned with maintaining the level of performance under the stated conditions (Table 11).

Table 11. Sub-Characteristics for Characteristic "Reliability" [12].

#	Sub-Characteristic	Description
C9.1	Error Detection	The ontology enables the users to detect faults
C9.2	Recoverability	The ontology can re-establish a specified performance level/data recovery in case of a failure
C9.3	Availability	The software component (*language, tools, ontology*) is operational and available when needed

3.3 Connecting Quality Characteristics and Metrics

As stated earlier, the unique feature of OQuaRE is the holistic view of quality. Not only quality metrics are proposed, but also an interpretation of which values are desirable. Further, [5] and the OQuaRE wiki [13] state how quality metrics influence the quality characteristics shown in the section above. The collected information on the resources is presented in Table 12.

By analyzing the results, one can see that some of the metrics are not described as influencing quality characteristics (cf., PROnto, POnto, NACOnto). All second versions of metrics (thus, end with a 2 like TMOnto2) are also not marked as influencing a quality characteristic. However, as they are concerned with similar aspects like their first versions, we assume they also influence the same quality characteristics (e.g., TMOnto2 also influences C1.6, C2.13).

Furthermore, while some metrics are connected to just two metrics (*TMOnto, CROnto*), others have much more diverse connections, like *AROnto, NOMOnto* (associated with eight metrics), *RROnto*, and *WMCOnto* (associated with eight metrics).

Further, OQuaRE describes two influences on quality characteristics that are not associated with a metric but are itself a sub-characteristic of the category *Structural*: *Formality (C1.1)* and *Consistency (C1.5)*. While we argue that there are metrics available that could be used to measure these two aspects, e.g., *Richness* and *Lawfulness* by Burton-Jones et al. [17], or *"Meta-Logical Adequacy"* and *"Generic Complexity"* by Gangemi et al. [1], the definition of new OQuaRE-metrics is beyond the scope of this paper. Further, it is noted that we did not analyze whether stated connections between quality characteristics and metrics or the proposed metric quality ranges are valid; they are merely collected out of the various resources.

Table 12. Metric Rating and Their Influence on the Quality Dimensions [12].

Metric	Influences	Best Worst				
		5	4	3	2	1
ANOnto	C1.3, C2.2, C2.4, C2.5, C2.6, C2.14	>0,8	0,6–0,8	0,6–0,4	0,4–0,2	<0,2
AROnto	C2.3, C2.4, C2.7, C2.8, C2.9, C2.12, C2.13, C2.14	>0,8	0,6–0,8	0,6–0,4	0,4–0,2	<0,2
CBOnto	C4.2, C5.2, C6.1, C6.2, C6.3, C6.4, 6.5	1–3	3–6	6–8	8–12	>12
CROnto	C2.9, C2.15	>0,8	0,6–0,8	0,6–0,4	0,4–0,2	<0,2
DITOnto	C3.1, C4.2, C6.2, C6.3, C6.4	1–2	2–4	4–6	6–8	>8
INROnto	C2.4, C2.8, C2.12, C2.13, C2.14	>0,8	0,6–0,8	0,6–0,4	0,4–0,2	<0,2
NACOnto		1–2	2–4	4–6	6–8	>8
NOCOnto	C3.1, C4.2, C5.2, C6.2, C6.4, C6.5	1–2	2–4	4–6	6–8	>8
NOMOnto	C2.5, C2.14, C3.1, C4.2, C5.2, C6.2, C6.3, C6.4	<=2	2–4	4–6	6–8	>8
LCOMOnto	C1.8, C2.14, C5.2, C6.3, C6.4, C6.5	<=2	2–4	4–6	6–8	>8
RFCOnto	C4.2, C5.2, C6.2, C6.3, C6.4, C6.5	1–3	3–6	6–8	8–12	>12
RROnto	C1.2, C2.3, C2.4, C2.5, C2.7, C2.8, C2.15	>0,8	0,6–0,8	0,6–0,4	0,4–0,2	<0,2
TMOnto	C1.6, C2.13	1–2	2–4	4–6	6–8	>8
TMOnto2		1–2	2–4	4–6	6–8	>8
WMCOnto	C3.1, C4.2, C5.2, C6.1, C6.2, C6.3, C6.4	1–2	2–4	4–6	6–8	>8
WMCOnto2		1–2	2–4	4–6	6–8	>8
PRONTO		>0,8	0,6–0,8	0,6–0,4	0,4–0,2	<0,2
PONTO		>0,8	0,6–0,8	0,6–0,4	0,4–0,2	<0,2
Formality	C2.2, C2.3, C2.4, C2.11, C2.14, C2.15					
Consistency	C2.2, C2.3, C2.14					

4   Evaluation of the OQuaRE Quality Ratings

The quality ratings presented in Table 12 are a unique feature of OQuaRE compared to other frameworks. In theory, they could enable the knowledge engineer to quickly interpret metric values and guide the development processes. Over the past years, the OQuaRE authors assessed some aspects of the framework on different ontologies. However, an analysis of whether the stated quality assumptions are valid for most real-world ontologies is missing.

Thus, this section is concerned with evaluating the stated quality assumptions. First, we recapitulate the evaluations performed for OQuaRE. Afterward, we use a large dataset to test whether these quality scores fit most of the analyzed ontologies.

4.1 Previous Evaluations of and with OQuaRE

The first and most cited OQuaRE paper [4] performed two evaluation case studies. For the first case study, the authors evaluated 11 ontologies for units of measurement manually and compared the results to the automatic evaluation of OQuaRE. The second case study regarded the assessment of two cell-type ontologies. Here, the first assessment was performed by eight master students, and the second one was automatically calculated with OQuaRE.

The second OQuaRE paper evaluated the framework's sub-characteristics and associated metrics [5]. Three professional ontology engineers in the biomedical field without affiliation to the OQuaRE framework first evaluated an ontology manually using the documentation and categories of the framework. Afterward, they repeated the evaluation using the automatic metric calculation. The results showed no significant results between the manual and automatic approaches.

[6] used OQuaRE to research the effect of training with the GoodOD development and evaluation guideline. It showed that students trained with GoodOD created ontologies with better values for most OQuaRE sub-characteristics.

[7] applied OQuaRE to an evolutional analysis of 16 versions of the EDAM, an ontology for managing bio-scientific data and data analysis. The authors argue that the OQuaRE framework can be used to detect changes systematically. [9] extended the research on ontology evolution and analyzed 408 ontology versions out of eight OBO foundry ontologies.

[10] analyzed 78 ontologies out of the AgroPortal and 119 OBO Foundry ontologies for correlations and clusters in these metrics and found that most ontologies of these repositories form stable groups.

Our evaluation is novel in manifold ways. First and foremost, it is the first OQuaRE evaluation applied by authors that are not affiliated with the creation of the framework itself. It, thus, provides a crucial outside perspective. Further, it regards not only biomedical ontologies but also other domains. At last, the previous authors always used the translated rating system from 1 (bad) to 5 (perfect) directly, without a critical analysis of the quality score mapping. Our evaluation assesses how this rating system performs on a large, diverse dataset.

4.2 Metric Data Origin Methodology

We gathered the data using the NEOntometrics application [11]. NEOntometrics calculates ontology metrics from online resources or git-based repositories. It implements measurements from various quality frameworks like OntoQA [3], oQual [1], or OQuaRE[2]. The software is available online and free to use. We performed the analysis using Project Jupyter, and the corresponding notebook file is available online[3].

NEOntometrics calculates values for all historical versions of git-based ontology repositories. For our analysis, we only selected the latest versions of an ontology with at least 100 axioms, resulting in a dataset of 4094 ontologies out of 154 git repositories.

---

[2] Except the value *LCOMOnto*.
[3] http://doi.org/10.5281/zenodo.7274815.

This dataset includes larger initiatives like the human disease ontology[4] and smaller ones like the vehicle signal specification ontology[5].

The resulting ontologies are as diverse as their origins. The smallest ontology has just 101 axioms, and the largest contains over 2 million. The mean number of axioms is 18,097.25, with a standard deviation of 76,156.42. This number of classes $(\sum_i c_i)$ represent this variety as well, which ranges from two to 178,398 classes with a mean of 1967.38 and a standard deviation of 7381.74.

4.3   The Distribution of OQuaRE Ratings on the Metric Dataset

One distinct feature of OQuaRE is the connection of ontology metrics with quality ratings. The framework describes recommended value ranges for the measured attributes, thus aiding their interpretation. Instead of a decimal value, where the meaning and especially its implications are often hard to understand, OQuaRE measures the performance of ontologies using a school-like grading system from 1 to 5 (cf. Table 12).

However, evaluations of whether these grades reflect the reality of the modeled ontologies are scarce. While the grading system does seem helpful and is a unique feature among the ontology quality frameworks, it is not yet validated whether this system does capture the variance of developed ontologies.

This analysis tests how the measured ontologies are sorted into the rating system and whether the OQuaRE grading system is suitable for an easy ontology evaluation. We accept the system as applicable to ontologies in general if the ontologies consistently fall in the given categories with a significant distribution between the rating grades. Further, quality typically follows a gaussian distribution. The presence of the typical bell curve in our graphs, thus, should be a clear indicator that OQuaRE indeed measures quality.

Figure 1 displays the quality ratings for each metric for the given ontologies. Some quality scores are more or less evenly distributed among the various ontologies. Taking the example of POnto, where 1018 of the 4094 ontologies fall in the worst category one, 269 in category two, 303 in category three, 548 in category four, and 1678 ontologies in the best category five. Even though POnto does not show a gaussian distribution, the values are evenly spread, and this metric seems to represent the available ontologies sufficiently. (Without making claims on whether the quality rating of worst to best applies to the ontologies from a users point of view).

However, the ontology metrics look less valid for many of the other measures. For *NOMOnto*, only 1.8% of the ontologies are better than the worst grade. *TMOnto, NACOnto, and CBOnto* results are even more imbalanced. *RFCOnto, INROno,* and *ANOnto,* are heavily imbalanced towards the upper end of the quality spectrum. Here, 81.9%, 85.2%, and 93.8% of the ontologies are given the best possible grade. Other metrics have concentrated results on only two measures: For *PROnto*, 96.2% of the ontologies are graded with 4 or 5. For *RROnto*, 94.9% of the ontologies are in category 1 or 2. *WMCOnto* paints a similar picture. Here, 97.2% of the ontologies are in metric groups 1 or 2.

---

[4] http://github.com/DiseaseOntology/HumanDiseaseOntology.

[5] github.com/klotzbenjamin/vss-ontology.

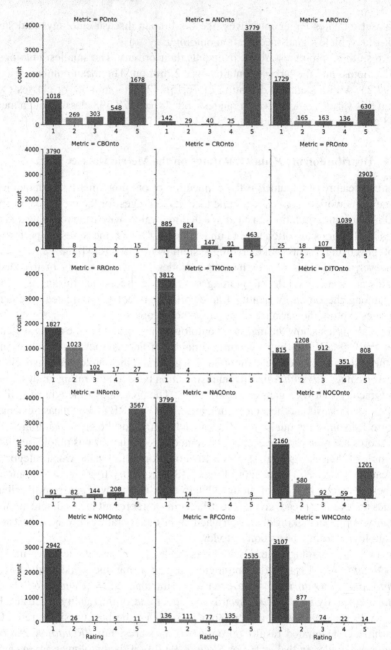

Fig. 1. The Distribution of the Various Quality Ratings for the Analyzed Dataset. Not all values can be calculated for all ontologies (e.g., division by zero). Thus, the values do not always add sup to 4094. At the point of the analysis, LCOMOnto is not yet part of NEOntometrics.

The disbalance is present in the majority of the measured ontologies. Of the 15 assessed ontologies, only four have metric results that are evenly spread out. The distributions are neither heavily left nor right-skewed for *AROnto, DITOnto, CROnto, and POnto*. Still, the value boundaries seem to not sufficiently cover the variety of the measurements as the most extreme measures take up 80.3% *(AROnto)*, 78.3% *(NOCOnto)*, and 58.5% *(POnto)*.

4.4 Implications of the Quality Evaluation

Awareness of an artifact's quality is crucial for making reusing decisions or identifying the need for improvement when developing ontologies. The structural assessments of ontologies are objective and can be calculated automatically. The results, however, are different to interpret, and their interpretation often remains arbitrary.It is difficult to understand the implications of given metric values without guidance. Knowing that an ontology has a *TMOnto* value of 2.3 does arguably help just little regarding deriving possible improvements without understanding if the value is desirable or not.

OQuaRE offers a translation of these measurements to a school-like grading system, thus providing an easy way to make sense of the data. For example, the tangledness measure *TMOnto* is best rated in the value between zero and two. We also know that *TMOnto* is defined as the number of superclasses per class, so we can derive improvement recommendations to reach the desired metric values.

Unfortunately, the proposed quality implications are not aligned with the reality of modeled ontologies. It is not easy to imagine that none of the 3816 ontologies with an associated *TMOnto* measure is sufficient. Moreover, this is an observation we can make not only on *TMOnto* but for many of the quality ratings of the framework.

As a result, we do not believe that the proposed quality ratings are fit for use in an actual evaluation context. The ratings do not sufficiently cover the variety of ontologies, and the conclusions made using the assessment results would likely be misleading. For example, by measuring *TMOnto* and applying the framework, one would likely conclude that heavy improvements and restructurings are necessary, as a poor evaluation score is likely. In reality, though, reaching a good evaluation score is unrealistic.

5 Discussion and Conclusion

Performing quality control on ontologies is crucial for developing and maintaining high-quality artifacts. Ontology metrics are a great way to provide reproducible, objective measures to assess the ontologies main attributes quickly. However, metrics alone do not offer guidance – the numbers need interpretation.

At first glance, OQuaRE offers a one-stop solution for evaluating ontologies. Not only does the framework propose metrics, but it also links them to quality characteristics. The authors further provide interpretation for the metrics through quality grades. In this regard, it is the most holistic ontology quality framework. It seems like OQuaRE is all a developer would need to ensure the creation of high-quality ontologies.

Conversely, during an implementation effort, significant heterogeneities and drawbacks of the framework came to light. Significant parts of the framework are primarily available on online web pages, putting the information at risk of being lost. The metrics are sometimes inconsistent in their definitions, making possible future applications cumbersome. The first part of the paper tackles these shortcomings by collecting information on the framework of the various online resources and homogenizing the metrics.

However, the identified shortcomings made it questionable whether OQuaRE can actually assess ontology quality. While measuring the attributes of ontologies can be performed using ontology metrics, measuring and ensuring quality is a much more difficult task. It is intriguing to think of a single, simple quality framework that connects these two worlds. That allows us to assess quality regarding the various dimensions quickly. Unfortunately, it does not seem that OQuaRE lives up to this promise. The quality scores seem to fail to capture the reality modeled in the ontologies. For some metrics, a good rating is not realistic to achieve; others are prone to achieve a perfect score. It is highly questionable whether a derived refactoring based on the calculated quality scores would lead to better ontologies.

These findings make it difficult to encourage the unfettered use of the framework, as the direct translation of metric results to quality does not yield promising results. That does not mean that the metrics can not be put into productive use, but they need to be carefully selected and interpreted, much like the measures of other frameworks, even though this negates the supposedly main advantage of the framework.

References

1. Gangemi, A., Catenacci, C., Ciaramita, M., Lehmann, J., Gil, R., Bolici, F.: Strignano onofrio: ontology evaluation and validation. An Integrated Formal Model for the Quality Diagnostic Task, Trentino, Italy (2005)
2. Yao, H., Orme, A.M., Etzkorn, L.: Cohesion metrics for ontology design and application. J. Comput. Sci. (2005).https://doi.org/10.3844/jcssp.2005.107.113
3. Tartir, S., Arpinar, I.B., Moore, M., Sheth, A.P., Aleman-Meza, B.: OntoQA: metric-based ontology quality analysis. In: Caragea, D., Honavar, V., Muslea, I., Ramakrishnan, R. (eds.) IEEE Workshop on Knowledge Acquisition from Distributed, Autonomous, Semantically Heterogeneous Data and Knowledge Sources, 27 November 2005, Houston (2005)
4. Duque-Ramos, A., Fernández-Breis, J.T., Stevens, R., Aussenac-Gilles, N.: OQuaRE: a square-based approach for evaluating the quality of ontologies. J. Res. Pract. Inf. Technol. **43**, 159–176 (2011)
5. Duque-Ramos, A., et al.: Evaluation of the OQuaRE framework for ontology quality. Expert Syst. Appl. (2013). https://doi.org/10.1016/j.eswa.2012.11.004
6. Duque-Ramos, A., Boeker, M., Jansen, L., Schulz, S., Iniesta, M., Fernández-Breis, J.T.: Evaluating the good ontology design guideline (GoodOD) with the ontology quality requirements and evaluation method and metrics (OQuaRE). PloS One (2014).https://doi.org/10.1371/journal.pone.0104463
7. Quesada-Martínez, M., Duque-Ramos, A., Fernández-Breis, J.T.: Analysis of the evolution of ontologies using OQuaRE: application to EDAM. In: CEUR Workshop Proceedings, vol. 1515 (2015)
8. Duque-Ramos, A., Quesada-Martínez, M., Iniesta-Moreno, M., Fernández-Breis, J.T., Stevens, R.: Supporting the analysis of ontology evolution processes through the combination of static and dynamic scaling functions in OQuaRE. J. Biomed. Semant. (2016).https://doi.org/10.1186/s13326-016-0091-z

9. Quesada-Martínez, M., Duque-Ramos, A., Iniesta-Moreno, M., Fernández-Breis, J.T.: Preliminary analysis of the OBO foundry ontologies and their evolution using OQuaRE. In: Randell, R., Cornet, R., McCowan, C., Peek, N., Scott, P.J. (eds.) Informatics for Health. Connected Citizen-Led Wellness and Population Health. Studies in Health Technology and Informatics, vol. 235, pp. 426–430. IOS Press, Amsterdam, Washington DC (2017)

10. Franco, M., Vivo, J.M., Quesada-Martínez, M., Duque-Ramos, A., Fernández-Breis, J.T.: Evaluation of ontology structural metrics based on public repository data. Brief. Bioinform. (2020). https://doi.org/10.1093/bib/bbz009

11. Reiz, A., Sandkuhl, K.: NEOntometrics – a public endpoint for calculating ontology metrics. In: Şimşek, U., Chaves-Fraga, D., Pellegrini, T., Vahdat, S. (eds.) Proceedings of Poster and Demo Track and Workshop Track of the 18th International Conference on Semantic Systems co-located with 18th International Conference on Semantic Systems (SEMANTiCS 2022), 13–15 September 2022. CEUR-WS, Vienna (2022)

12. Reiz, A., Sandkuhl, K.: Harmonizing the OQuaRE quality framework. In: Proceedings of the 24th International Conference on Enterprise Information Systems, Online, 25–27 April 2022, pp. 148–158 (2022)

13. OQuaRE Wiki (2016). http://miuras.inf.um.es/oquarewiki. Accessed 9 Nov 2021

14. OQuaRE: A SQuaRE based Quality evaluation framework for Ontologies. http://miuras.inf. um.es/evaluation/oquare/. Accessed 9 Nov 2021

15. Fernandez-Breis, J.T., Franco, M., Vivo, J.M., Quesada-Martinez, M., Duque-Ramos, A., Bernabe-Diaz, J.A.: OQuaRE calculation web service. Universidad de Murcia (2018)

16. Tibaut, A.: ontology-evaluation (2018). https://freesoft.dev/program/124565495. Accessed 25 Nov 2021

17. Burton-Jones, A., Storey, V.C., Sugumaran, V., Ahluwalia, P.: A semiotic metrics suite for assessing the quality of ontologies. Data Knowl. Eng. (2005). https://doi.org/10.1016/j.datak. 2004.11.010

Successful Practices in Industry-Academy Collaboration in the Context of Software Agility: A Systematic Literature Review

Denis de Gois Marques[1]([⊠]) [iD], Tamara D. Dallegrave[2] [iD],
Cleyton Mario de Oliveira[1] [iD], and Wylliams Barbosa Santos[1] [iD]

[1] University of Pernambuco, POLI/UPE, Recife, Brazil
{denis.marques,cleyton.rodrigues,wbs}@upe.br
[2] Polytechnic School of Pernambuco, POLI/UPE, Recife, Brazil
tldad@ecomp.poli.br

Abstract. Research and development (R&D) are the main factors for inserting new technologies and knowledge in the industry, starting from research developments established through collaborations between industry and academia (IAC) within software development organizations. These collaborative practices between these environments enhance the exchange of knowledge and experiences, favoring both communities involved. One of the main processes applied within software development is agile methodologies that present low effort and reduced failure rates in software development. The purpose of this article is to present an exploratory and empirical study of IAC practices in the context of Agile Software Development (ASD), exploring and characterizing solutions and practices, the challenges encountered in the application of IAC, and collaboration. A Systematic Literature Review (SLR) and a Snowballing were conducted in five databases, evaluating/analyzing 8460 articles following the defined criteria. As a result, ten categories of challenges and fourteen categories of good practices for application in collaborative projects were described, in addition to describing seven collaboration models.

Keywords: Industry-academia collaboration · Agile software development · Systematic literature review · Software engineering

1 Introduction

In the context of Software Engineering (SE), the Industry and Academia communities have a considerable volume of participants, which are formed by contexts and members with very heterogeneous profiles, also presenting different growth factors [12]. Due to the importance of these two communities, the cooperation between them is of paramount importance for the improvement of software development. However, these two areas are portrayed as disconnected, with few researchers and practitioners collaborating synergistically with the other community [13, 17].

These disconnections between academic and industrial knowledge have always permeated the discussions that "research in Software Engineering is divorced from the

J. Filipe et al. (Eds.): ICEIS 2022, LNBIP 487, pp. 292–310, 2023.
https://doi.org/10.1007/978-3-031-39386-0_14

problems of the real world". Or even that "many research results are hidden behind academic paywalls, which make them inaccessible to professionals who are not willing to bet $40 on the possibility that an article contains something useful". These are passages portrayed by the technology blog called "It Will Never Work in Theory" which presents itself as a bridge between researchers and professionals, highlighting "useful results of studies" for application in the industry [30].

Collaboration practices between Industry and Academia significantly impact higher education, as the Communities can identify each other's needs and develop strategies for cooperation to address these demands. Sharing knowledge and experiences among organizations (Government, Academy, and Industry) has become increasingly employed, both for the qualification of education professionals (academics) through applied research and for the application and development of scientific solutions within organizations (government and industry). Several articles and projects already address this theme [1,13,33] and reaffirm the importance and values of collaborations between industry and academia, as well as best practices within the organization and the need for more research, demonstrating that both communities benefit from these iterations.

Several workshops and panels are organized to highlight the importance and impact of IACs. In the Brazilian Congress Software (CBSoft), through the track "Industry Talks", also in the panel called "What Industry wants from Research", in the 33rd International Conference on Software Engineering (ICSE, 2011), and since 2015 has an industry-oriented "track" panel, Industry Forum. Also, there are incentives and proposals from government agencies in Brazil for inserting researchers into the industry, Also, there are incentives and proposals from government agencies in Brazil for inserting researchers into the industry.

Undoubtedly, there are considerable challenges in the insertion of collaborative projects between industry and academia, from the industry's perspective, compared to academia. While the industry focuses on building and selling products, academia is usually oriented to constructing new knowledge, developing academic articles, and fundraising [16,31]. However, the collaborations between industry and academia (with the participation of governments) present skills that empower both communities. The insertions of research applied to the industry tend to assist in the training of researchers and make businesses more prone to technologies, and new and better processes [15,33].

Thus, with the growth of collaborations between the industry-academy, mainly targeting aspects of Software Engineering and Agile Software Development, it is essential to summarize what is being developed and applied in the literature, describing state of the art in the research domain [13,21]. The analyzes and descriptions of these collaborative projects in the literature employ and enable benefits for both communities. From the practices analyzed, it is possible to identify potential risks and possible solutions and propose ways of risk mitigation to intervene with the best decisions before, during, and after the collaborative projects.

Furthermore, the collaboration between industry and academia starts from a problem that is not well defined, it works in an environment of constant change, and iterative steps plan the objectives during the execution of the project. The similarities discussed present good evidence for observing agile methodologies practices in a collaborative environment between industry and academia. The objective of this study is to carry out a Systematic Literature Review (SLR), together with a Snowballing, of articles that present perspectives of collaboration between industry and academia in context in Agile Software Development (ASD), focusing on in the identification of models, practices, challenges and correlations between the challenges and good practices of mitigation in these IAC projects.

2   Theoretical Reference

2.1   Industry-Academy Collaboration (IAC)

The disconnection and lack of interactivity between industry and academia knowledge directly damage both communities: one of the significant challenges is the convergence of objectives between the communities, where the industry focuses on building and selling the product and academia on the development of new knowledge [28,29,31]. Glass and DeMarco [17] devoted two chapters of their book to reporting the misalignments of knowledge and practice between industry and academia ("theory versus practice" and "industry versus academia"), categorizing and describing this disconnect as "disturbing" and describing that "Software engineering research is so divorced from real-world problems that it has nothing of value to offer them".

Even before Glass and DeMarco [17] present their book criticizing these disconnections between communities, Beckman [4] constructs and highlights the gap between communities, mainly aimed at the academic community, where industries need software engineers, and academia provides computer scientists. The challenges of integration between communities extend to the present, where several articles describe the difficulties in hiring new graduates in software engineering due to misalignments of competencies required by the industry and presented in academic curricula. Also, the lack of opportunity to build scientific research involving both communities [3,8,13,15,28].

In parallel with the criticisms directed at teaching practices, applying the IAC in the context of research utilized in the industry is one of the main points to reduce the distance between the fields, which is the central point of this work. In this way, industry-academy collaboration is a robust mechanism for innovation and knowledge transfer for industry software development [5,15]. To establish and conduct an efficient IAC, it is necessary that the collaborating partners (researchers and professionals) understand the motivations, goals, and needs to be involved in the project of research, with these objectives being aligned across communities [15,29,33].

The critical points of the collaborations are the analytical skills, case studies, and development of solutions employed by the academy, together with the practices and experiences in project development. The applications of this knowledge generate impact projects in industry, based on practical solutions, and in academia with funding and scientific production [2].

Thus, several articles present the perspective of inclusion of the researcher in the industry, and that demonstrate positive results in this practice [3,23,26]. As an example, Petersen [23] applies two action-research at Ericsson, following five cyclical development steps of action research, with the following steps: i) Diagnosis; ii) Planning and Design of action; iii) Taking Action; iv) Evolution; and v) Learning Specification. Among the results, the action research cycles underwent refinements in each of the interventions applied, with action research being a powerful tool for collaborative research and the success of the method depends a lot on cooperation and teamwork.

In the perspective of models for applications of collaborative projects, there are several types, and examples for the application, one of these models is the one proposed by Philbin [25]. Another model, one of the main ones in the literature, is the one proposed by Wohlin [34], described as the "maturity model" or "proximity model" by Garousi [13,15], as shown in Fig. 1, featuring five levels of "closeness" in collaborations between industry and academia.

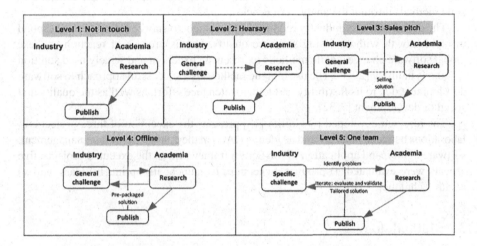

Fig. 1. Five levels of proximity between industry and academia [33].

The five levels proposed by the model can be evaluated increasingly, where level 1 presents significant weaknesses in its execution. In contrast, level 5 is the case considered ideas, where both communities are integrated as a team. Practices in collaborative projects classified as level 4 and 5 are defined as IAC, while at level 1 (no contact), level 2 (I heard), and level 3 (Sales Discussion), the link between communities is weak or non-existent [15].

The work presented by Marijan and Sen [27], arising from 14 years of collaborative projects between researchers in software engineering and the European software industry. Observation methods and interviews with communities were used, highlighting 28 best practices and recurring problems to be avoided, in addition to describing and exemplifying patterns and anti-patterns in collaborative projects. Following this same theme, the work developed by [14] presents the execution of opinion polls as a way to collect information from researchers and professionals in collaborative projects, aiming to classify challenges, patterns, and anti-patterns in international projects.

2.2 Agile Software Development (ASD)

According to Dingsøyr et al. [9], the articulations and construction of the manifesto agile in 2001 brought unprecedented changes to the field of software engineering, valuing process iterations, valuing individuals, software working with summarized documentation and rapid acceptance of proposed design changes. In that way, agility implies the ability to create and respond quickly and flexibly to changes in domains business [19].

In the article by Dyba and Dingsøyr [11], empirical studies for agile software development were reviewed to evaluate four themes: Introduction and adoption, social factors, perceptions about agile methods, and comparative studies. From this review, 1996 documents were identified, of which 36 were identified as empirical and analyzed studies. The review investigated the benefits, limitations, and strengths of the evidence of agile methods through the analysis of articles.

The global software industry requires improvement to stay competitive and respond to rapid growth without losing software quality. In this process of reaction to these global software changes, agile methods represent an excellent and widely used solution in today's industry [20]. O The use of agile methods is a successful approach to software development due to its flexibility and low maintenance effort, as well as the quality and speed of development [7,32].

The research developed by Guillot [18] presents the execution of three project collaborations between industry and academia (IAC) in the context of project management, software agility, and application of the Scrum framework. In the executed projects, five aspects were evaluated: i) product ownership, ii) launch; iii) sprint, iv) team, and v) health technique.

3 Methodology

A systematic literature review is a mean of identifying, evaluating and interpreting all available research on a particular phenomenon of interest, research question or area of interest [21]. The main reasons for applying an SLR are: 1) Residing facts and evidence regarding an existing topic; 2) Identify gaps current research and generate new topics for future investigations; And 3) Provide a framework to properly position new research activities [21]

Kitchenham [21,22] describes that to perform an SLR it is necessary to build a protocol that aims to identify, evaluate and interpret the evidence available in the data obtained, such evidence being interconnected with the research questions of the study. This protocol building process is shown in Fig. 2.

The purpose of this study is to provide an insight into the challenges and practices carried out in the articles analysed, with a focus on agile practices. Thus, the following research questions were applied:

RQ1. What challenges to the application of IACs in ASD were raised?
RQ2. What are the proposed practices for improving IAC in the context of ASD?
RQ3. What types of IAC models have been proposed in the context of ASD?

Some other questions were raised in isolated ways, with the objective of complementing and supplementing the SLR, described externally to the paper, such as: 1) "What were the success and failure criteria elicited by the articles?"; 2) "What are the direct benefits/impacts of the applications of these collaborations, from the perspectives

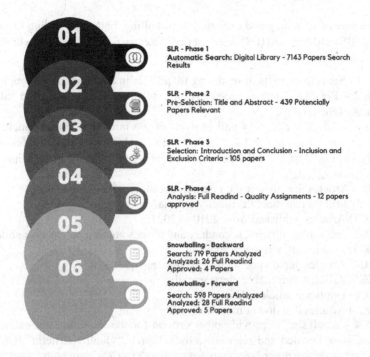

Fig. 2. SLR Driving and Snowballing.

of industry and academia?"; 3) "What are the most used agile methodologies and practices in these projects?"; And 4) "What are the maturity levels of the articles, according to the Wohlin scale, of the analyzed articles?".

Following the SLR guidelines proposed by Kitchenham [21] and Petersen [24], for searches of relevant studies, five main search engines were used, such as: ACM Digital Library, IEEE Xplore, Science Direct, Scopus and Springer. And for the accomplishment of Snowballing Google Scholar was used. The final search string was based on two seminal works in the context of IAC [13], and Agile Software Development [9], described in Table 1.

Table 1. String Search.

IAC: ("Industry" OR "Practice" OR "University" OR "Academia" OR "Theory" OR "Collaboration" OR "Relationship" OR "Relation")
AGILE: ("Agile Development" OR "Agile Methodologies" OR "Agile Software Development" OR "Agile Methods" OR "Agile Projects" OR "Agile Project Management" OR "Scrum" OR "ScrumBan" OR "Extreme Programming" OR "Lean Software Development" OR "Lean Development" OR "Kanban")

From the Snowballing perspective, the first step is to find the seed set (initial set) for carrying out the snowballing steps. In this case, the seed set is the results found in the

SLR. As a way of searching and executing snowballing, both snowballing techniques, Backward (BS) and Forward (FS) were applied. To carry out SLR and Snowballing, the protocols were structured in four phases, these are:

- **Phase 1 - Search:** consists of searching for articles in databases and through seeds;
- **Phase 2 - Pre-Selection:** is pre-selection, which involves reading the titles and abstracts of the articles;
- **Phase 3 - Selection:** reading and analysis of the introduction and conclusion of the articles is carried out and the application of inclusion (IC) and exclusion (EC) criteria, such as "Articles with a publication date before 2010 and greater than 2021". These criteria are described as:
    * IC1) Articles that answer the research questions;
    * IC2) Articles that meet the quality criteria.
    * EC1) Articles published from 2010 to 2021;
    * EC2) Incomplete articles, secondary and tertiary studies, abstracts or slideshow;
    * EC3) Articles not written in English;
    * EC4) Articles that do not answer the research questions;
    * EC5) Articles unavailable electronically;
    * EC6) Duplicate articles; And
    * EC7) Articles that do not meet the quality criteria.
- **Phase 4 - Analysis:** To provide other criteria for the inclusion and exclusion of articles, more detailed and relevant to the research, "Quality Criteria" (QC) were developed for evaluation of the analyzed studies. The CQs were built based on three directives: i) criteria aimed at the objectives of the articles, if in fact, there is synchrony with the objective of the research (OBJ); ii) criteria aimed at the quality of articles and their applied methodologies (WED); and iii) criteria aimed at IAC practices, where "minimum criteria" were defined that the article can be considered as IAC. In Table 5, it is possible to observe all the criteria for quality applied in the study [10, 16, 33], which are detailed in Table 2.

Table 2. Quality Criteria.

COD	Description	Directives
QC01	Is the research related to Agile Software Development?	OBJ
QC02	Is the article based on research/action (or just considerations/opinions)?	OBJ
QC03	Is the research design adequate for the research objectives?	OBJ
QC04	Are the objectives clearly defined?	WED
QC05	Is the methodology clearly defined?	WED
QC06	Are the results clearly defined?	WED
QC07	Is the IAC theme addressed directly?	IAC
QC08	Is there a case analysis in the company (Environment study)?	IAC
QC09	Is it a real project? With a real "customer"?	IAC
QC10	Is the researcher inserted in the context, acting on the spot?	IAC
QC11	Are there changes in the environment with the insertion of the researcher? IAC	IAC
QC12	Do you clearly answer the research questions (related to IAC)?	IAC

The execution of snowballing refers to the analysis and use of article lists, where citations and article references are the sources of information in order to benefit the study literature [35]. Thus, analyzes of article references (backward) and article citations (forward) were performed. The seeds (start set or seeds) used in snowballing resulted from the finalization of the Systematic Review and the finalization of the snowballing iterations, where with the approval of the article also converted the seed. To evaluate the articles detected in the snowballing, the same criteria of SLR were used, where they need to be validated by the same inclusion, exclusion and quality criteria.

4   Search Results and Discussions

In this section, the results found in the data analysis are presented. In this way, the data are presented according to the research questions. The number of articles searched and the final results are shown in Table 3. Following the research protocols described by Wohlin [35] regarding snowballing, forward (FS) and backward (BS) searches were performed.

Table 3. Number of Articles Analyzed.

Database/Snowballing	Phase 1	Phase 2	Phase 3	Approved
IEEE	730	47	18	2
ACM	97	19	5	1
Science Direct	613	72	10	1
Scopus	3516	142	29	3
Springer	2187	159	43	5
BS	719	80	26	6
FS	598	68	28	3
Total	**8460**	**587**	**159**	**21**

From the primary studies found in SLR and Snowballing (available in the Appendix), the process of coding and categorizing the data began. For analysis of the results, electronic spreadsheets were used, for the construction of simplified results, and for coding and categorization, the support software, Atlas TI, was used. The analyzes and coding are built following the principles of Charmaz [6], where Open and Axial coding were used to develop subcategories and categories in the data analysis.

Within these analyses, 10 categories were created to which the challenges and impediments of collaborations are highlighted (RQ1), another 14 categories were created highlighting the best practices and standards regarding the industry-academy collaboration process (RQ2), in addition to presenting 7 models of collaboration for the practice of IAC (RQ3). In addition to the analysis of correlations between the challenges and best practices for mitigating the execution of these projects.

In the link (shorturl.at/DLOU1) it is possible to observe the research protocols, the approved articles, the demographic data and the list of categories and subcategories of each of the research questions.

4.1   A. (RQ1) What Challenges to the Application of IACs in ASD Were Raised?

This section analyzed the challenges and impediments found in the researched articles. As a result, 10 categories and 37 subcategories (labeled by CI) stood out. Category 01 represents the incompatibility between industry and academia, where they represent the contrasts of reality on both sides. This category describes the different perceptions of the collaboration fields, where they demonstrate other interests and objectives (CI06) and perceptions of challenges (IC15) of different solutions and results (CI25). In addition, they represent differences between the availability of industry and academia resources (IC04) and difficulty in synchronizing schedules (IC03). This category is represented by 10 subcategories.

Category 03 is represented by the lack of training, experience and skills of project employees. Thus, the lack of skill and experience in Software Engineering (CI21) and Agile (CI20) are factors that hinder these collaborations, as well as the difficulties in collaborating with the researchers' solutions (CI16), where training in both contexts are necessary (CI20 and CI21). This category is made up of only 3 subcategories.

The lack of proper communication between the members of a collaboration is one of the main challenges to leverage collaborations. In Category 05, 3 subcategories are presented that represent this knowledge. The insertion and availability of the researcher at the collaboration site (CI12) and the communication gaps between communities (CI26) are challenges for the practice of collaboration. Another form of communication challenge is concealment or difficulty in obtaining information from the team (CI37).

Agile practices, in Category 07, is the main point of the research project, so agile practices also present difficulties in applications in collaborative projects. The difficulties involve the difficulty of following the iterations (CI09), the interruption of the iterations (CI10) and short sprints (CI11) of the project. In addition, some projects are resistant to changes from traditional approaches to agile (CI33). This category is made up of only 3 subcategories.

In this section, 4/10 categories were presented that represent the challenges and impediments for a collaborative project, in addition to the representations of the subcategories. The other categories and subcategories are exposed in the link (available in the results section).

4.2   B. (RQ2) What Are the Proposed Practices for Improving IAC in the Context of ASD?

The purpose of the question is to analyze the good practices that were developed and represented in the analyzed articles. 14 categories and 76 subcategories (labeled by GP) of practices were constructed in the development of an IAC. For example, in Category 01 (Knowledge Management), good knowledge management practices of collaborating teams are described, such as communication skills, training and social and research skills. Some that were described in this category are: holding seminars and workshops (GP31), conducting training (GP54), promoting satisfaction with learning (GP24) and developing social and managerial skills (GP42). In addition to these four practices, four more practices were also developed, totaling 8 subcategories.

Category 02 (Project Management and Engagement Assurance) focuses on the collaboration and involvement of project participants. Thus, it is necessary to meet the needs of the industry (GP35), ensuring the involvement of the industry (GP11), the transfer of knowledge (GP16) between employees and the development and encouragement of the industry (GP03). This category is represented by 8 subcategories.

The agile aspects is the central point of the research, where the observations and analyzes of these practices are very suitable for the context of the IAC. In Category 06, the Software contexts Agility focused on collaborations were presented. This category is represented by 15 subcategories. The inclusion of agile practices in collaborative projects (GP32) acts in a very positive way, as in the management of expectations [11] and in the conversion of large projects into smaller projects (GP30). In addition, the use of short iterations, short cycles and sprints (GP26) are one of the main factors in the insertion of agile methodologies. The practices Sprint Retrospective (GP62), Planning Ceremony, Sprint Review (GP40), Daily Meeting and/or Planning Meeting (GP07) and Daily Standup (GP69) are well applied in the studied contexts.

Agile skills such as Spike Solution (GP56), Pair Programming (GP57), Planning Poker (GP61) and User Story (GP21) were also reported in the development of collaborations, these skills being related to team training, project execution, skill leveling (described in Category 01, GP59) or in the construction of the research/development proposal.

In this section, 3/14 categories were presented, with the filing of some subcategories that represent the knowledge of the respective category. The other categories and subcategories are presented in the link (available at the top of the results).

4.3   C. (RQ3) What Types of IAC Models Have Been Proposed in the Context of ASD?

Some collaboration models involving the construction of IAC projects were presented in the articles by Garousi [13] and by Marijan and Gotlieb [26]. However, it addressed a limited number of models and methodologies. In the article, the case study and action-research methodologies were widely applied, directly and indirectly, in the analyzed reports, as well as some IAC collaboration models, such as the "Certus Model" and "Spiral Model".

Collaboration models offer necessary elements for collaboration between industry and academia, determining a practical framework that details the roles and responsibilities of project participants [26]. Thus, this review discovered seven models of collaboration between industry and academia.

The article by [ID06] presents the **"technology transfer model"**, which is directly related to the structure of a Minimum Viable Product (MVP) in academia and, after construction, the transfer to the domain of industry. Among the steps used to apply the model are: 1) Identify problems in the industry through process evaluation or case study; 2) Formulate a solution to the problem in the industry, with the cooperation of the industry; 3) Perform validations of these solutions; and 4) Release step by step for implementation in the industry.

In this collaboration, the integration and exchange of information and hypothesis testing are crucial for the benefit of the model. For example, *"In the Software Factory laboratory, a company brings the vision about a new product or service and an initial value hypothesis. The company is responsible for testing the value hypothesis whereas the Software Factory is responsible for creating the MVP"*.

In [ID7] the **Cooperative Method Development (CMD)** model is presented, whose main characteristics are the principles of action research, with a more extraordinary qualitative approach, which also combines problem-oriented methods and techniques and improvements of the processes. Thus, *"the researchers, who are motivated to understand how software developers face the daily challenges in software development, can combine technical innovation, and method and process improvement within one CMD cycle"*. The CMD application cycles are: 1) Understanding the practice; 2) Deliberate improvements; And 3) Implementing and observing improvements.

The article by [ID12] brings the junction of Cooperative Method Development (CMD) with Design Science Research (DSR) for the application of collaboration. The intersection of these principles created the methodology "SoftCoDer". The work of [ID11] it is presented the **"SoftCoDer UserX Story: Incorporating UX Aspects"** approach that brings the CMD foundation, with Design Science Research (DSR) guidelines, such as *"design artifacts of value based on both real need of industry (relevance) and scientific knowledge (rigor of research)"*.

The work of [ID15] **Dialogical Action Research (DAR)** presented an emerging and engaged approach in both research and practice. This approach was designed to promote an understanding of applications of practical phenomena. As a form of collaboration, *"DAR presents a researcher/practitioner partnership that allows for a reflective dialog to explore and shape change in an organization and also specify learning for a scientific community"*. O DAR proposes an interactive engagement of practices to address a practitioner's problems through dialogues, action planning, decision-making, and evaluations.

In the article by [ID13] the **Collaborative Practice Research (CPR)** presented a model where the concepts of practitioners ("insiders") and researchers ("outsiders") working in close collaboration are applied, where they take advantage of bringing their knowledge in joint identification, analysis, and interpretation in the project. The "Insider" is an active partner of the industry and gives an in-depth view of the needs of the industry. However, the "outsider" brings the knowledge of experiences as a researcher, with the vision of several projects analyzed.

In articles [ID22] and [ID23] the **ARN model** is presented, which proposes to *"seek to work closely with agile practitioners in their workplaces, investigating challenges identified by professionals"*. The model uses short cycles of iterative surveys, where the deadlines are defined by the model using 12 weeks of project effort. The approach presents four distinct phases of execution: i) Start of collaboration, ii) investigation; iii) Implementation; and iv) Evaluation. The model shows a timebox, and the ARN researchers work closely with the industry team during each phase.

Another model presented is the **Certus Model**, in the article [ID26], which describes the elements of a process of collaborative projects between researchers and professionals in software engineering. This study focus on the principles of co-creation of knowledge in research and commitment to solving problems. The model structure has seven phases, namely: i) Problem Scope; ii) Conception of Knowledge; iii) Development; iv) Transfer; v) Exploration; vi) Organizational Adoption; and vii) Market Research.

4.4   Correlations Between Challenges and Good Practices

As a way of building the correlations, a new analysis linked these challenges to good practices to minimize the impacts of the problems and propose solutions, thus, in Table 4 the challenges and good practices described in the literature are shown, related to the categories of Incompatibility, Research Methods and Lack of Training.

In Table 5, issues related to Human and Organizational Factors, Issues related to Project Management and related to Agile Practices were elicited.

All the related factors, through the correlations, are extracted from the practical executions demonstrated in the analyzed articles. By observing and analyzing risks in collaborative projects, researchers and industry professionals can use the explicit survey to plan the risk management of their projects and add good practices to their projects.

4.5   Discussions and Threats to Validity

The research focused on descriptions of factors that influence collaborative research. Through the analysis and explanation of these factors, it is possible to contingency possible failures during the execution of collaborations. Some research also presents, as a foundation, the analysis of problems and the practice of solutions developed in collaborations, as a way of mitigating and guiding partnerships between industry and academia ([3, 13, 15, 18, 33]). Differently from these researches, the objective of this RSL is the analysis and evaluation of IAC in contexts of Agile Software Development, in an applied way, in a direct way, for the inclusion of researchers in the industry.

One of the main practices, based on the number of citations in the analysis, was the insertion of the researcher in the industrial context, carrying out an assessment of the work environment, leveling knowledge and carrying out the exchange of knowledge. The researcher in the inserted context is free to collect evidence and experiences, feelings and difficulties of the participants, to build insights, hypotheses and solutions for the environment. All these elements must be discussed openly between the collaborating parties, through Workshops, Lectures and Training, which has been demonstrated as a good practice for collaborations. The process of internal and external dissemination of the results is fundamental for the continuation of the research project, as well as allowing new collaborations, even in other contexts.

For a better interposition of the researchers in the context of the industry and a collaboration successful among communities, an excellent and frequent practice in data analysis it is the realization of training and leveling of knowledge between practitioners and researchers. However, challenges such as lack of time and incompatibility of schedule for building new insights can be found in conducting the collaboration. Given the incompatibilities of schedules, interruptions of iterations and time windows for research, practices agile solutions present a good way out of these problems. Agile practices, such as short iterations, sprint review and planning meetings, react very well to changes in the environment.

The models described in RQ3 are important ways to run an IAC on some context. From these models, it is possible to develop a research project following a methodology already exemplified and described in the bibliography, repressing risks in the accomplishments of future research. As presented in RQ3, there are several models that

Table 4. Relational List between Challenges and Good Practices. Part 1.

Challenges	Good Practices
Category 01: Incompatibility between Industry and Academia	
(DI36) Different expectations about the collaboration - manage expectations	- (BP26) Use iterations and Cycles - through deliverables incremental [ID09] - (BP32) Use agile methods - through deliveries incremental [ID09] - (BP06) Planning meetings, daily meeting or periodic [ID09]
(DI06) Different interests and goals	- (BP39) Establish common goals between the industry and academia [ID13] - (BP38) Construction of collaboration agreement [ID13] [ID22] - (BP02) Socialization and Demonstration of the proposal in the organization [13] - (BP40) Sprint Planning Ceremony and Sprint Review [ID13] - (BP04) Proper presentation and communication of researchers [ID13]
(DI03) Difficulty in time synchronization	- (BP76) Pilot Construction [ID24]
(DI04) Different schedules between industry and academia	- (BP76) Pilot Construction [ID24]
Category 02: Research Method	
(DI27) Difficulties in integration of new solutions and of improvements	- (BP06) Planning Meetings or Meetings Daily or Periodic [ID04] - (BP03) Create and Encourage side membership and support industry [ID04] - (BP66) Bring in experts - own industry [ID21]
(DI34) Execute a project with research method flexible	- (BP17) Synchronization of Activities [22] - (BP33) Detailed schedule creation [ID22] - (BP09) Co-located researcher [ID22]
Category 03: Lack of Training, Experience and Skills	
(DI21) Lack of Training, Experience and Skills (General)	- (BP54) Need and/or Performance of Training [ID07], [ID12] - (BP31) Conducting a Seminar and Workshop [ID07] - (BP24) Satisfaction with learning [ID07] - (BP66) Bring in specialists - in this case, senior [ID10] [ID21] - (BP55) Use of Coach and Mentorship practices - as facilitators of activities and communication [ID10]
(DI20) Lack of training, experience and skills (Agile)	- (BP07) Daily Meetings and/or Meetings of Planning - rework in difficulties in sprints [ID09] - (BP60) Use of Sprint Backlog [ID09] - (BP54) Need and/or Performance of Training [ID14]
(DI16) Ability of professionals to work with the solution of researchers	- (BP03) Create and encourage buy-in and side support industry [ID14] - (BP66) Bring in experts - external professional of the project [ID14]

Table 5. Relational List between Challenges and Good Practices. Part 2.

Challenges	Good Practices
Category 05: Problems related to Communication	
(DI26) Gaps of communication between researchers and professionals	- (BP55) Employ or make a representative the researcher [ID10] - (BP66) Bring in experts - form a industry representative [ID10] - (BP09) Researcher co-located and being present in the industry [ID10] - (BP41) Sympathy, Empathy and Respect [10] - (BP17) Activity Synchronization [ID13]
Category 06: Human and Organizational Factors	
(DI32) Resistance to Change and Inflexibility of practices	- (BP29) Formation of development groups in process improvement [ID04] - (BP27) Show your partner the benefits of the solution [ID04], [11] - (BP11) Ensure the involvement of professionals [ID11] - (BP51) Use of Coach practice and Mentorships [ID14] - (BP53) Provide training in the use of academic tools [ID14] - (BP02) Socialization and Demonstration of research proposal [ID21] - (BP01) Get feedback on expectations and results [ID21]
(DI23) Difficulty after the training	- (BP03) Create and encourage buy-in and support on the industry side [ID14] - (BP66) Bring in experts - professional project external [ID14]
Category 07: Questions related to Agile Practices	
(DI33) Barriers against development approaches non-traditional	- (BP23) Visible process improvement [10] - (BP27) Show your partner the solution benefits [ID10]
Category 08: Issues related to Management	
(DI17) Output of project participants	- (BP54) Need and/or Achievement training [ID14]

propose the execution of collaborative practices, such as the Certus Model [ID17], spiral model [ID30], collaborative research aimed at agile research ([ID10], [ID26]) and, the one that has been gaining a lot of space in research, the technology transfer model ([ID31], [ID32]).

Threats to Validity. The stages of systematic literature review and snowballing were well defined and the The research process was carried out in a systematic and precise manner, seeking to expose transparency and reliability to the research. However, some threats to the validity of the study are presented, such as:

* **External Partner.** There was no intervention or support from an external partner in the evaluations the protocol or the validity of the data obtained;

* **Search String.** With search limitations exposed by the String, it is possible that some articles have not been captured by the search process in the databases;
* **Manual Bases.** The search was not carried out in manual bases, which could have more data for analysis;
* **Quality Criteria.** The quality ratings of the articles were based on the "five levels of closeness" proposed by Wholin [33] and described by Garousi [13]. In this way, stricter criteria were placed as defined by authors on IAC.
* **Secondary and Tertiary Articles.** The article selections for the RSL were objective to obtain primary articles, which actually represent a collaboration, excluding secondary and tertiary articles of the analyses, missing possible information about the topic of research;
* **Snowballing.** Search performed only on Google Scholar, disregarding searches in other databases;
* **Snowballing Forward.** Search carried out in March (2022), articles published later to that date were not computed.

5   Conclusion

Collaborative projects between industry, academia, and government have the power to enhance all environments in a very positive way. By including researchers in these environments, industries tend to have better processes and technological improvement, and researchers become more qualified, receive funding, and put into practice the knowledge acquired in academia.

In collaborative projects, there are also challenges in the planning and execution of the project, such as interest to participate, development time, and distinction in research objectives, among many others. One of the best risk mitigation practices is using agile practices, demonstrating results throughout the research process.

The research aims to carry out an exploratory review of articles that present a perspective of industry and academia collaboration (IAC) from a software agility perspective. A Systematic Literature Review (SLR) was carried out as a form of research construction with the five main search engines. In addition, Snowballing practices were applied.

From the number of articles approved and with the performance of qualitative analyzes of these articles, 10 categories and 37 subcategories were developed on the challenges and impediments of the execution of the IAC (RQ1), and 14 categories and 76 subcategories that express good practices in the execution of collaborations between industry and academia (RQ2). In addition to these results, 7 models for performing IAC were described, in the context of ASD (RQ3), exposed in [28,29].

Agile practices in collaborative projects (IAC) present good practices due to the need for these projects to demonstrate fast and qualified results, together with some practices related to agility, such as the use of short iterations, quick meetings, and short sprints.

For future research, our next step is to validate the results with professionals in the field and design an ontology model relating the challenges with good practices within the scope of the IAC. Therefore, this conceptual model can organize and structure the domain, through which professionals can perform reasoning tasks to infer a set of practices to be introduced within a specific IAC project based on the identified challenges.

Acknowledgements. This study was financed in part by the Coordenacão de Áperfeiçoamento de Pessoal de Nível Superior - Brasil (CAPES) - Finance Code 001.

Appendix A. Primary Studies Included in SLR

ID01 - Ardito, C., Baldassarre, M. T., Caivano, D., & Lanzilotti, R. (2017). Integrating a SCRUM-Based Process with Human Centred Design: An Experience from an Action Research Study. doi: 10.1109/cesi.2017.7

ID04 - Pino, F. J., Pedreira, O., García, F., Luaces, M. R., & Piattini, M. (2010). Using Scrum to guide the execution of software process improvement in small organizations. doi: 10.1016/j.jss.2010.03.077

ID06 - Münch, J., Fagerholm, F., Johnson, P., Pirttilahti, J., Torkkel, J., & Jäarvinen, J. (2013). Creating Minimum Viable Products in Industry-Academia Collaborations. doi:10.1007/978-3-642-44930-7_9

ID07 - Choma, J., Zaina, L. A. M., & Beraldo, D. (2015). Communication of Design Decisions and Usability Issues: A Protocol Based on Personas and Nielsen's Heuristics. doi: 10.1007/978-3-319-20901-2_15

ID09 - Santos, N., Fernandes, J. M., Carvalho, M. S., Silva, P. V., Fernandes, F. A., Rebelo, M. P., ... Machado, R. J. (2016). Using Scrum Together with UML Models: A Collaborative University-Industry R&D Software Project. doi:10.1007/978-3-319-42089-9_34

ID10 - Wen, Melissa, Meirelles, Paulo, Siqueira, Rodrigo & Kon, Fabio. (2018). FLOSS Project Management in Government-Academia Collaboration. 10.1007/978-3-319-92375-8_2.

ID11 - Choma, J., Zaina, L. A. M., & Beraldo, D. (2016). UserX Story: Incorporating UX Aspects into User Stories Elaboration. doi:10.1007/978-3-319-39510-4_13

ID12 - Choma, J., Zaina, L. A. M., & Silva, T. S. D. (2015). Towards an Approach Matching CMD and DSR to Improve the Academia-Industry Software Development Partnership: A Case of Agile and UX Integration. doi: 10.1109/sbes.2015.18

ID13 - Sandberg, A. B., & Crnkovic, I. (2017). Meeting Industry-Academia Research Collaboration Challenges with Agile Methodologies. doi:10.1109/icse-seip.2017.20

ID14 - Sousa, T. L. de, Venson, E., Figueiredo, R. M. da C., Kosloski, R. A., & Junior, L. C. M. R. (2016). Using Scrum in Outsourced Government Projects: An Action Research. doi:10.1109/hicss.2016.672

ID15 - Babb, J. S., Hoda, R., & Nørbjerg, J. (2014). XP in a Small Software Development Business: Adapting to Local Constraints. doi:10.1007/978-3-319-09546-2_2

ID17 - Nalepa, G., Fontana, R. M., Reinehr, S. & Malucelli, A. (2019). Using Agile Approaches to Drive Software Process Improvement Initiatives. doi: 10.1007/978-3-030-28005-5_38

Appendix B. Primary Studies Included in Snowballing

ID18 - Ardito, C., Buono, P., Caivano, D., Costabile, M. F., & Lanzilotti, R. (2014). Investigating and promoting UX practice in industry: An experimental study. International Journal of Human-Computer Studies, 72(6), 542–551. doi:10.1016/j.ijhcs.2013.10.004

ID19 - Petersen, K., Gencel, C., Asghari, N., Baca, D., & Betz, S. (2014). Action research as a model for industry-academia collaboration in the software engineering context. Proceedings of the 2014 International Workshop on Long-Term Industrial Collaboration on Software Engineering - WISE, 14. doi: 10.1145/2647648.2647656

ID20 - Costello, G., Donnellan, B., & Conboy, K. (2011). Dialogical Action Research as Engaged Scholarship: An Empirical Study. Proceedings of the International Conference on Information Systems, ICIS 2011, Shanghai, China, December 4–7, 2011.

ID21 - Meirelles P., Wen, M., Terceiro, A., Siqueira, R., Kanashiro, L. & Neri, H. (2017). Brazilian Public Software Portal: an integrated platform for collaborative development. In Proceedings of the 13th International Symposium on Open Collaboration (OpenSym '17).

ID22 - Barroca, L., Sharp, H., Salah, D., Taylor, K., & Gregory, P. (2015). Bridging the gap between research and agile practice: an evolutionary model. International Journal of System Assurance Engineering and Management, 9(2), 323–334. doi:10.1007/s13198-015-0355-5

ID23 - Gregory, P., Plonka, L., Sharp, H., & Taylor, K. (2014). Bridging the gap between research and practice: The agile research network. In ECRM 2014, 13th European conference on research methodology for business and management studies. London, 16–17 June, 2014. Reading: Academic Conferences and Publishing International Limited.

ID24 - Choma, J., Zaina, L.A.M. & da Silva, T.S. (2016). SoftCoDeR approach: promoting Software Engineering Academia-Industry partnership using CMD, DSR and ESE. Journal of Software Engineering Research and Development. 4, (Nov. 2016), 8:1 - 8:21.

ID25 - Persson, J. S., Nørbjerg, J. & Nielsen, P. A. (2016). Improving ISD Agility in Fast-moving Software Organizations. ECIS 2016 Proceedings., 96, Association for Information Systems. AIS Electronic Library (AISeL), Atlanta, GA, Proceedings of the European Conference on Information Systems, 24th European Conference on Information Systems, ECIS 2016, Istanbul, Turkey.

ID26 - Marijan, D., & Gotlieb, A. (2020). Industry-Academia Research Collaboration in Software Engineering: The Certus Model. Information and Software Technology, 106473. doi:10.1016/j.infsof.2020.106473

References

1. Ankrah S., Al-Tabbaa, O.: Universities-industry collaboration: a systematic review. Scand. J. Manag. 31(3), 387–408 (2015). ISSN 0956–5221, https://www.sciencedirect.com/science/article/pii/S0956522115000238

2. Banal-Estañol, A., Macho-Stadler, I., Pérezcastrillo, D.: Research output from university-industry collaborative projects. Econ. Dev. Q. **27**(1), 71–81. SAGE Publications (2013)
3. Barbosa, A., et al.: Fostering industry-academia collaboration in software engineering using action research: a case study. In: Anais do XIX Simpósio Brasileiro de Qualidade de Software. Porto Alegre, RS, Brasil: SBC, pp. 411–419 (2020). https://sol.sbc.org.br/index.php/sbqs/article/view/14239
4. Beckman, K., et al.: Collaborations: closing the industry-academia gap. IEEE Softw. **14**(6), 49–57. (IEEE) (1997). https://doi.org/10.1109/52.636668
5. Carver, J.C., Prikladnicki, R.: Industry-academia collaboration in software engineering. IEEE Softw. **35**(5), 120–124 (2018)
6. Charmaz, K.: Constructing Grounded Theory: A Practical Guide Through Qualitative Analysis, vol. 1 (2006)
7. Chookittikul, W., Kourik, J.L., Maher, P.E.: Reducing the gap between academia and industry: the case for agile methods in Thailand. In: 2011 Eighth International Conference on Information Technology: New Generations, pp. 239–244 (2011)
8. Dallegrave, T., et al.: Are we ready? Identifying the gap between academia and software industry in the context of agile methodologies. In: Anais do XXXII Simpósio Brasileiro de Informática na Educação (SBIE 2021). [S.l.]: Sociedade Brasileira de Computação - SBC (2021)
9. Dingsøyr, T., Nerur, S., Balijepally, V., Moe, N.: A decade of agile methodologies: towards explaining agile software development. J. Syst. Softw. **85**, 1213–1221 (2012)
10. Dyba, T., Dingsoyr, T., Hanssen, G.K.: Applying systematic reviews to diverse study types: an experience report. In: Proceedings of the First International Symposium on Empirical Software Engineering and Measurement, ESEM 2007, pp. 225–234. IEEE Computer Society, USA (2007)
11. Dyba, T., Dingsøyr, T.: Empirical studies of agile software development: a systematic review. Inf. Softw. Technol. **50**(9–10), 833–859 (2008)
12. MCTI, S.: Indústria de software e serviços de tic no brasil: Caracterização e trajetória recente. In: [s.n.], p. 46. (2022). https://softex.br/estudoindustriatics/
13. Garousi, V., Petersen, K., Ozkan, B.: Challenges and best practices in industry-academia collaborations in software engineering: a systematic literature review. Inf. Softw. Technol. **79**, 106–127 (2016). ISSN 0950–5849. https://www.sciencedirect.com/science/article/pii/S0950584916301203
14. Garousi, V., et al.: Industry-academia collaborations in software engineering. In: Proceedings of the 21st International Conference on Evaluation and Assessment in Software Engineering. ACM, New York, NY, USA (2017)
15. Garousi, V., et al.: Characterizing industry-academia collaborations in software engineering: evidence from 101 projects. Empir. Softw. Eng. **24**(4), 2540–2602 (2019). https://doi.org/10.1007/s10664-019-09711-y
16. Garousi, V., Shepherd, D.C., Herkiloglu, K.: Successful engagement of practitioners and software engineering researchers: evidence from 26 international industry-academia collaborative projects. IEEE Softw. **37**(6), 65–75 (2020)
17. Glass, R., Demarco, T.: Software Creativity 2.0. Developer. Books, 2006. (Online access: EBSCO Computers & Applied Sciences Complete). ISBN 9780977213313. https://books.google.com.br/books?id=DozsD0zxb5wC
18. Guillot, I., Paulmani, G., Kumar, V., Fraser, S.N.: Case studies of industry-academia research collaborations for software development with agile. In: Gutwin, C., Ochoa, S.F., Vassileva, J., Inoue, T. (eds.) CRIWG 2017. LNCS, vol. 10391, pp. 196–212. Springer, Cham (2017). https://doi.org/10.1007/978-3-319-63874-4_15
19. Highsmith, J., Cockburn, A.: Agile software development: the business of innovation. Computer **34**(9), 120–127 (2001)

20. Kamei, F.K., Pinto, G., Cartaxo, B., Vasconcelos, A.: On the benefits/limitations of agile software development: an interview study with Brazilian companies (2017)

21. Kitchenham, B.A., Charters, S.: Guidelines for performing Systematic Literature Reviews in Software Engineering. [S.l.], (2007). https://www.elsevier.com/__data/promis_misc/525444systematicreviewsguide.pdf

22. Kitchenham, B.A., Budgen, D., Brereton, P.: Evidencebased software engineering and systematic reviews. Philadelphia, PA: Chapman & Hall/CRC, 2015. (Chapman & Hall/CRC Innovations in Software Engineering and Software Development Series)

23. Petersen, K., et al.: Action research as a model for industry-academia collaboration in the software engineering context. In: Proceedings of the 2014 International Workshop on Long-term Industrial Collaboration on Software Engineering. ACM, New York, NY, USA (2014)

24. Petersen, K., Vakkalanka, S., Kuzniarz, L.: Guidelines for conducting systematic mapping studies in software engineering: an update. Inf. Softw. Technol. **64**, 1–18 (2015)

25. Philbin, S.: Process model for university-industry research collaboration. Eur. J. Innov. Manag. **11**(4), 488–521. Emerald (2008)

26. Marijan, D., Gotlieb, A.: Industry-Academia research collaboration in software engineering: the Certus model. Inf. Softw. Technol. **132**(106473), 106473. Elsevier BV (2021)

27. Marijan, D., Sen, S.: Industry-academia research collaboration and knowledge co-creation: patterns and anti-patterns. ACM Trans. Softw. Eng. Methodol. **31**(3), 1–52. Association for Computing Machinery, New York, NY, USA, (2022). ISSN 1049–331X, https://doi.org/10.1145/3494519

28. Marques, D., et al.: Industry-academy collaboration in agile methodology: preliminary findings of a systematic literature review. In: Proceedings of the 24th International Conference on Enterprise Information Systems. SCITEPRESS - Science and Technology Publications (2022). https://doi.org/10.5220/0011115900003179

29. de Marques, D.G., et al.: Industry-academy collaboration in agile methodology: a systematic literature review. In: 2022 17th Iberian Conference on Information Systems and Technologies (CISTI). IEEE (2022). https://doi.org/10.23919/cisti54924.2022.9820166

30. Neverworkintheory. It Will Never Work in Theory (2022). https://neverworkintheory.org/

31. Sandberg, A., Pareto, L., Arts, T.: Agile collaborative research: action principles for industry-academia collaboration. IEEE Softw. **28**(4), 74–83 (2011)

32. Santos, W.B., Cunha, A., Moura, H., Margaria, T.: Towards a theory of simplicity in agile software development: a qualitative study. In: 43rd Euromicro Conference on Software Engineering and Advanced Applications, pp. 40–43 (2017)

33. Wohlin, C., et al.: The success factors powering industry-academia collaboration. IEEE Softw. **29**(2), 67–73. (IEEE) (2012)

34. Wohlin, C.: Software engineering research under the lamppost. In: Cordeiro, J., Marca, D.A., van Sinderen, M. (eds.). ICSOFT 2013 - Proceedings of the 8th International Joint Conference on Software Technologies, Reykjavík, Iceland, 29–31 July 2013. [S.l.]: SciTePress, p. IS-11 (2013)

35. Wohlin, C.: Second-generation systematic literature studies using snowballing. In: Proceedings of the 20th International Conference on Evaluation and Assessment in Software Engineering. Association for Computing Machinery, New York, NY, USA (2016). (EASE 2016). ISBN 9781450336918. https://doi.org/10.1145/2915970.2916006

Human-Computer Interaction

Distance Digital Learning for Adult Learners: Self-paced e-Learning on Business Information Systems

Anke Schüll[✉] [iD] and Laura Brocksieper

University of Siegen, Siegen, Germany
anke.schuell@uni-siegen.de

Abstract. This paper contributes to ongoing research by providing a selection of design elements to enrich the learning experience for adult learners and to actively involve them in their learning process. To supplement the courses, a web-based interactive mixed-media approach was selected, providing access to the learning content anywhere and anytime. Within this paper we extend previous research and report on experiences with several self-paced e-learning courses on business information systems at a German University. The self-paced e-learning environments presented here were developed in a repetitive scheme, following the same pattern. A stencil was developed for this pattern to facilitate the design of further e-learning environments. To evaluate the acceptance of the concept within the target group of adult learners, qualitative and quantitative evaluations were conducted in real-life situations. This evaluation confirmed a positive impact on flow, interest, motivation, and perceived usefulness regarding the learning efficiency.

Keywords: Self-paced e-Learning · Adult learning · Business information system

1 From Distance Education to Self-paced e-Learning

Distance education dates back to the time of postal correspondence and covers a process, involving sender and recipients: distance teaching plus distance learning [23]. In an effort to reach out to geographically dispersed students almost every technological achievement was employed, from written content to television-based distance education [4, 6, 40, 55], radio broadcasts [44] and computer/web based learning materials [1, 22, 27]. Distance education can be used not only to teach the students, but also to teach the teachers [35]. To delineate the field of distance education, Keegan (1980) identified six elements of distance education [23]:

- Separation of teacher and learner,
- media usage to provide the content,
- provision of means for two-way communication,

© The Author(s), under exclusive license to Springer Nature Switzerland AG 2023
J. Filipe et al. (Eds.): ICEIS 2022, LNBIP 487, pp. 313–337, 2023.
https://doi.org/10.1007/978-3-031-39386-0_15

- the possibility of occasional meetings (for didactic and social reasons),
- influence of an educational organization on the learning process, and
- an industrialized form of education.

The influence of an educational organization on the learning process is not a core element, but rather a side effect of organizations specializing in distance education. These institutions represent the most industrialized form of education [23]. With increased maturity of the concepts, the last elements gain importance.

The term "tele-learning" rooted in research on distance learning or distance education [32]. Even though still referred to (e.g. [17, 51]), the majority of research on the term "tele-learning" reached its prime in 1999 (e.g. [15]) as a form of distance learning, supported by the technological means of that time. The term TelE-Learning [34] led the way towards e-learning and with the passage towards this millennium, research embraced the term e-learning and e-education, bringing computer-based/web-based learning and teaching into the focus of attention. The broad acceptance of the internet offered students access to [19]:

- Self-paced interactive mixed media learning environments,
- (self-)assessment of knowledge and skills,
- additional materials to improve performance or motivation, and
- online communication.

In 1999, efforts on distance learning were perceived as poor alternatives to traditional campus based programs, a necessary evil and a compromise to circumstances [54]. During the pandemic social distancing created a learning situation similar to geographical dispersion and the terms "tele-learning" and "distance learning" were revived (e.g. [51]). And again, almost any available technology was employed to make it work: e.g. online-lectures, live skill demonstrations and virtual reality for anatomy and surgical training [10]. Still, the perception hasn't changed: distance learning is still perceived as an inferior teaching/learning method, even though the variety of available digital means would now allow true student-centered learning experiences and active involvement of students in their learning process.

Self-paced learning is a structured way to encourage students to learn individually at their own pace [3, 5]. They can stop, repeat, slow down or accelerate speed according to their individual learning process [9]. Self-paced learning provides students with all necessary resources and allows them to organize their learning independently from the teacher [3], autonomous and self-regulated [3]. This can accelerate the learning rate, as waiting time within the learning process until other students catch up or receive individual support is eliminated [9]. On the other hand, those requiring some extra time can take this without slowing down the pace for the class. As students interact with the environment, the level of participation increases, leading to a higher degree of cognitive engagement and, in consequence, improved effectiveness [9]. By this students should develop an intrinsic motivation, improve their time-management [3] and become prepared for lifelong learning [5].

With this paper we report on a pilot and several self-paced e-learning environments developed in a repetitive scheme with the potential to mature into a process of developing e-learning environments. Various design elements were combined to enrich the learning experience for students attending specific courses at our university. As a critical discourse and a collaborative exchange remain essential for in-depth and meaningful learning [18], regular video-meetings accompanied every course for didactical as well as for social reasons. With the pandemic subsiding, video-meetings turned into on-premises courses, supplemented by these self-paced e-learning environments.

The main target group of the courses discussed here are students of BA Business Administration, Business Information Systems and Business Law. These students are adults, which is paramount when considering their learning needs. Due to higher physical, cognitive, and mental abilities, learning approaches for adult learners need to be different from those for the education of children. Adult learning theory focusses on the involvement of the learner in the learning process [11]. In their book on adult learning, Thorndike et al. (1928) voted for changes in the quantity and nature of educating students between fifteen and forty-five, especially above twenty-five years [50]. They attributed adult learners with [50]:

- higher senso-motoric abilities,
- better observation and reaction to details,
- easy adoption of simple and more elaborate habits,
- better memorizing, and
- in more complex functions.

But not only are the abilities of adult learners different. Lindeman (1926) pointed out that it should be recognized that adult learners find themselves in specific situations, thus adult learning shouldn't be oriented on subjects, but on situations and on solving problems [28]. Also, the intellectual aspirations of adult learners require teachers to take on a different role, providing guidance instead of authority, and discourse instead of one-directional sharing of knowledge [28]. Based on the literature of that time, the foundation stones of adult learning theories can be summarized as [25]:

- Motivation: Adults understand the necessity to learn, to satisfy their needs, interests, and aspirations.
- Life-centered: Adults' approach to learning is life-centered. Adult teaching thus should be more organized towards life situations and problem solving, less around subjects.
- Experience: Learning from experience should be at the core of adult learning and teaching.
- Self-directed: Adults are not children and seldom appreciate being treated as such. They require guidance, but do not respond well to authoritarian teaching.
- Individuality: Differences between individuals increase in age. Style, time, place, and pace of learning should take these differences into account.

This tendency to self-directed learning represents a fundamental difference between children and adults [30]. But sure, self-directed learning has its limits. As Garrison (2009) points out, a high degree of self-direction requires a certain extent of skills and motivation. Lacking these, self-directed learning "becomes a sink or swim approach" [18].

The students addressed within this approach have decided to take courses at their faculty, attend at free will and most of them come with a high intrinsic motivation and some basic skills. It therefore seemed suitable to opt for an e-learning environment heading in that direction.

2 Design Elements of the Self-paced e-Learning Environments

Learning is a personal experience, thus should be student-centered, not teacher-centered [9]. As teachers are confronted with a heterogeneous group of students, with different backgrounds, different experiences, different personality types, and different learning styles, a classroom course will be too fast for some, too slow for others, too complex for one part of the course, too trivial for another etc. [9]. A curriculum for self-paced learning isn't predetermined by the teacher, but should be based on the capabilities of the students [26]. This can make e-learning more targeted to their individual needs [9]. When dealing with adult learners, this should be taken into consideration.

All courses discussed here are related to Business Information Systems. The target group is undergraduate students of Business Administration, Business Law, and Business Information Systems. That successful digitalization goes beyond a mere electrification of existing processes towards a radical rethinking of this process was a frequently repeated mantra in most of these courses. When considering the challenge of a digital transformation of learning/teaching, a good taste of our own medicine seemed adequate. After a pilot supplementing an 8h-repetition course was well received by the students, several self-paced learning environments were developed following the same pattern in an almost industrialized process. To avoid any technical barriers, and to facilitate the access, a roadmap of the course was made accessible on a password protected website. The roadmap gave an overview of the content of the course with the look and feel of a game board, with a starting point, and arrows pointing from step to step up to a finish-flag and each chapter of the course accentuated by a cloud or plate and labelled correspondingly. A playful, "non-pedagogical", appearance was decided on, something with a less earnest and serious look and feel. The aim was to create a more relaxing environment, unrelated to "learning". As mixed media-based e-learning generates a higher user concentration (Liu et al. 2005), different media and tools were combined to dissolve monotony.

As the attention span in digital media is limited, videos have a maximum of 20–25 min. Explanatory videos were enriched by external videos to provide variety in explanatory style. As well organized, meaningful vivid and personal content can increase retention [9], voiceover presentations were enriched by introducing videos filmed at selected locations, to keep a personal connection. Links were included, providing examples, software applications and case studies without infringement of copyrights. Mini quizzes and learning cards were implemented to get the students actively involved and to trigger their engagement. The icons on these roadmaps are largely self-explanatory (Table 1).

Straightening the arrows was discussed as an option but dismissed to keep a playful appearance. To keep the association with a game board, the background is colored, the colors of the arrows harmonized with the background. Background colors are muted for a good contrast.

Table 1. Elements of the Roadmap.

Icon	Function
	Each panel corresponds to a chapter of the lecture.
	Arrows point the way along the roadmap, from start to finish. The color is harmonized with the background of the roadmap.
	Start
	Finish
	Explanatory videos
	External videos
	Introduction videos
	URLs link to examples or other practical clarifications of the content.
	This icon is linked with an pdf document containing tasks and assignments, to deepen the knowledge or to self-research some content.
	Case study
	Sample solution
	The set of slides for each chapter is linked to this icon as a pdf document.
	Mini-quizzes or short evaluations. The icon is linked with a URL, so that even without scanning the QR code, access to the quiz is granted.
	Electronic learning cards
	Feedback option/evaluation
	Forum in our learning management system, upload of documents
	Literature recommendations beyond those already given for the course
	Software recommendations or hands-on exercises
	Visually supported audio files

All materials were put online at the beginning of the semester. The roadmap was accessible on a website. Each chapter was visually distinguished from the others, the sequence clear. Since students could track their progress on the roadmap, no additional display was necessary. Even though the learning environment covered the relevant content, it's a supplement instead of a substitute. As critical discourse deemed irreplaceable, each course was accompanied by weekly video meetings to discuss the content and the results of the assignments. With the subsiding of the pandemic, video meetings were replaced by on-premises courses. A time schedule was provided to assist in time management and to narrow down the topic of discussion of the regular meetings.

The layout of the roadmaps, the structure of the content and the icons sedimented into a notation that could be applied on all other courses. To facilitate the development of further e-learning-environments, a stencil for Microsoft Visio was designed (Fig. 1).

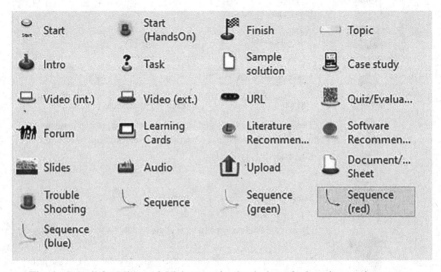

Fig. 1. Stencil for Microsoft Visio to assist the design of e-learning-environments.

3 Pilot and Case Studies

3.1 Pilot

With the spreading of Covid-19 in spring 2020, an 8h-repetition course for business executives on Business Information Systems was cancelled on short notice. An asynchronous learning environment was agreed on, two weeks ahead of the scheduled time. The roadmap of this course was simple. Each step was visualized by design elements that were linked to mixed media content: voice-over presentations, external videos, documents, or examples. To assist in time management, information on the duration of each video was provided. Quizzes, exercises, and case studies were included, to actively engage the students in their learning process. Arrows showed the sequence, leading from start to finish. Tasks belonging to the same chapter of the course were embraced by a cloud. The design was user friendly, but for the target group of business executives, the layout was a bit of a risk.

With everyone aware of the situation, the concept of a wrap-up video-meeting complemented by a self-paced e-learning environment was received well by the students. Even though a bit heavily built, this pilot proved the concept. This concept was rolled out onto several courses afterwards. With students' feedback on each iteration, the courses could evolve. Media production improved in quality, due to an upgrade in hardware and software, but of course, within strict financial and temporal limitations.

3.2 Course "Modelling and Design of Business Information Systems"

Following the pattern of the pilot, a course on modelling and design of business information systems was digitalized in autumn 2020. The target group of this course studied BA Business Information Systems. At the core of this course were different approaches on the modelling and design of organizations, business processes and information systems including e.g., Business Process Model and Notation 2.0, the House of Business Engineering, the Supply Chain Operations Reference Model, and some other methods evolving around Business Intelligence. To understand the advantages of each approach and to get familiar with them, a considerable part of the course involved case studies. Providing (digital) support for students working on these case studies was the major challenge of this course.

In reference to the main topic of the course, the content was modelled like a learning process (Fig. 2). Chapters were highlighted by geometrical shape instead of a cloud to provide a clearer and less confusing layout. Each chapter of the course contained different tasks, to actively involve the students. Each type of task was visualized with a design element corresponding to the media used.

Explanatory videos were kept to a length of 20–25 min. Information about the duration of the videos was exchanged by short information about the content for external videos. Several external videos and links to websites were included to provide further information and explanatory variety. External videos were marked in a different color. Mandatory content was part of the process, optional content for those students interested in additional information was available alongside the process. A pdf-file with a collection of all slides was included for each chapter, to facilitate commenting and/or printing the slides. To prepare students for their professional careers, recommendations of software-tools for modeling and design were included in each chapter.

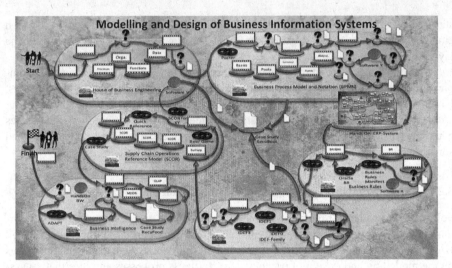

Fig. 2. Roadmap of the course on "Modelling and Design of Business Information Systems".

Question marks provided access to case studies. The complexity of the case studies increased with the progress in each chapter. With the knowledge of the explanatory videos, students could manage to solve each case on their own and upload their solutions into the learning management system used at our faculty. There was corrective feedback for each uploaded solution and weekly videoconferences to discuss the content of the course and the case studies.

Students were free to work with the learning environment anytime, but there was a schedule for finishing the case studies and for the discussion within the videoconferences. Sample solutions were provided to allow self-control, but only after the corresponding video meetings. During the semester, explanatory videos for each sample solution were added that addressed the difficulties becoming obvious within the uploaded solutions or were voiced within the discussions.

A similar environment was developed in summer 2022 for Master students on the topic "Data Warehousing". The course was on-premises, and in contrast to the pilot and the course on "Modeling and Design on Business Information Systems", where all materials were online at the beginning of the semester, the material of this course was developed along the course of the semester.

3.3 Hands-On-Training on SAP ERP-Systems (SAP S/4 HANA)

Within the same semester, a course on Business Information Systems was digitalized. The course consisted of two parts that were synchronized: a series of lectures on different business information systems and a hands-on training in a specific ERP-System to gain practical experience with these systems. Within this course, business processes about sales and production of bikes and the procurement of the necessary raw materials are executed within the ERP-System SAP S/4 HANA. Most students were novices to the topic. A main challenge for a remote course on this system was in dissipating anxiety to work with the system on their own. Encouraging students was therefore an important aspect. As this course was presented within previous research [46], we will not elaborate further on this.

Observations of students, their reactions and feedback pointed towards adjustments and improvements to the self-paced learning environment (Fig. 3). Videos were exchanged, content and quality improved. For the autumn semester 2022/23 e.g., visually supported audio files gave explanations for several terms that students within their first year are not familiar with (e.g., MRP-Run, minimum inventory level, or purchase requisition). In addition to a Trouble Shooting Portal, emails with questions and answers for each domain were anonymized and made accessible in the learning management system.

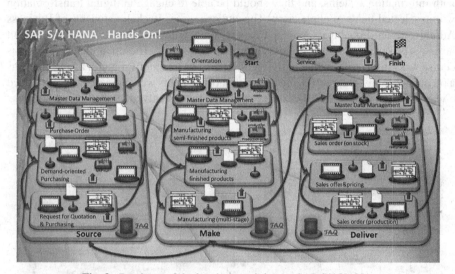

Fig. 3. Roadmap of the hands-on training on SAP S/4 HANA.

Students can work with this environment anywhere and anytime, but they are also supported in on-premises courses. As a supplement to the on-premises courses, this environment showed:

- Students proceed in these scenarios at different speed. Some are quick, while others take their time or make mistakes that cause them to fall behind. Explanations about the "why" and "how" thus come too late for one group of students, while too early for the other, limiting the attention both groups pay. Within the self-learning environment every student can get explanation exactly when needed.
- The self-learning environment encourages students to learn autonomously, with support provided whenever needed.
- Limitations caused by a certain number of computers in a computer lab diminish, increasing flexibility for the students.
- With some students working autonomously from home, all students can participate but there are less on-premises groups necessary, thus reducing the costs for the faculty while improving the service for our students.

- The self-paced learning environment ensures a constant explanatory quality in all groups and avoids discrepancy between the groups or between students working on premise or remote.

3.4   Course "Business Information Systems"

The Hands-On Training on Business Processes in the ERP-System SAP S/4 HANA was accompanied by lectures on Business Information Systems that have been the topic of a previous article [45]. Within this course, business process models were the starting points for the development or evaluation and customizing of business information systems. At the end of the course, students should be aware of the penetration of businesses with information systems, and they should be able to engage in digital transformation within selected fields of application. The audience were Bachelor students of Business Administration, Economics, and/or Business Law, without previous knowledge about business processes, and/or business information systems. The main challenge of this course was thus in raising their curiosity, their interest and motivation for a topic that was a bit outside their core interest and to which most students were novices. The self-paced e-learning environment for this course supplemented weekly video conferences.

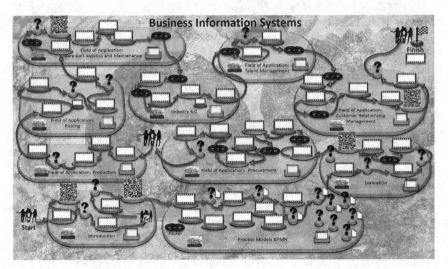

Fig. 4. Roadmap of the course on "Business Information Systems" [45].

The playful layout was kept, but with a slightly less obtrusive background and arrows colored in harmony with the background (Fig. 4). Each panel corresponds to a chapter of the course. All learning materials were put online at the beginning of the semester. Students could learn anywhere and anytime, but there was a schedule for the topics of the weekly video-meetings. Quizzes, mini-evaluations and learning cards with the most relevant terms and their explanation were included to actively involve the students in the learning activity. The benefits of learning cards are disputable, but students appreciated these and voiced concern of falling behind when a software update made these learning cards temporarily unavailable.

3.5   Course "Management and Control of IT"

Within the summer semester, a course on Management & Control of IT was supplemented by another e-learning platform (Fig. 5). The target group of the course were students within the 3rd or 5th semester of BA Business Administration [45]. Within the summer semester 2022, some adjustments were made to the concept. Additional literature recommendations were added for those students ready for a deeper understanding. A new symbol for case studies was added and each task and case study had an accompanying link to an upload area in our learning management system, to improve connectivity.

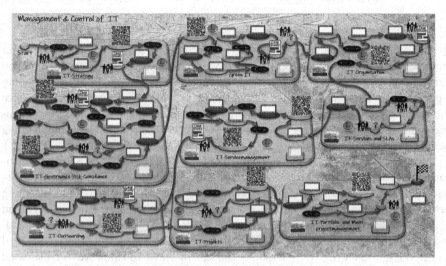

Fig. 5. Roadmap of the course on "Management & Control of IT".

4   Research Model and Data Collection

The main goal of these self-paced e-learning environments was providing the best possible learning environment for students during the drastic changes in the teaching/and learning situation enforced by the pandemic within the financial, technical, and temporal restrictions of a small teaching unit. Observations of the students, their feedback, their results in the exams and the teaching evaluations gave evidence of a positive perception of this environment. To assess the potential of the concept for courses in the aftermath of the pandemic, a systematic evaluation was conducted, that relied mainly on the Unified Theory of Acceptance and Use of Technology (UTAUT) (Venkatesh et al. 2003) and on the Technology Acceptance Model (TAM) (Davis 1989), slightly adopted to the context. This UTAUT model is frequently used to assess the mechanisms of e-learning acceptance (Table 2).

Table 2. Recent Literature on the Acceptance of e-Learning, grounded on UTAUT.

Reference	Topic	Research method
[2]	Mobile Learnings system of the university	Online survey
[8]	Mobile Learning systems in universities	Online survey
[36]	D-Learning, Generation Z Behavior	Online survey
[43]	E-learning systems	Online survey
[38]	Learning management systems, social isolation within COVID-19 pandemic	Online Survey

Expectancies regarding the usefulness of a technology for a specific purpose differ from person to person. Within UTAUT, the subjective expectancy to enhance the job performance is assessed with the construct Perceived Usefulness (PU) [52]. Commuting students regard attending university as such: more like a job [47]. E-learning can enhance the learning performance and can improve the accomplishment of the assigned learning goals. It supports learning activities and improves educational skills and learning performance [43].

Technical and organizational infrastructure to support the use of an e-learning environment is addressed with the construct Facilitating Conditions (FC) [43]. Three factors have a particularly high influence on system use, user satisfaction and self-learning behavior within an e-learning environment: System quality, information quality, and computer-related self-efficacy, which is also influenced by the service quality surrounding the e-learning offerings [42]. The perceived ease of use of a technology is measured by Effort Expectancy (EE) [53]. Usability and ease of use should improve with increased maturity of e-learning environments [43].

E-learning environments should recognize their users as learners and should follow a design philosophy dedicated to building up the learners concentration [29]. The term "flow" describes a state in which individuals become so totally immersed in an activity, that they lose awareness of their surroundings or their sense of time [12]. They focus entirely on this activity, experience an intense level of mental and physical engagement and act free of doubt or worries regarding the outcome or completion of the performed activity [13, 14, 33]. When learners reach this psychological state, they focus entirely on their learning behavior without getting distracted by their environment [29]. Achieving the state of "flow" during a learning activity can increase learning motivation and accelerate academic success [33]. This psychological state can thus play a major role in learning behavior [29].

The Behavioral Intention (BI) to use an e-learning environment is thus also positively correlated with its perceived usefulness [43]. Perceived usefulness has a positive impact on the intention to use a technology [53].

The impact of social influence on perceived usefulness is indisputable, but a hypothesis on the social influence on perceived usefulness would be biased by the exceptional circumstances and be neither comparable to previous literature nor a robust indicator for the time after the pandemic. Omitting the hypothesis that social influence has a significant and positive impact on perceived usefulness deemed consequential to the fact that during the Covid-19-pandemic, social distancing diminished the impact of social influence on the acceptance of this e-learning system. The hypotheses of the research model thus are [45]:

H1: Performance Expectancy (PE) has a significant and positive influence on Perceived Usefulness (PU).

H2: Facilitating Conditions (FC) have a positive impact on Perceived Usefulness of e-learning environments.

H3: The degree of ease (Effort Expectancy = EE) related to the use of an e-learning environment has a positive effect on its usage to achieve the assigned learning goals.

H4: Flow (F) has a positive influence on accomplishing the assigned learning goals (PU).

H5: PU has a positive effect on the Behavioral Intention (BI) to use the e-learning environment.

To evaluate the concept, an anonymous online survey amongst the participants of the courses was conducted. These students are the target group and possess the familiarity with the concept for a sound assessment.

Within the faculty of Business Administration, Business Law and Business Information Systems at the University of Siegen (Germany), students of two different courses were invited to participate in an anonymous online survey: "Management and Control of IT" (3.5) and "Business Information Systems" (3.4). The student groups did not overlap. All students were undergraduate, of about the same age, their gender not relevant for this analysis. Demographic items were therefore omitted from the questionnaire. No personal data was gathered, and the analysis strictly limited to gaining insights on the evaluation of the learning environment and its elements. Questions were derived from literature, some modified to the context, some questions were inverted (Table 3). A five-point Likert scale was applied.

Students were invited to participate in the survey on a voluntary basis, without any incentives, gifts, or credit points. 361 students accepted the invitation, 139 data sets were incomplete and had to be omitted, leaving 222 data sets for further analysis.

The research model was evaluated using Partial-Least Square Structural Equation Modeling (PLS-SEM) as this approach is widely used in IS research. The evaluation was conducted in SmartPLS (v3.3.3) [39]. Reliability and validity of the constructs was evaluated first, followed by an evaluation of the structural model.

Table 3. Items in questionnaire, Mean and Standard Deviation (SD) (PE - Performance Expectancy, FC – Facilitating Conditions, EE – Effort Expectancy, F – Flow, PU – Perceived Usefulness, BI – Behavioral Intention) [45].

Code	Item	Mean	SD
Performance Expectancy (PE)			
PE1	I like learning the content on my own	3.410	1.106
PE2	I like being able to proceed at my own pace	3.973	1.301
PE3	I like learning anytime	4.284	1.324
PE4	I like learning anywhere	4.104	1.357
Facilitating Conditions (FC)			
FC1	This is a convenient way to learn the content	3.649	1.140
FC2	It is convenient, not needing to gather the learning material myself	4.068	1.325
FC3	I get exactly the information I need in a learning situation	3.721	1.246
FC4	It is more time consuming to use the learning environment (inverted)	3.671	1.327
Flow (F)			
F1	Time flies when using the environment	3.387	1.050
F2	When using the environment, I won't be distracted by noises	3.189	1.212
F3	I feel good using the environment	3.482	0.985
F4	Using the environment is fun	3.667	1.064
F5	The environment triggers my curiosity	3.554	1.067
Effort Expectancy (EE)			
EE1	Orientation in the learning environment is easy	4.086	1.335
EE2	Interactions with the environment are easy and understandable	4.050	1.299
EE3	It's getting easier to navigate within the environment	4.036	1.252
EE4	The environment and its elements are easy to use	4.135	1.273
Perceived Usefulness (PU)			
PU1	I have learned more about the content	3.757	1.172
PU2	My understanding of the content has improved	3.770	1.149
PU3	My learning productivity is higher than in lectures	3.604	1.247
PU4	I learn the content more effectively than in lectures	3.536	1.176
PU5	This is a more convenient way to learn the content than in lectures	3.586	1.139
Behavioral Intention to Use (BI)			
BI1	I would continue in this learning environment	4.018	1.298
BI2	I would use this way of learning voluntarily	3.806	1.298
BI3	I would recommend this course to other students	3.959	1.275
BI4	I would wish for more self-learning courses	3.383	1.363

Cronbach's alpha indicates internal consistency among the items and should be above 0.7 [49] (Table 4). With a Cronbach's alpha above 0.7 for all constructs, this criterion is met. For the constructs EE and PE, the Cronbach's alpha exceeds the value of 0.9, which is almost too high. A Composite Reliability (CR) above 0.7 indicates reliability of the results [20], a criterion that is also fulfilled. The Average Variance Extracted (AVE) indicates the convergence validity and satisfies the nominal value of 0.5 for all constructs [16]. With all criteria met, reliability and validity could be confirmed.

Table 4. Construct Reliability (PE - Performance Expectancy, FC – Facilitating Conditions, EE – Effort Expectancy, F – Flow, PU – Perceived Usefulness, BI – Behavioral Intention) [45].

	Cronbach's Alpha	Composite Reliability	Average Variance Extracted
BI	0.888	0.923	0.751
FC	0.871	0.913	0.725
EE	0.975	0.983	0.952
F	0.868	0.906	0.661
PU	0.933	0.949	0.788
PE	0.890	0.925	0.756

The loading of each item on its construct should be above 0.7 and it should load higher on its own constructs than on the others. Cross-loadings (Table 5) confirm that this criterion is satisfied.

Discriminant validity can be confirmed, if the AVE square root is above 0.5 and higher that on the other constructs [16]. Table 6 shows that this Fornell-Larcker criterion is fulfilled, thus discriminant validity could be confirmed.

To test the significance of item loadings, a bootstrap procedure was applied with 1000 subsamples and a significance level of 0.05, confirming all hypotheses (Table 7).

Individuals perceive the expected usefulness of a technology to enhance their learning performance differently. In this context, PU is defined as the degree of perceived usefulness for accomplishing the assigned learning goals. That PE has a significant and positive influence on PU is in line with previous research on e-learning [31]. PE covered several items addressing the self-paced characteristics of the learning environment, something several students underlined as beneficial for their learning progress (Table 8).

With growing age, differences between individuals increase [25]. Requirements concerning time, place and pace differ as well as the perception of a specific style. The self-paced e-learning concept thus did not work well for all of them. One student commented on being ok with the setting as a compromise to circumstances: *"[...] I would have preferred learning a bit more dynamically – directly from the lecturer and at the university. Nonetheless, I like having the option to do everything from home."* [45]. Even though weekly discussions spotlighted crucial aspects, prioritizing the content was not easy for all of them. One student commented: *"Not bad, but in some articles and videos it is hard to figure out what to take out of it. They were not uninteresting, but you didn't really learn a lot. I have no clue what I should have learned out of all these parts."* [45].

Table 5. Cross-Loadings (PE - Performance Expectancy, FC – Facilitating Conditions, EE – Effort Expectancy, F – Flow, PU – Perceived Usefulness, BI – Behavioral Intention) [45].

	BI	FC	F	EE	PE	PU
BI1	**0.857**	0.686	0.628	0.677	0.731	0.705
BI2	**0.931**	0.743	0.730	0.612	0.780	0.784
BI3	**0.907**	0.783	0.661	0.714	0.784	0.756
BI4	**0.763**	0.511	0.581	0.341	0.610	0.589
FC1	0.687	**0.883**	0.619	0.617	0.692	0.675
FC2	0.736	**0.885**	0.568	0.812	0.788	0.693
FC3	0.714	**0.904**	0.612	0.670	0.684	0.677
FC4	0.549	**0.721**	0.464	0.430	0.490	0.507
F1	0.561	0.515	**0.786**	0.365	0.538	0.568
F2	0.346	0.292	**0.625**	0.206	0.306	0.386
F4	0.692	0.633	**0.877**	0.450	0.599	0.638
F5	0.700	0.638	**0.882**	0.593	0.631	0.645
F6	0.685	0.565	**0.866**	0.492	0.576	0.626
EE1	0.658	0.734	0.509	**0.982**	0.727	0.659
EE2	0.688	0.755	0.555	**0.970**	0.752	0.679
EE4	0.665	0.720	0.505	**0.975**	0.719	0.663
PE1	0.653	0.587	0.618	0.402	**0.750**	0.605
PE2	0.711	0.692	0.549	0.682	**0.902**	0.665
PE3	0.803	0.759	0.600	0.799	**0.925**	0.741
PE4	0.751	0.692	0.561	0.693	**0.890**	0.696
PU1	0.745	0.706	0.625	0.676	0.725	**0.885**
PU2	0.741	0.738	0.662	0.670	0.746	**0.884**
PU3	0.727	0.599	0.638	0.576	0.656	**0.881**
PU4	0.706	0.602	0.617	0.498	0.659	**0.901**
PU5	0.725	0.688	0.632	0.597	0.670	**0.887**

Within this study it could be confirmed that facilitating conditions have a significant positive influence on perceived usefulness, which is in line with recent research [31]. Accessibility and availability of the content anywhere and anytime is important (Table 9), but so is quality. Students can benefit from e-learning when the media content provided via e-learning is "likely to be instructionally superior to the low-quality media normally available in traditional teaching environments" [9]. Due to technical issues at the beginning of the pandemic, the sound quality of the first videos was poor, an issue several students commented on. This was a serious issue as the choice of the medium

Table 6. Fornell-Larcker-Criterion (PE - Performance Expectancy, FC – Facilitating Conditions, EE – Effort Expectancy, F – Flow, PU – Perceived Usefulness, BI – Behavioral Intention) [45].

	BI	FC	EE	F	PU	PE
BI	**0.867**					
FC	0.794	**0.852**				
EE	0.687	0.755	**0.976**			
F	0.752	0.668	0.536	**0.813**		
PU	0.822	0.755	0.684	0.716	**0.888**	
PE	0.842	0.788	0.751	0.668	0.781	**0.869**

Table 7. Path coefficients (PE - Performance Expectancy, FC – Facilitating Conditions, EE – Effort Expectancy, F – Flow, PU – Perceived Usefulness, BI – Behavioral Intention) [45].

Hypothesis	Path Coefficients	T Statistics (|O/STDEV|)	P Values
H1: PE ->PU	0.332	4.298	0.000
H2: FC ->PU	0.206	2.727	0.007
H3: EE ->PU	0.123	1.992	0.047
H4: F ->PU	0.290	5.280	0.000
H5: PU ->BI	0.822	32.183	0.000

Table 8. Sample statements on Self-Paced Learning.

Statements

I personally found learning via the self-study platform very convenient, it was fun, and the content was well presented. What I liked most was that I could decide for myself whether and how much I wanted to learn in a week, even if there was a schedule that you tried to stick to

A nice alternative in the current situation to learn independently and on your own

Overall, a very effective learning environment to learn independently and regardless of time and place

Through the learning platform, the lecture material, and the associated tasks, you can easily find your way into the topics and can work on and understand them at your own pace. I also find the effort behind this idea very remarkable, and it makes the start for freshmen but also for already "experienced" students easier

might be secondary, but the quality is not: Content that can't be listened to properly is not adequately accessible.

Table 9. Sample Statements on Facilitating Conditions.

Statements
As a student, you know exactly what to do and what to learn. You can complete the tasks at your own pace, and are also flexible for other lectures."
As a foreigner, I usually have a hard time understanding the professors during lectures, either because of pronunciation or speed of speech. And I can always go back to the videos if I encounter the difficulties mentioned above
Very convenient self-learning, you do not have the feeling of being aimless but always work "from one point to the next"
You get enough information on each topic and the learning material is gathered in one place. The self-learning environment is clearly arranged and structured. You get a good overview of all the topics right from the start
In lectures you sometimes can't keep up or your mind was somewhere else. With this option, this problem is completely avoided. I am very satisfied, and it is fun
I especially like the fact that I can watch the videos [...]as often as I want. This helps especially if confusion arises, or I was not attentive enough the first time listening
It is very convenient to learn with this self-learning environment, you have a good overview and through the many small videos etc., you are more motivated to continue than with a lengthy script because it is more varied and interesting
[...] you can work on everything at your own pace and thus also flexibly adapt your schedule or learning plan to different situations and circumstances during the semester

That Effort Expectancy has a positive impact on the perceived usefulness of the learning environment is in accordance with recent literature [31]. It could also be confirmed that flow has a significant positive influence on the perceived usefulness of the learning environment, an aspect several students commented on (Table 10). When students reach a state of flow and get absorbed by their learning activity, they concentrate entirely on this learning process, which could be beneficial for the learning outcome. Liu et al. (2005) recommended that the design philosophy of e-learning environments should be dedicated to build up students' concentration [29].

That Perceived Usefulness has a significant positive impact on the behavioral intention is in accordance to literature [29] und is reflected by the students' comments (Table 11).

The comments of the students underlined the results from the quantitative analysis and revealed several aspects not covered by the questionnaire e.g., statements on raised interest and motivation and on "not feeling left alone" with the learning content (Table 12). This last item points towards the relevance of connectivity that would be worth further exploration.

Table 10. Sample statements on Flow.

Statements
It's really fun to learn like this, and some facts are easier to understand. [45]
It is fun to work with the learning environment. As a result, you learn more readily and with more joy and curiosity
So far, I am very enthusiastic about the self-learning environment, especially about the computer lab. It's a pleasure to playfully gather the information and learn that way. I completed more than half of the self-learning environment within three days, not because I want to finish the course quickly, but because it's hard to stop once you're really into it. [45]
It was really an interesting experience to work with such a learning platform. I especially liked the quizzes and surveys because they gave a playful feeling. I also liked the fact that the content of the instructional videos was presented in a short and concise manner
I find the learning platform very well done & in general I find it a great idea because it motivates in a playful way
"The playful roadmap arouses my curiosity". At the beginning, I thought Business Information Systems would not interest me at all, but my curiosity grew with every panel." [45]

Table 11. Sample statements on Perceived Usefulness.

Statements
I enjoyed the course a lot. The topics don't drag too long and for every topic there are examples from reality. The practical examples make everything understandable and anchors the knowledge in the brain.
Overall, I find this learning concept very helpful. It is very varied and through the learning cards you can also check your current knowledge very well. In my opinion, this variety also makes you want to continue learning
The many short videos are more motivating to proceed than a [...] script, because of the variation, and it is more interesting." [45]
Very well organized, it says what would be best to work on until when, so that you can cope with the material well!
[...] it is really fun to learn with it and it is also easier to understand certain facts. In addition, it is also better to learn with this learning environment than to read through completely monotonous scripts
Easy handling of the learning environment, thus better understanding of the teaching content
So far, I am more motivated to study for the Business Informatics course than for courses in which you are "only" provided with lecture slides
The playful environment arouses my curiosity and motivates me to keep working. With the ability to set my own pace, learning is much easier, and I can adapt my process to achieve my best performance

Table 12. Sample statements on Connectivity.

Statements
Through the learning environment, you can determine your own learning speed and you still get everything explained very well, based on your recordings and the links. This means that you don't just have a script in front of you, which is monotonous in the long run. Through the zoom meetings there is the possibility to ask questions and you get them answered directly. In addition, you don't feel left alone with the material due to this possibility
Great videos, compact, everything important is well explained, fun to use the environment and you do not feel "left alone" with the material
Personally, I was a little worried that I wouldn't be able to handle either Business Information Systems or the computer lab since the subject was a bit out of the scope of my degree program. In retrospect, this was completely without reason, as you "took us by the hand" with everything

5   Conclusions

The main goal to develop the environments presented within this paper was coping with the teaching and learning situation during the pandemic. The University was locked down, access forbidden or unadvised, face-to-face-contact with students almost impossible. Teaching and learning had to work from a distance. With the pandemic subsiding, the learning and teaching setting changed again into a combination of the self-learning platform with on-premises courses.

Within this paper, a web-based interactive mixed-media approach was presented, providing access to the elements anywhere and anytime, allowing students to proceed along the learning process at their own speed. This paper extends previous research [45] by providing additional case studies with different characteristics and challenges. A self-paced e-learning environment was explored within a pilot study that emerged into a cornerstone for a sequence of self-centered e-learning environments. To drag students out of their learning routine, a playful, non-pedagogical look and feel was decided on for each learning environment. Associations with non-learning activities were fueled and different tools were mixed to raise their curiosity and to dissolve monotony in the learning process. Explanatory videos were kept short (max. 25 min), external videos provided examples or a practical perspective. As the students were not allowed onto the campus, each chapter began with a small intro video shot on the campus or in the city for a more personal contact and to connect them to the university. Quizzes and mini evaluations were included to actively involve the students in the learning activity. All courses were accompanied by regular video or face-to-face-meetings for didactic and for social reasons.

The concept initiated with the pilot evolved into a pattern of icons. This pattern allowed a time-efficient development of e-learning environments and was slightly adopted to the specifics of the corresponding course. The elements of these roadmaps sedimented into a symbolic language, for which a stencil for MS Visio was designed, to facilitate the development of further e-learning environments. Quantitative and qualitative evaluations in real-life-situations based on UTAUT confirmed a positive impact on interest, motivation, and perceived usefulness regarding the learning efficiency. The importance of "flow" for the acceptance of these learning concepts was underlined and examples of design elements provided that can be combined to actively involve students in their learning process and to enrich their learning experiences.

Even though initiated as a compromise to circumstances, the students' feedback on the pilot and the case studies presented here gives encouragement to continue this path. But the research presented here is not without its limitations. The focus of the different courses was on "making things work" for the students, not on the rigor of its evaluation. As capacities are limited, most of the effort went into the design elements and the content. The design of the elements as well as the questionnaire should be grounded deeper in existing research and additional research should be conducted for further confirmation of the results. The evaluation was conducted within a real-life situation, there was no control group to compare the results with and it took place under the influence of the pandemic. Further research would be necessary to evaluate the concept in the aftermath of the pandemic.

The participants of the survey were students from our target group, the results are thus not representative on a broader scale. And it needs to be pointed out that the anonymous online survey was embedded in the self-paced-e-learning environments at the very end. As only students completing the course got this far, the results will be biased. Evaluations earlier in the course would be necessary to cover the feedback of those students struggling with the concept. Expanding the evaluation of the e-learning platform on the learning outcomes would also be recommended.

Literature clearly assigns technology an important role in learning and teaching, but the challenge paramount to the introduction of these new technologies is in determining its value for educational purposes within the specific context of deployment [48]. And the value students draw out of a learning experience "is not dependent upon the medium used to deliver the learning, but actually upon the 'orchestration' from the teacher." [24]. Whilst several studies on the acceptance of e-learning point towards more pros than cons from the students' perspective, the acceptance of teachers and lecturers remains underexplored. To spur digital teaching, research on factors influencing the lecturers' acceptance of digital teaching would be recommended.

Facilitation Conditions and Ease of Use are expected to influence lecturers' acceptance of e-teaching. To facilitate the development of e-learning environments, a stencil for MS-Visio was presented within this paper, covering the shapes and icons used within the self-paced-e-learning environments. An evaluation of the acceptance of these might be worth further exploration.

Amongst the factors that have an impact on lecturers' acceptance of digital teaching is their tendency to comply with the norms and policies of their faculty. The influence of an educational organization on the learning process, and an industrialized form of education are thus two elements paramount for a diffusion of digital teaching of learning [23]. Very early research voiced concerns for a lack of interaction between lecturers and students and the fear of some faculties that e-learning environments would make lecturers obsolete [56]. But there are several reasons why academic institutions should embrace the deployment of information and communications technology for education [21]:

- Students' demands for improved educational services.
- Political pressure to modernize and improve efficiency.
- Neglect of the potential could cause a competitive disadvantage.

Students are adult learners, with different interests, skills, and motivation. Self-paced-e-learning environments can support their learning activities according to their specific needs. Students using digital technologies within their studies begin to master these e-learning/leaching tools and can pass their skills on to the next generation of students [41]. They get hands-on experience with technologies that allows them to assess the benefits and limits of these technologies more critically [41]. Students of Business Administration and Business Law could profit from these experiences later in their professional career. We thus strongly advise to assess the chances and risks of e-learning environments from a strategic perspective. A strategic agenda for e-education could include [7]:

- Identification of the competitive forces [37] to anticipate the development of the educational market and secure the competitive position.
- Analysis of the educational institution's strengths and weaknesses and its external opportunities and threats (SWOT).
- Consideration of strategic alliances to provide e-educational services.

References

1. Abrami, P.C., Bures, E.M.: Computer-supported collaborative learning and distance education. Am. J. Dist. Educ. **10**(2), 37–42 (1996). https://doi.org/10.1080/08923649609526920
2. Almaiah, M.A., Alamri, M.M., Al-Rahmi, W.: Applying the UTAUT model to explain the students' acceptance of mobile learning system in higher education. IEEE Access **7**, 174673–174686 (2019). https://doi.org/10.1109/ACCESS.2019.2957206
3. Anurugwo, A.O.: ICT tools for promoting self-paced learning among sandwich students in a Nigerian university. Eur. J. Open Educ. E-learn. Stud. **5**(1) (2020)
4. Baldwin, L.V.: An electronic university: engineers across the United States can earn master's degrees by enrolling in a new television-based institution. IEEE Spectr. **21**(11), 108–110 (1984). https://doi.org/10.1109/MSPEC.1984.6370348
5. Bautista, R.G.: Optimizing classroom instruction through self-paced learning prototype. J. Technol. Sci. Educ. **5**(3), 183–194 (2015). https://doi.org/10.3926/jotse.162
6. Buck, G.H.: Far seeing and far reaching: how the promise of a distance technology transformed a faculty of education. J. E-Learn. Dist. Learn. **23**(3), 117–132 (2009)
7. Chan, P.S., Welebir, B.: Strategies for e-education. Ind. Commer. Train. **35**(5), 196–202 (2003). https://doi.org/10.1108/00197850310487331
8. Chao, C.-M.: Factors determining the behavioral intention to use mobile learning: an application and extension of the UTAUT model. Front. Psychol. **10**, 1652 (2019). https://doi.org/10.3389/fpsyg.2019.01652
9. Clark, D.: Psychological myths in e-learning. Med. Teach. **24**(6), 598–604 (2002). https://doi.org/10.1080/0142159021000063916
10. Co, M., Cheung, K.Y.C., Cheung, W.S., Fok, H.M., Fong, K.H., Kwok, O.Y., et al.: Distance education for anatomy and surgical training - a systematic review. Surgeon : J. Royal Colleges of Surgeons of Edinburgh and Ireland (2021). https://doi.org/10.1016/j.surge.2021.08.001

11. Cook, D.A., Dupras, D.M.: A practical guide to developing effective web-based learning. J. Gen. Intern. Med. **19**(6), 698–707 (2004). https://doi.org/10.1111/j.1525-1497.2004.30029.x

12. Csikszentmihalyi, M.: Flow: The Psychology of Optimal Experience, 1st edn. Harper & Row, New York (1990)

13. Csikszentmihalyi, M. (ed.): Flow and the Foundations of Positive Psychology. Springer, Dordrecht (2014). https://doi.org/10.1007/978-94-017-9088-8

14. Csikszentmihalyi, M., Abuhamdeh, S., Nakamura, J.: Flow. In: Csikszentmihalyi, M. (ed.) Flow and the Foundations of Positive Psychology, pp. 227–238. Springer, Dordrecht (2014). https://doi.org/10.1007/978-94-017-9088-8_15

15. Finley, M.J.: Tele-learning: the "Killer App"? [Guest Editorial]. IEEE Commun. Mag. **37**(3), 80–81 (1999). https://doi.org/10.1109/MCOM.1999.751499

16. Fornell, C., Larcker, D.F.: Evaluating structural equation models with unobservable variables and measurement error. J. Mark. Res. **18**(1), 39–50 (1981). https://doi.org/10.1177/002224 378101800104

17. Galvao, J.R., Lopes, J.M., Gomes, M.R.: How the WWW environment and tele-learning build teaching system architecture? In: IEEE Computer Society Conference on Frontiers in Education, San Juan, Puerto Rico, 10–13 Nov. 1999 (12A3/22). Stripes Publishing L.L.C. (1999). https://doi.org/10.1109/fie.1999.839267

18. Garrison, R.: Implications of online and blended learning for the conceptual development and practice of distance education. Int. J. E-Learn. Dist. Educ. **23**(2), 93–104 (2009)

19. Gunasekaran, A., McNeil, R.D., Shaul, D.: E-learning: research and applications. Ind. Commer. Train. **34**(2), 44–53 (2002). https://doi.org/10.1108/00197850210417528

20. Hair, J.F., Black, W.C., Babin, B.J., Anderson, R.E.: Multivariate Data Analysis (Pearson Custom Library). Pearson, Harlow (2014)

21. Heilesen, S.B.: What is the academic efficacy of podcasting? Comput. Educ. **55**(3), 1063–1068 (2010). https://doi.org/10.1016/j.compedu.2010.05.002

22. Jonassen, D., Davidson, M., Collins, M., Campbell, J., Haag, B.B.: Constructivism and computer-mediated communication in distance education. Am. J. Dist. Educ. **9**(2), 7–26 (1995). https://doi.org/10.1080/08923649509526885

23. Keegan, D.J.: On defining distance education. Distance Educ. **1**(1), 13–36 (1980). https://doi.org/10.1080/0158791800010102

24. Kidd, W.: Utilising podcasts for learning and teaching: a review and ways forward for e-Learning cultures. Manag. Educ. **26**(2), 52–57 (2012). https://doi.org/10.1177/089202061 2438031

25. Knowles, M.S.: Andragogy: adult learning theory in perspective. Community Coll. Rev. **5**(3), 9–20 (1978). https://doi.org/10.1177/009155217800500302

26. Kumar, M.P., Packer, B., Koller, D.: Self-paced learning for latent variable models. In: Proceedings of the 24th Annual Conference on Neural Information Processing Systems, pp. 1189–1197 (2010)

27. Lauzon, A.C., Moore, G.A.B.: Computer: a fourth generation distance education system: integrating computer-assisted learning and computer conferencing. Am. J. Dist. Educ. **3**(1), 38–49 (1989). https://doi.org/10.1080/08923648909526649

28. Lindeman, E.C.: The Meaning of Adult Education. New Republic Inc., New York (1926)

29. Liu, S.-H., Liao, H.-L., Peng, C.-J.: Applying the technology acceptance model and flow theory to online e-learning users's acceptance behavior. Issues Inf. Syst. **IV**(2), 175–181 (2005). https://doi.org/10.48009/2_iis_2005_175-181

30. Loeng, S.: Self-Directed learning: a core concept in adult education. Educ. Res. Int. **2020**, 1–12 (2020). https://doi.org/10.1155/2020/3816132

31. Mahande, R.D., Malago, J.D.: An E-learning acceptance evaluation through UTAUT model in a postgraduate program. J. Educ. Online **16**(2), n2 (2019)

32. Mood, T.A.: Distance Education: An Annotated Bibliography/Terry Ann Mood. Libraries Unlimited, Englewood (1995)
33. Özhan, ŞÇ., Kocadere, S.A.: The effects of flow, emotional engagement, and motivation on success in a gamified online learning environment. J. Educ. Comput. Res. **57**(8), 2006–2031 (2020). https://doi.org/10.1177/0735633118823159
34. Passey, D., Kendall, M.: TelE-learning: the challenge for the third millennium. In: Passey, D., Kendall, M. (eds.) IFIP 17th World Computer Congress: TC3 Stream on telE-learning. IFIP, 25–30 August, 2002, Montréal, Québec, Canada, vol. 102. Kluwer Academic, Boston (2002)
35. Perraton, H.: Distance Education for Teacher Training. Routledge, London (2002)
36. Persada, S.F., Miraja, B.A., Nadlifatin, R.: Understanding the generation Z behavior on D-learning: a unified theory of acceptance and use of technology (UTAUT) approach. Int. J. Emerg. Technol. Learn. (iJET) **14**(05), 20 (2019). https://doi.org/10.3991/ijet.v14i05.9993
37. Porter, M.E.: Competitive Strategy: Techniques for Analyzing Industries and Competitors. Free, New York (2004)
38. Raza, S.A., Qazi, W., Khan, K.A., Salam, J.: Social isolation and acceptance of the learning management system (LMS) in the time of COVID-19 pandemic: an expansion of the UTAUT model. J. Educ. Comput. Res. **59**(2), 183–208 (2021). https://doi.org/10.1177/073563312096 0421
39. Ringle, C.M., Wende, S., Becker, J.-M.: SmartPLS 3. SmartPLS GmbH (2015)
40. Rovai, A.P., Lucking, R.: Sense of community in a higher education television-based distance education program. Educ. Tech. Res. Dev. **51**(2), 5–16 (2003). https://doi.org/10.1007/BF0 2504523
41. Royalty, R.M.: Ancient christianity in cyberspace: a digital media lab for students. Teach. Theol. Relig. **5**(1), 42–48 (2002). https://doi.org/10.1111/1467-9647.00117
42. Saba, T.: Implications of E-learning systems and self-efficiency on students outcomes: a model approach. Human-Centric Comput. Inf. Sci. **2**(1), 6 (2012). https://doi.org/10.1186/2192-196 2-2-6
43. Salloum, S.A., Shaalan, K.: Factors affecting students' acceptance of E-learning system in higher education using UTAUT and structural equation modeling approaches. In: Hassanien, A.E., Tolba, M.F., Shaalan, K., Azar, A.T. (eds.) AISI 2018. AISC, vol. 845, pp. 469–480. Springer, Cham (2019). https://doi.org/10.1007/978-3-319-99010-1_43
44. Sasidhar, P., Suvedi, M., Vijayaraghavan, K., Singh, B., Babu, S.: Evaluation of a distance education radio farm school programme in India: implications for scaling up. Outlook Agric. **40**(1), 89–96 (2011). https://doi.org/10.5367/oa.2011.0033
45. Schüll, A., Brocksieper, L.: In the flow: a case study on self-paced digital distance learning on business information systems. In: 24th International Conference on Enterprise Information Systems, Online Streaming,---Select a Country---,25.04.2022–27.04.2022, pp. 332–339. SCITEPRESS - Science and Technology Publications (2022). https://doi.org/10.5220/001106 8000003179
46. Schüll, A., Brocksieper, L., Rössel, J.: Gamified hands-on-training in business information systems: an educational design experiment. In: 18th International Conference on e-Business, Online Streaming,---Select a Country---,07.07.2021–09.07.2021, pp. 165–171. SCITEPRESS - Science and Technology Publications (2021). https://doi.org/10.5220/001 0618401650171
47. Stalmirska, A.M., Mellon, V.: "It feels like a job …" understanding commuter students: motivations, engagement, and learning experiences. J. Hosp. Leis. Sport Tour. Educ. **30**, 100368 (2022). https://doi.org/10.1016/j.jhlste.2021.100368
48. Sutton-Brady, C., Scott, K.M., Taylor, L., Carabetta, G., Clark, S.: The value of using short-format podcasts to enhance learning and teaching. ALT-J **17**(3), 219–232 (2009). https://doi.org/10.1080/09687760903247609

49. Tabachnick, B.G., Fidell, L.S.: Using Multivariate Statistics (Pearson Custom Library). Pearson, Harlow (2014)
50. Thorndike, E.L., Bregman, E.O., Tilton, J.W., Woodyard, E.: Adult Learning. The Macmillan Company, New York (1928)
51. Tomlinson, J., Shaw, T., Munro, A., Johnson, R., Madden, D.L., Phillips, R., et al.: How does tele-learning compare with other forms of education delivery? A systematic review of tele-learning educational outcomes for health professionals. N. S. W. Public Health Bull. **24**(2), 70–75 (2013). https://doi.org/10.1071/NB12076
52. Venkatesh, V., Davis, F.D.: A theoretical extension of the technology acceptance model: four longitudinal field studies. Manage. Sci. **46**(2), 186–204 (2000). https://doi.org/10.1287/mnsc.46.2.186.11926
53. Venkatesh, V., Morris, M.G., Davis, G.B., Davis, F.D.: User acceptance of information technology: toward a unified view. MIS Q. **27**(3), 425–478 (2003)
54. Weiser, M., Wilson, R.L.: Using video streaming on the internet for a graduate IT course: a case study. J. Comput. Inf. Syst. **39**(3), 38–43 (1999)
55. Whittington, N.: Is instructional television educationally effective? A research review. Am. J. Dist. Educ. **1**(1), 47–57 (1987). https://doi.org/10.1080/08923648709526572
56. Wilson, R.L., Weiser, M.: Adoption of asynchronous learning tools by traditional full-time students: a pilot study. Inf. Technol. Manage. **2**(4), 363–375 (2001). https://doi.org/10.1023/A:1011446516889

Human-Computer Interaction: Ethical Perspectives on Technology and Its (Mis)uses

Maria Cernat[1]([✉]) [iD], Dumitru Borțun[1] [iD], and Corina Matei[1,2] [iD]

[1] National School of Political and Administrative Studies, Bucharest, Romania
maria.cernat@comunicare.ro
[2] Titu Maiorescu University, Bucharest, Romania

Abstract. The developments in the human-computer interaction are astonishing. The Neuralink experiment showed a monkey literally moving a mouse on a computer screen „with its mind". The neuro device inserted in its brain was able to pick up the impulses and translate them into computer commands. Not very long ago, researchers presented an experiment in which a human patient diagnosed with severe depression had a neuro chip inserted in her brain to regulate her emotions. The fact that these major discoveries are not accompanied by equally well funded research to investigate the social, political and ethical implications that such results may have reveals the importance of this debate. The ethical challenges of human-computer interaction discoveries are themselves at the beginning since it is very difficult to keep up with the latest results. Nevertheless, it is vital that we start asking the important questions and put forward challenging hypotheses, since technology will not only be part = of our lives, abut also of our bodies.

Keywords: Ethics · Technology · Human-computer interaction · Deep brain stimulation

1 Introduction

Our paper is focused on recent developments in human-computer interaction. Our aim is not to provide conclusive answers, but rather to ask important questions related to the political economy of these scientific discoveries. While most discussions and research hypotheses are focused on the ethical challenges, researchers ask whether it is morally acceptable to implant chips into depression patients, for example. In this paper, we shall address more mundane, yet more troubling questions mostly related to power and money: who finances these researches, who controls the results, who makes sure the results are available for the general public, who is going to manage the impact of these discoveries? One of the most important books recently written about the Neuralink experiment belongs to Slavoj Žižek. He is among the few who dared to ask the question: who controls the computers? Of course, one might think that it is far too soon to ask such questions, assuming that, at some point, the human-computer interaction will allow us to "download" our very consciousness into a giant computer and thus achieve what Žižek calls "singularity" – a common consciousness. Yet, this question still remains unanswered: who pulls the plug of that computer? [1].

© The Author(s), under exclusive license to Springer Nature Switzerland AG 2023
J. Filipe et al. (Eds.): ICEIS 2022, LNBIP 487, pp. 338–349, 2023.
https://doi.org/10.1007/978-3-031-39386-0_16

The political economy of scientific research, in general, and of the human-computer interaction, in particular, is not just one but many steps behind the results put forward by scientists. In this paper we shall focus our attention on the Neuralink experiment and on the severe depression patient treated with a "deep brain stimulation device" [2]. Of course, our analysis could not be conducted without a broader perspective on ethics and ontology, since the results of these experiments also alter the limits of what is considered acceptable and what it means to have a human identity.

2 The Neuralink Experiment and the Financial Power to Support Scientific Research

When presenting a draft of our initial paper [38], we hinted at the serious dangers of the concentration of financial power in the field of scientific research. The Neuralink organization was set up to solve some of the most important health problems humanity faces. Elon Musk, the billionaire behind it, wanted to produce revolutionary technologies that would ultimately lead to a better communication between the human brain and the computer. Human paralysis, for example, may be cured with the aid of such technologies [3, 4]. Apparently, to this day, nobody actually challenged a fundamental issue: if too much power is concentrated in just one person problems related to access and transparency may soon arise. As researchers show, there are several problems related to the private model of financing technology-based ventures [5]. We asked, at that time, rhetorically, what would happen if a billionaire like Elon Musk would simply deny access to the results of the scientific discoveries he is financing to categories of individuals? At the time of the presentation, this idea might have seemed too far-fetched, but recent developments in the geopolitical area show just how much power he actually holds. Offering and then threatening to deny Internet access to a country involved in a military conflict was a sad confirmation of our predicament: vital research programs financed by only one person may lead to arbitrary decisions [6].

Still, in spite of these recent developments, there is little inquiry and debate related to this model of funding scientific research that would allow a closer interaction with the computers through brain stimulation devices and AI. The prospect of having a cure for paralysis obscures very important questions and challenges that, in our view, need to be faced. First of all, there is little debate on the fact that the public discussion on these results which facilitate a closer human computer interaction shifted from curing paralysis to providing pleasure on-demand [7]. The uses and the resulting financial gains are beyond the public's control. In an era where innovation depends so much on private funding [8], the political economy of scientific research is still considered sometimes mere speculation. Yet, the consequences of ignoring questions related to the public's ability to shape, control and decide on what is being researched, who might have access to the results of cutting-edge technologies are very real and reveal a serious gap between the have and the have-nots: "Unreliability of support is not the only or even the greatest, limitation on drug's company ability to help the world's poor. More fundamental are the demands of the corporate from itself (…).That's because – says Cohen – the 80 percent of the world's population that lives in developing countries represents only 20 percent of the global market for drugs. (The entire African continent represents only 1.3

percent of the world market.) Conversely, the 20 percent of the world's population who live in North America, Europe, and Japan account for 80 percent of the drug market. Predictably, of the 1,400 new drugs developed between 1975 and 1999, only 13 were designed to treat or prevent tropical diseases and 3 to treat tuberculosis" [9].

As in many situations before, the prospect of cutting-edge technologies helping us cure severe diseases crippling the lives of so many of us acts as a major deterrent preventing us from asking these difficult questions: who is it going to benefit from a brain stimulation device developed by Neuralink? Since major drug companies are investing their money in treating baldness and impotence because it is highly profitable, while completely ignoring diseases killing millions on the African continent, it is logical to infer that a market for brain stimulation devices providing pleasure on-demand would exceed in terms of profit the one of paralysis patients. This is why the course of the investment will follow the money as the ultimate business imperative. Even if the researchers advocating for open innovation politicies [8] do not criticize the current model of financing research, centralization of (financial) power could lead to authoritarianism.

Some of the critics of deconstructivism [10, 11] simply fail to acknowledge that deconstruction does not mean that the scientific truths are relative, but that there is a political economy of knowledge. What gets to be researched, who gets to be selected to become a researcher, how are the results of scientific discoveries made available to the public or kept at the disposal of a small elite, these are questions related to deconstruction. The idea that science evolves on a linear ascendant trajectory where scientists are perfectly rational beings completely separated from mundane asperities of life such as money, hierarchy and power is simply an ideal that, if we are to believe Thomas Kuhn's seminal research, will never become reality. [12] Strictly confined to some academic journals in a highly polarized political intellectual environment, deconstructivism fails to provide us with the necessary tools to question in a very efficient way the current business model of financing research in technology. But, nonetheless, it is the theoretical ground offering us the guidelines to analyze, compare and challenge the current business models in technological research.

Researchers focus on the evolution and social organization of research teams [13] and on the model of the so-called angels investing in technology ventures [5], yet no real challenges have been put forward so far to this business model. The anti-trust legislation that should prevent consolidation and monopoly formation was weakened during the Reagan administration [14], and it went without saying that as far as technology was concerned there were almost no legislative barrier preventing billionaires from financing, owning and patenting technological discoveries. This resulted in arbitrary decisions with regard to the hypotheses that need to be analyzed, the uses of technology, the transparency of it and the general public's access to it. While there are many attempts, especially by the EU, to regulate the digital giants [15], there is a need explore a similar effort to regulate the way scientific research programs are financed by the very owners of these companies. If digital monopoly proves to be a danger for the public sphere [16] there is no reason to infer that it represents no danger for the technological research sphere. Transparency of data, protection of scientists who may have findings important for the public, but contrary to the financial interest of the company supporting the research, the tempering of data to obscure possible negative effects of the discovered products and prevent public scrutiny [17], all of these problems and many more are to be explored thoroughly in order to suggest alternative regulation policies with regard to the way technological research is financed.

Apart from these very serious structural problems resulting from a financing model that is neither public nor democratic, the Neuralink experiment paves the way for many philosophical interpretations since its implications exceed by far the sphere of technology.

3 Treating Depression with a Deep Stimulation Device

The treatment of psychiatric illness through electrical stimulation techniques is not new. As researchers showed [4], attempts to control and modulate neurological disorders were used since ancient times. The physician of Roman emperor Claudius notes in a text that was preserved from that period (46 AD) that applying an electric ray on the surface of the skull could alleviate headache [4]. The medical history contains references to the use of electric fish to treat seizures, depression or pain in the eighteenth century [4, 18, 19]. Important observations on the topography of the brain were made in the late 19th century, but it was in 1950 that fundamental studies allowed researchers to have a clear representation of the brain functions [4].

The attempts to control and to treat antisocial behaviors through direct intervention on the human brain had been dominated, especially at the beginning of the 20th century, by extremely unethical and misogynistic behaviors. It was especially women suffering from depression or other type of mental illnesses that were subjected to a so-called miraculous treatment: lobotomy [20]. The research conducted by Egaz Moniz resulting in this type of highly controversial treatment consisting in severing the frontal lobes was hailed at that time and Moniz was even awarded the Nobel Prize in 1949. This was a direct intervention through surgical means aimed at altering human (especially female) behavior [20, 35].

As for the electrical stimulation, the nonetheless controversial technique of electroshocks was introduced by Ugo Cerletti in 1938 for the treatment of psychosis [4]. According to researchers, there are three types of techniques where electrical stimulation is used as a therapeutic solution: transcranial magnetic stimulation (it can stimulate only a particular area through the production of a magnetic field to modulate the excitability of the cortex), cortical brain stimulation (it involves the insertion in the brain of electrodes connected to a generator with a battery located in the chest) and deep brain stimulation (it involves implanting microelectrodes precisely in some brain areas that would send, through an external subcutaneous pacemaker, impulses to the brain [4].

The Neuralink experiment showed a monkey moving the arrow of the computer mouse on the screen with "its mind". In 1952, the Spanish neuroscientist Jose M. Delago started a series of ground-breaking experiments that proved the opposite was possible too. This time around it was not the monkey expressing its will with the help of deep brain stimulation devices connected to a computer. It was the researcher that controlled the behavior of the monkey using radio waves to control the microelectrodes inserted into the monkey's brain. In his seminal book Physical Control of the Mind: Toward a Psychocivilized Society, [21] the researcher presents the remarkable results in the control of motor functions, inhibition of maternal emotions or aggressive behavior, the ability to evoke emotions or memories, to induce anxiety, to alter the mood. Of course, doctor Delago did not have at his disposal a super computer that could perform the operation, as in the Neuralink experiment. He relied on rather primitive devices that were seen on the skull of his patients. But he was fascinated with the results of his own research and even claimed that "some women have shown their feminine adaptability (sic!) to circumstances by wearing attractive hats or wigs to conceal their electrical headgear" [21]. But the spectacular experiments Delago put forward remain a landmark in terms of deep brain stimulation. Amongst such experiments, one caught our attention. Ali was a violent chief of a monkey colony. He often mistreated his fellows and exerted dictatorial power. It made it the perfect target of doctor Delago who wanted to inhibit intraspecies violent behavior. After inserting electrodes in its brain, the doctor used radio stimulation to block his aggressive behavior. Apparently, the radio stimulation made Ali very obedient and prevented it from attacking other monkeys. In a second stage of the experiment, a lever that controlled the radio stimulation was placed inside the cage. Pressing the lever would have resulted in inhibiting Ali's aggressive behavior. Shortly after, a female monkey, Elsa, discovered it and pressed it whenever Ali threatened her. She failed to climb the social ladder, but, as the experiment showed, she was able to maintain the peaceful coexistence in the whole colony. As doctor Delago puts it: The old dream of an individual overpowering the strength of a dictator by remote control has been fulfilled, at least in our monkey colonies, by a combination of neurosurgery and electronics, demonstrating the possibility of intraspecies instrumental manipulation of hierarchical organization [21]. One could only speculate that this can once be achieved in human society. As we showed in the previous chapter, the business model for technology research sometimes favors individuals accused of unethical behavior [22]. This is part of the reason why the ideal of reaching social harmony by inhibiting intraspecies violence is not possible since the very persons financing the research may be the ones accused of controversial behavior. Nevertheless, the experiments conducted by doctor Delago were quite spectacular [4]. Such an experiment was conducted in 1962. CNN presented in 1985 the video recording of the moment where an aggressive bull was prevented from charging as a response to radio stimulation exerted through an electrode inserted into animal's brain [22].

However, it is worth noting that doctor Delago did not put great emphasis on the ability of radio stimulation to produce behavior. He was of the opinion that much of what we call personal identity or mind could not be self-sufficient Education, especially in Western cultures, is, to a great extent, based on the belief that individual personality is a self-contained and relatively independent entity with its own destiny, well differentiated from the surroundings, and able to function by itself even when isolated from Earth and traveling in an orbiting capsule. A more detailed analysis of reality, however, shows that cerebral activity is essentially dependent on sensory inputs from the environment not only at birth but also throughout life. Normal mental functions cannot be preserved in the absence of a stream of information coming from the outside world. The mature brain, with all its wealth of past experience and acquired skills, is not capable of maintaining the thinking process or even normal awareness and reactivity in a vacuum of sensory deprivation: The individual mind is not self-sufficient [21]. This is why he emphasizes the limitations of deep brain stimulation: it can stimulate certain responses, it can evoke certain experiences and memories, but it cannot create these manifestations of the cerebral activity.

With regard to the ethical challenges of such research, he pinpoints the rights of the individual as well as the controversial practice of using healthy individuals for the experiments since they often have limited choices – most of them are prisoners or severely impaired patients whose ability to consent is questionable [21].

The therapeutic value of deep brain stimulation is nowadays widely recognized. As researchers show, chronic pain, tremor, medically refractory Parkinson's disease began to be treated using deep brain stimulation and in the USA it was approved in 1989 by the Food and Drug Administration and, in the last decades, we witness a growing interest in this type of neurosurgical technique [4].

The journal Nature Medicine published in October 2021 an article about a severely depressed patient who was treated using a chip implanted in her brain. The novelty of this experiment resides in the fact that it was for the first time when a disease disputed by psychology and psychiatry has been treated through mechanistic methods. "We developed an approach that first used multi-day intracranial electrophysiology and focal electrical stimulation to identify a personalized symptom-specific biomarker and a treatment location were stimulation improved symptoms. We then implanted a chronic deep brain sensing and stimulation device and implemented a biomarker-driven closed-loop therapy in an individual with depression. Future work is required to determine if the results and approach of this n-of-1 study generalize to a broader population" [2].

The case is controversial since the psychologist, the psychiatrist and the neurologist might differ in their opinions regarding the triggering factors and the best therapeutic trajectory for severely depressed patients. Important research was conducted showing that there is a growing and concerning tendency of overmedicalizing mental illness [25–27]. Equally troubling is the attempt to solve psychological problems created by a more and more precarious, hyper-competitive, unequal and polarized society through the means of diagnosis and psychiatric treatment [28, 29]. With the use of deep brain stimulation techniques to cure severe depression, we move a step closer down the path of the medicalization of psychological suffering patients and of a mechanistic perspective on human suffering that could potentially be inhibited as the aggressive behavior of Ali, the violent monkey boss. "In this study, we established proof-of-concept for a new powerful treatment approach for neuropsychiatric disorders. The new framework presented in this article could advance biomarker-based neural interfaces and enhance the mechanistic understanding and treatment of a broad range of neuropsychiatric conditions" [2]. We could speculate that one might indulge themselves into believing that this is a rather tempting perspective, but then again we have to go back to Žižek's question: who controls the technology? [1] As previously showed, the current model of financing technological research does not look very promising and, as such, the figure of a human equivalent of Ali as the ruler and financier of this type of technology might come to haunt us.

And, for the time being, a closer look to the experiment aimed to be a proof-of-concept used the ethical guidelines available for the treatment of epilepsy and it was accepted that they are extendable to cases of severe depression. The Food and Drug

Administration's standards are said to have been applied to NeuroPace RSN (responsive cortical stimulation for the treatment of refractory partial epilepsy). In other words, their research was guided by the standards already available for technology used in epilepsy [30].

This only shows that the ability of technology to develop and advance new therapeutic solutions advances at a rapid pace and it is very difficult to incorporate discussions about the political, social and ethical aspects into future decisions related to such experiments. Metaphorically speaking, the technological advances seem to resemble a train that some-times seems to run on its own. The new experiments, the financing models exceed our capacity to incorporate all the voices of those concerned and potentially affected by such developments. And, since in some cases the financial support is no longer made public, this makes it even more difficult to accommodate the ideal of democratic pluralism with that of technological progress.

4   Neuralink, Mental Illness and the Use of Human-Computer Interaction to Exert Social Control

The idea of mentally programming individuals to perform tasks in a robot-like fashion was a long-lasting fantasy depicted in books, movies and other cultural manifestations. The way of perfecting tools and techniques to remotely control individuals is not new and it displayed several sordid signs in our recent history. Psychiatrist Donald Cameron is a sinister figure of psychiatry. In the 1950s, in the peak of the Cold War, he conducted a series of incredibly cruel experiments on his patients. The concern of the Intelligence divisions was that the Russians were probably programming ordinary citizens to carry out certain commands, like, for instance, to kill someone, without them even being aware that they were programed to act in this manner [31]. McGill's Allan Memorial Institute and doctor Donald Cameron that received funding from the CIA became, from 1957 to 1964, the place of the most horrific, inhumane and morally apprehensive psychiatric treatments ever to be conducted in such a systematic manner over such a long period of time. These extremely harsh treatments implied: electric shocks, huge doses of psychotropic drugs, torture techniques, sleep deprivation, sexual abuse, endless repetition of sound sequences. Those patients were thus meant to be re-programmed. It is only recently that 55 families of the former patients were granted the right to a fair trial [32] and it happened by serendipity: the documents that proved the existence of the experiment were not destroyed, they had been misplaced in the financial department. This is why in 1977 the team of Doctor Cameron was brought to a US Congress hearing [31].

This sordid episode in the history of modern psychiatry should definitely serve as a cautionary tale. The idea of using science and technology to exert social control proves to be central at least in some parts of the Intelligence apparatus. As we move closer to a time where technology will no longer be just part of our lives but of our beings [33], there is a growing concern related to the (mis)uses of technology in the human-computer interaction. The possibility of having our personality tampered with and making us unknowingly act against our will determines certain scientists to insist on banning this type of research: As doctor Delago pointed out back in 1969 "The possibility of scientific annihilation of personal identity, or even worse, its purposeful control, has sometimes been considered a future threat more awful than atomic holocaust. Even physicians have expressed doubts about the propriety of physical tampering with the psyche, maintaining that personal identity should be inviolable, that any attempt to modify individual behavior is unethical, and that method and related research -which can influence the human brain should be banned" [21].

However, he further argues that it is unrealistic to expect technological progress to simply stop. Since public institutions are no longer the sole providers of financial support for major scientific research programs, such a prohibition seems even less likely nowadays. The only possible soothing answer is that "a knife is neither good nor bad; but it may be used by either a surgeon or an assassin" [21] But that opens the debate our paper tries to initiate: what kind of technology is being researched, what research projects are financed, who provides the financial support, what kind of regulations could prevent an entity or person from abusing its power, who has access to the research process, who has access to the newly discovered technology. As scholars showed, there is always a facilitator between the scientist and the public, there is always an entity – the state or a private corporation – that makes the innovation available to the public [34].This is the place where regulation policies could shape an ethical research environment.

While the debate is still on its outset, international research programs aimed at providing a general framework for future regulation were advanced by the European Commission in 2020. The SIENNA Project [36] is an international project coordinated by The University of Twente, the Netherlands. The so-called Human Enhanced Technologies refer to all aspects of technological devices used to make the human computer interaction stronger than before. The scientists in this project put forward some of the most important challenges that future regulation policies have to acknowledge as key elements in the development of useful and comprehensive regulation:

- Right to privacy; the intrusion and the monitoring of man's life by HET, including the physiological intimacy is quite likely in the future;
- Freedom to be "imperfect" refers to the charm and the uniqueness of someone's personality which could be rendered by his/her own imperfection;
- Addiction to emerging technologies affects individuals regardless of their age, but HET and Artificial Intelligence could only deepen this trend, invalidating man as social being, as professional and even as rational being.
- Social inequalities and discrimination would deepen the differences in social status, financial power, inequality of chances between various social categories.
- Misuse of emerging technologies by transforming them into tools or weapons in order to achieve immoral or illegal purposes (theft, violence, illegal surveillance, espionage, etc.).
- Ownership and censorship of expensive HET and AI devices; these issues are linked to the economic and financial problems posed by the sale or loan of technologies to users, which would allow manufacturers to enforce certain restrictive conditions and increase addiction to their products.
- Issues of security, safety and liability refer to the risk that HET and AI would trigger multiple types of errors and accidents, especially where they are supposed to interact with humans or replace humans endowed with reason, intuition, instinct and wisdom.

- Possible weaponization of enhancements; It is known that, throughout history, the military field was the first to benefit from technological innovations and inventions which, after their wear and tear, were upgraded and disseminated to the general public.

5   Conclusions

In the light of the above, we can draw a first obvious conclusion. Debates resulting in regulation are vital. It is not only practical, feasible, but also in line with the common sense widely shared by large social and civil categories. Furthermore, if we raise awareness on the seriousness of the new technological revolution, a suitable legislative framework is likely to be designed, preventing future abuses. The conclusion is that the strategies and structures of research, funding and implementation, maintenance and results control belonging to these top fields ought to be transparent and confirmable. This should happen all over the world, either they belong to the private or governmental structures, since their importance and impact on the life of the entire humankind are overwhelming and exceed by far business, negotiation and market considerations. An ethical and philosophical matter of civilizational importance cannot be dealt with by business-driven decision-takers.

It is very foreseeable that all areas of HET and AI research will be affected by the lack of transparency and the concentration in the hands of a few financially powerful individuals of huge decision-making power regarding the use of new devices. Even the direction of research will suffer deviations to the extent that the general public's interest in technological achievements does not fit the interests of certain publics. It is predictable that the interests of more politically and economically influential publics will prevail and that, instead of registering progress in treating serious diseases that challenge humanity, funds could be assigned to new types of entertainment or to weaponization. Furthermore, in the absence of ethical regulations of the new fields, researchers themselves could lose their freedom of choice and would be prevented from becoming whistle blowers. Therefore, we believe that it is time for extensive public debates on the transparency of funding sources in research and production in these revolutionary fields, as well as on the ethical regulations crucial in the prevention of the misuses thereof.

We shall further develop other broader conclusions related to the current human civilization.

We are the only self-referential species in the known universe, i.e. capable of retrospectively analyzing its own evolution - therefore, we are also the only species capable of problematizing its future. The acceleration of historical evolution that we witness nowadays makes this problematization even more difficult: the evolution curve becomes asymptotic, while the economic, technological and political systems enter the "far from equilibrium" state [36], i.e. they become fundamentally unpredictable, not just due to human ignorance (deficiency of knowledge or limited ability to synthesize it).

We live in a time deprived of past anchors to guide us. The future is more uncertain than ever, as the new technological revolution impatiently pushes us into a realm yet uncharted from an ethical perspective. Even when we do anticipate the emergence of a new phenomenon, we cannot know whether it is "good" or "bad" and, in some cases, the two notions simply become irrelevant.

The four forces that got us here are the revolutions in genetics, robotics, nanotechnology and artificial intelligence. The first will probably make possible eugenics, the shaping of the human genome according to dominant social values – which could be shared by the vast majority of the population or only by the elites. The second consists in the transition from robots capable of performing only repetitive operations to machines capable of "emotional" reactions. The integration of technologies into the human body will make it possible for the human being to evolve based on technological progress – which will raise a new ethical issue: what is the legitimate point up to which a bio-technological being can evolve physically and intellectually, aesthetically and functionally? Nanotechnology will is likely to make it possible to implement micro-robots at cellular level; in conjunction with artificial intelligence, they will change the reality inside the body (the blood, endocrine, immune and brain systems), but also the one outside it; and man will be able to be treated as a product more or less useful, more or less expensive, more or less desirable. The assessment criteria will depend, once again, on the dominant values - dominant either statistically or by their enforcement by a dominating elite.

These changes call into question the accepted criteria for defining the human being as Homo sapiens, as well as the set of values on which our civilization is founded. One of these values is freedom of choice, but these changes refer to the dominance, control, ruthless ambition, consumerism, competitiveness and – as a consequence of the latter – competition; and they can impact on the fabric of humanity, can lead to its degeneration. In any case, a change is emerging in the meaning of evolution: the transition from one degree of freedom to another, greater degree of freedom. At least since the Enlightenment, the idea of progress has been confused with the idea of the emancipation of the human being, of increasing his margin of integrity, freedom and autonomy. As researchers showed, the human identity is not fixed, it is the result of constant interaction with the environment. It is precisely for this reason that an profit driven environment will incentivize un-ethical behavior that coupled with the centralized power offered by the current business model in research could lead to serious damage.

Either as an effect of the technological revolution (which calls into question the centrality of the human being), or as an effect of globalization (which calls into question the centrality of the European civilization), or as a result of a combination between the two, today we witness a double obsolescence: of anthropocentrism and of Eurocentrism. We shall not rule out the fact that the two were mere ideological illusions, useful in their time, which collapse as a result of decentration - a process that Jean Piaget explained by the multiplication of interactions and by the increasing distances at which they are established. This would be tantamount to returning to Hegel in the Phenomenology of Spirit (B. IV. A, # 148), who cynically argued: "Self-consciousness exists in itself and for itself, in that, and by the fact that it exists for another self-consciousness; that is to say, it is only by being acknowledged or recognized" [37]. (Hegel 1965, p. 107). Consequently, he claims that the real engine of history is the "struggle for recognition," as Hegel defines it in the Phenomenology of Spirit (in particular, paragraph IV.A.3., "Master and Slave"). A return to Hegel would mean accepting the fact that we have only two choices: to be masters or to be slaves; it would mean accepting the struggle for power as the only meaning a human life can have.

It is common knowledge that the world of money has nothing to do with empathy and solidarity, but it is also marked, even tarnished by competition and competitiveness. If the technological revolution is governed by the logic of profit maximization, Homo sapiens will not stand a chance, as the machines of the future will probably be incomparably smarter, more powerful and more economically efficient. The people in charge and capable of great decisions will have to decide now, before it is too late: do we leave these processes in the "invisible hand" that Adam Smith believed fine-tunes the free market (but which does not regulate it), or debate on and decide how far one can go on the path of redefining the human being?

References

1. Žižek, S.: Hegel in a Wired Brain. Bloomsbury Academic, London (2020)
2. Scangos, K.W., Khambhati, A.N., Daly, P.M., et al.: Closed-loop neuromodulation in an individual with treatment-resistant depression. Nat. Med. **27**, 1696–1700 (2021). https://doi.org/10.1038/s41591-021-01480-w
3. Larson, P.S.: Deep brain stimulation for movement disorders. Neurotherapeutics **11**(3), 465–474 (2014). https://doi.org/10.1007/s13311-014-0274-1
4. Sironi, V.A.: Origin and evolution of deep brain stimulation. Front. Integr. Neurosci. **5**, 42 (2011). https://doi.org/10.3389/fnint.2011.00042
5. Freear, J., Sohl, J.E., Wetzel, W.: Angles on angels: Financing technology-based ventures - a historical perspective. Ventur. Cap. **4**(4), 275–287 (2002). https://doi.org/10.1080/136910 6022000024923
6. Marquardt, A.: Exclusive: Musk's SpaceX says it can no longer pay for critical satellite services in Ukraine, asks Pentagon to pick up the tab, CNN, Updated 6:38 PM EDT, Fri October 14 (2022)
7. Yanakieva, N.: Can Musk's Neuralink Promise us Orgasm on-Demand? DevStuleR (2022). https://devstyler.io/blog/2022/02/07/can-musk-s-neuralink-promise-us-orgasm-on-demand/
8. Chesbrough, H.: Open Innovation. The New Imperative for Creating and Profiting from Technology. Harvard Business School Press, Boston (2003)
9. Bakan, J.: The Corporation. The Pathological Pursuit of Profit and Power. Free Press, New York (2004)
10. McIntyre, L.: Post Truth. MIT Press, Massachusetts (2018)
11. Pluckrose, H., Lindsay, J.: Cynical Theories: How Activist Scholarship Made Everything about Race, Gender, and Identity—and Why This Harms Everybody. In Pitchstone Publishing (US&CA) (2020)
12. Kuhn, T.: The Structure of Scientific Revolutions. Chicago University Press, Chicago (1996)
13. Milojević, S.: Principles of scientific research team formation and evolution. PNAS **111**(11), 3984–3989 (2014). https://doi.org/10.1073/pnas.1309723111
14. Correia, E., Budeiri, P.: Antitrust legislation in the Reagan era. Antitrust Bull. **33**(2), 361–393 (1988). https://doi.org/10.1177/0003603X8803300206
15. Clayton, J.: Europe agrees new law to curb Big Tech dominance, BBC (2022) https://www.bbc.com/news/technology-60870287. Accessed 29 Oct 2022
16. Fuchs, C.: The digital commons and the digital public sphere: how to advance digital democracy today. Westminster Papers Commun. Cult. **16**(1), 9–26 (2021). https://doi.org/10.16997/wpcc.917
17. Oreskes, N., Conway Erik, M.: Merchants of Doubt: How a Handful of Scientists Obscured the Truth on Issues from Tobacco Smoke to Global Warming. Bloomsbury Press, New York (2010)

18. Kellaway, P.: The part played by the electrical fish in the early history of bioelectricity and electrotherapy. Bull. Hist. Med. **20**, 112–137 (1946)
19. Schwalb, J.M., Hamani, C.: The history and future of deep brain stimulation. Neurotherapeutics **5**, 3–13 (2008). https://doi.org/10.1016/j.nurt.2007.11.003
20. Tone, A., Koziol, M.: (F)ailing women in psychiatry: lessons from a painful past. CMAJ. **190**(20), E624–E625 (2018). https://doi.org/10.1503/cmaj.171277. Epub 2018 May 21. PMID: 30991349; PMCID: PMC5962395
21. Delago, J.M.: Physical Control of the Mind: Toward a Psychocivilized Society. Harper Collins, New York (1969)
22. McHugh, R.: A SpaceX flight attendant said Elon Musk exposed himself and propositioned her for sex, documents show. The company paid $250,000 for her silence. Business Insider (2022). https://www.businessinsider.com/spacex-paid-250000-to-a-flight-attendant-who-acc used-elon-musk-of-sexual-misconduct-2022-5
23. CNN Special Report on Electromagnetic Mind Control featuring José Delgado and his famous 1965 experiment with an implanted bull. https://www.youtube.com/watch?v=23pXqY3X6c8
24. Szasz, T.: The Medicalization of Everyday Life. Syracuse University Press, New York (2007)
25. Conrad, P.: The Medicalization of Society: On the Transformation of Human Conditions into Treatable Disorders. John Hopkins University Press, Baltimore
26. Kaczmarek, E.: How to distinguish medicalization from over-medicalization? Med. Health Care Philos. **22**(1), 119–128 (2018). https://doi.org/10.1007/s11019-018-9850-1
27. Davis, J.E.: Medicalization social control and the relief of suffering. In: Cockerham, W.C. (ed.) The New Blackwell Companion to Medical Sociology. Wiley, Oxford (2010)
28. Raplay, M., Moncrieff, J., Jaqui, D.: De-Medicalizing Misery Psychiatry. Psychology and the Human Condition. Palgrave McMillian, New York (2011)
29. Sun, F.T., Morrell, M.J.: The RNS System: responsive cortical stimulation for the treatment of refractory partial epilepsy. Expert Rev. Med. Devices **11**, 563–572 (2014)
30. Collins, A.: In the Sleep Room: The Story of the CIA Brainwashing Experiments in Canada. Key Porter Books, Toronto (2001)
31. CBS News: Class action suit by families of those brainwashed in Montreal medical experiments gets go-ahead (2022). https://www.cbc.ca/news/canada/montreal/class-action-lawsuit-families-montreal-brainwashing-mk-ultra-1.6371416
32. Hayles, K.: How We Became Posthuman. Virtual Bodies in Cybernetics, Literature and Informatics. University of Chicago Press, Chicago (1999)
33. Wu, T.: The Master Switch: The Rise and Fall of Information Empires. Atlantic, London (2011)
34. Jansson, B.: Controversial Psychosurgery Resulted in a Nobel Prize. The Nobel Prize. https://www.nobelprize.org/prizes/medicine/1949/moniz/article/
35. SIENNA Report, 12.05.2020, sectionD3.4: "Ethical Analysis of Human Enhancement Technologies -WP3 Human Enhancement", pp. 77–88 (2020). file:///C:/Users/user/Downloads/Attachment_0%20(1).pdf
36. Prigogine, I., Stengers, I.: Temps et Éternité. Editions Fayard, Paris (1988)
37. Hegel, G.W.F.: Fenomenologia spiritului. The Phenomenology of Spirit. Academiei Press, Bucharest (1965)
38. Cernat, M., Bortun, D., Matei, C.: Remote Controlled Individuals? The Future of Neuralink: Ethical Perspectives on the Human-Computer, Proceedings of ICEIS, ISBN: 978-989-758-569-2 (2022). https://doi.org/10.5220/0011086700003179

Enterprise Architecture

Meet2Map: A Framework to Support BPM Projects Implementation Collaboratively

Danilo Lima Dutra[1], Carla Ribeiro[1], Simone C. dos Santos[1(⊠)], and Flávia M. Santoro[2]

[1] Centro de Informática, Universidade Federal de Pernambuco, Recife, Brazil
{cemmgrb,scs}@cin.ufpe.br
[2] Institute of Technology and Leadership, São Paulo, Brazil
flavia@inteli.edu.br

Abstract. Process understanding is essential to the Business Process Management (BPM) lifecycle for diverse strategic organizational projects. However, in many companies, the processes are still informal and unstructured. In this scenario, no process models are available, and all the knowledge about the processes remains on the people involved in their execution. Thus, it is necessary to engage stakeholders to share their expertise in current processes. Additionally, in the context of complex processes involving different sectors of the organization, stakeholder participation represents one more challenge. In this context, this research argues the use of the framework called Meet2Map that aims to promote identifying, discovering, and modeling processes involving participants based on collaborative interactions and human-centered design principles. This study is an evolution of the results published at the ICEIS 2022 conference, focusing on the conception of Meet2Map and the first case study concerning an ERP project. From the second case study in an authentic organizational environment, the results revealed the adequacy of the proposal, with very positive evaluations regarding the planned objectives. From both studies, a list of good practices could be identified, helping managers and process analysts apply Meet2Map in their BPM projects.

Keywords: BPM projects · As-is modeling · Collaboration · Design principles

1 Introduction

Organizations usually face challenges when implementing initiatives to change their business processes. The initial stages of the Business Process Management (BPM) life cycle [5], which include the identification, discovery, and analysis of business processes are especially hard to conduct due to the low level of BPM maturity in many organizations. Process analysts conduct these steps by interviewing many stakeholders to identify business processes, detail their tasks, and produce process models that clarify how to work procedures are performed (the so-called as-is models) and how they can be improved (the so-called to-be models). Two fundamental problems emerge from this scenario: 1) the lack of information from outdated, inconsistent, or non-existent process documentation [12] and, 2) the need for a holistic view of processes from different

J. Filipe et al. (Eds.): ICEIS 2022, LNBIP 487, pp. 353–378, 2023.
https://doi.org/10.1007/978-3-031-39386-0_17

stakeholders' perspectives. So, in these organizations, the primary source of information about processes is the people involved in their execution - the domain experts (e.g., users, owners, managers) [21].

A typical example of this scenario is the implementation of business management systems, better known as ERP (Enterprise Resource Planning). These systems support the key activities of any company. ERP implementantion is a project that usually brings about changes in critical processes in sectors such as production, services, HR, finance, logistics, distribution, among others, and, therefore, involving all employees who make the company, from operational to a strategic level [24, 27]. Many ERP projects have been renewed recently, mainly because of cloud computing and big data technologies emergence, aiming at service-based platforms to reduce costs and increase business efficiency through greater integration between the company and its value chain [26]. Thus, we observe that ERP solutions are constantly evolving, demanding the implementation of new ERP projects [14, 15].

Another common scenario is BPM projects based on the need to optimize complex business processes, which are not always supported by IT as they should be. Unlike ERP projects that bring with them efficient processes embedded in technological solutions, these projects have the main characteristic of understanding the process improvement points from various sectors of the organization and people involved, identifying possibilities for efficiency, greater productivity, higher quality of services and IT support suitable for these improvements, generally associated with integration needs between systems and data internal and external to the organization.

In both scenarios, profound transformations in organizations might directly impacting all employees' activities, at all levels, conducted for years in the same way and with the same people. Thus, it becomes clear how essential the commitment of the interested people is in organizational changes and how these changes need to be understood effectively, not only with messages and presentations, which are almost always unidirectional. BPM can be used as a strategy for engaging employees with future changes, based on the exposure of their current processes and, together with them, understanding the changes that these processes will undergo with the implementation of the new ERP or the new customized systems.

The discovery of processes through interactions between process analysts and domain experts could be a hard task [16]. Process analysts usually collect information from domain experts and documents and build process models according to their understanding. However, domain experts (like users, owners, managers) are generally unfamiliar with process thinking and notations, failing to provide meaningful feedback. Another widespread problem during process modeling is that the process analysts could misinterpret information because they are not knowledgeable with the process, generating an incomplete and/or incorrect model. Nonetheless, errors in a process model cause rework, costs, and compromise stakeholder confidence.

Considering this context, a central question is formulated in this study: "How to engage stakeholders to participate and share knowledge and experience in BPM projects?". To answer the question, this research describes a proposal to identify, discover and model business processes from interactions of people involved in those processes, called the "Meet-to-Map" Framework (Meet2Map, for short). Based on collaborative interactions and human-centered design principles, this approach might foster new ways of process modeling by engaging participants. For this, the Meet2Map framework comprises an application process, a design tool, and artifacts to guide the process analyst in modeling the As-Is processes together with domain experts. Literature presents some related works [1, 10, 16], but none of them provides guidance aimed at a collaborative process modeling able to support the analyst in this task.

The contributions of this work are twofold: (i) a detailed step-by-step procedure on how to perform the As-Is modeling stage of the BPM life cycle; and (ii) a collaborative approach supported by a tool based on design principles. Following a qualitative methodological approach based on action-research, the proposal was evaluated in two real cases, showing the viability and adequacy of this proposal. Meet2Map was first presented in Dutra et al. [6]. The current paper extends and consolidates the results from its usage by presenting a new case study that evidences its adequacy and viability, besides providing a list of good practices to apply Meet2Map.

This paper is organized as follows. Section 2 describes the methodological approach adopted. Section 3 presents background knowledge on BPM and design principles. Section 4 discusses related works. Section 5 describes Meet2Map Framework and its components. Section 6 presents the evaluation of the proposal through two case studies. Section 7 discusses the results obtained as well as lessons learned. Finally, Sect. 8 concludes the paper with future perspectives of this research.

2 Methodological Approach

This research is regarded as qualitative, with an emphasis on action research. According to Patton [20], research is said qualitative when it aims to investigate what people do, know, think, and feel through data collection techniques such as observation, interviews, questionnaires, document analysis, among others. On the other side, Merriam & Tisdell [17] explain that action research is an approach intended to solve a problem in practice, contributing to the research process itself, addressing a specific problem in a real environment such as an organization. This research was conducted in three stages: 1) Research delimitation; 2) Action-research cycles; and 3) Consolidation of results.

In the first stage, a literature review was carried out, which supported the construction of the theoretical background on the themes related to BPM, process modeling, and benefits of design principles (described in Section "Background knowledge"). In the second stage, we specified the problem, research questions (presented in the section "Case Studies") and planned the 2 case studies. To guarantee the reliability of the results, we defined a protocol for each one, which described the procedures of data collection, selection of the participants, analysis methods, and documents used.

The first case study had an exploratory nature, aiming to come up with the main challenges related to process modeling with the usage of [16]. The results of this exploratory study provided an additional basis for the design of the Meet2Map Framework in [6]. The data collected allowed the identification of the problems related to the process elicitation and modeling phase as well as the use of the collaborative tool. The problems were categorized into 10 types of failures and these failures were associated with 9 distinct types of consequences arising from them. The categorization of failures and consequences allowed the proposition of four critical success factors for discovering and modeling activity, using the collaborative modeling tool. The goal of the second case study was to evaluate the framework's application and provide its users with valuable guidelines. Section "Case Studies" describes that case study in detail. In the third stage, we consolidated the results proving a discussion and lessons learned from the findings.

3 Background Knowledge

Two main knowledge areas and related concepts lie behind the proposal of this research: the characteristics of the discovery and process modeling step of the BPM cycle and the principles of human-centered design. They are presented in following sections.

3.1 Business Process Discovering and Modeling

The first step in any BPM project is to understand the process and identify its problems, and as so, to model the As-Is process [5, 8, 12]. It is necessary to understand the current process to identify improvement points and avoid repeating past mistakes [25]. Another essential factor is the acceptance of the model by its participants. The imposition of a new process, not considering the participants' experience, contributes considerably to negative results.

Baldam et al. [3] emphasize the relevance of the following steps on the identification and modeling of the As-Is process: i) Project planning, understanding of scope (which process will be modeled, goals, metrics, strategic alignment, deadlines, deliverables), team composition, schedule, the infrastructure needed, access to documentation related to the process (laws, regulations, references); ii) Data collection with users (business experts and facilitators), including interviews (open or structured), the joint definition of activities, information about the process, meeting records; iii) Process documentation, building the process model, according to the methodology adopted. Additionally, some information could also be registered, such as publications, references, scope, etc. In this phase, it is expected the use software to support modeling; iv) Process validation, evaluating the model in an authentic process instance to check for coherence; v) Model adjustment, in which any perceived distortions during validation should be corrected.

Together with critical issues extracted from the BPM CBOK[1], these steps grounded the definition of the Meet2Map framework. The first issue considered is communication. All the stakeholders must fully understand the externalized knowledge. It is essential to establish a clear goal to support defining priorities and needs. The second issue is the need to identify the process at a high level before describing its activities. Going straight to the details can lead to constructing a large and challenging to understand model. A decomposition approach of "understanding the whole - understanding the parts - choosing one part" assists in the conception and comprehension of the model.

The third issue is managing expectations about modeling goals. Process modeling is not an end but a means to achieve a goal. Thus, making clear what is expected of the modeling and validating it by those who will use it should be done. Finally, the

---

[1] https://www.abpmp.org/page/guide_bpm_cbok.

last issue is related to a confusion between understanding and standardization. Process modeling does not represent an institutionalization of the process as a standard. These critical factors served as a reference to define the method for the process modeling phase proposed in this paper.

3.2 Design Principles to Foster Business Processes Discovering

Although the design thinking approach has been used more frequently in the context of process innovation, some studies suggest that alternatively, organizations could use design principles [4] and practices to address some issues faced by dealing with BPM in a general way, considering these issues are essentially human-centered [13].

In the context of "As-Is" process modeling, three phases can be considered: identification, to achieve a full understanding of the process, analyzing it both from a rational and experienced perspective; discovering, discussing processes, and proposing potential solutions that represent them; as-is modeling, in which a process is designed and evaluated by people involved, in a real scenario, getting feedback from them. As in the design of new product or processes, some principles also need to be adopted in the discovering and modelling of processes, encouraging the participation and engagement of domain analysts, users, and other stakeholders. In this context, Brown [4] defines seven design principles (DP) that can be related to processes, as shown in Table 1.

Table 1. Design principles in process modeling (Source: [6]).

ID	Design Principle	Process Modeling
DP1	Human-centered	People who participate in the modeling process are the most significant resource and should be valued
DP2	Collaboration	Stakeholders should be able to model process collaboratively, making the participants owner of the process
DP3	Experimentation	Look for different ways of modeling processes and breaking down barriers, allowing domain experts to model the process
DP4	Reflective	Observe the process from different perspectives to allow more discussion on the process
DP5	Multidisciplinary	Involve people from different areas and functions related to the process
DP6	Process thinking	Ensure that process stakeholders can analyze their processes, think about the problems faced, and propose possible improvement
DP7	New ways of thinking	In this context, about the processes

These principles were used in the Meet2Map conception to define its method and guidelines on how to apply it as described in Sects. 5.

4 Related Work

According to Forster et al. [11], since various domain experts and engineers participate in collaboratively creating of process models, the frontier between elicitation and representation is blurred. The authors argue that the distinction between those two phases disappears and is replaced by an iterative process of performing them. So, despite the most traditional method for process discovery and modeling be the structured interview [5, 9], some research on collaborative process modeling has been proposed. In this sense, Adamides and Karacapilidis [1] presented a framework for collaborative business process modeling, in which during a collaborative modeling session, a group participates in four interrelated activities: construction, presentation and understanding, critique, and intervention on the model. Furthermore, Antunes et al. [2] developed a modeling approach and a collaborative tool to support end-users in modeling business process based on a composition of scenes that form storyboards. None of these proposals provide a structured method for the modeling phase of business processes.

On the other hand, although some companies are already starting to use design thinking in BPM (e.g., SAP1), and this approach is mentioned in white papers from consulting groups (e.g., [19]), the academic literature has not explored this topic in depth yet. Based on design science and action research, Rittgen [23] proposes a six-step method to guide groups in workshops and a software to support applying this method, also advocating the use of collaborative methods for process modeling [22]. This work is focused on the dynamics of the workshop itself. Cereja et al. [10] described the use of design within the BPM cycle. A real case in a company has motivated that work when a traditional method of requirements elicitation was not successful in a specific project, and moreover, the users did not perceive the process that the system should support. So, a consulting company was hired to apply design thinking. They did not focus on the modeling phase and neither used a collaborative application to support the approach, as the framework presented here. Besides, Miron et al. [18] described the DIGITRANS Project, which is intended for an automated transformation of haptic storyboards into diagrammatic models. They are focused on the development of an app that delivers a design-based method and a set of tools to support the digital transformation process of Small and Medium Enterprises.

Luebbe and Weske [16] proposed a tool to support business process modeling called TBPM (Tangible Business Process Modeling). They aim to model the process during the discovering phase, so the domain specialist is no longer only interviewed, but he/she models together with the analyst. This proposal is based on design principles. The tool comprises a set of acrylic artifacts based on BPMN notation (task, event, gateway, and data object), which can be managed by the team to create the model. According to the authors, the semantics associated with the artifacts impel participants to use the concept of control flow, data flow, and resources. Thus, participants can create, delete, organize, and rearrange objects. Luebbe and Weske [16] pointed out the benefits of the tool: people talk and think more about their processes; people review their processes more often and apply more corrections during the discovery session; more fun and new insights during process modeling; active creation involves the participants and increases the engagement to complete the task. In view of these results, we propose this work as the initial stage of our research methodology, as described in the next section.

5 Proposal: The Meet2map Framework

Meet2map is a framework proposed to support a business analyst to conduct the discovery and modeling phases of a BPM life-cycle in a collaboratively manner [5]. Therefore, the framework is composed of a collaborative modeling method, a tool for collaborative modeling processes, and group of artifacts, as shown in Fig. 1 [5]. Each component is detailed in the following sections.

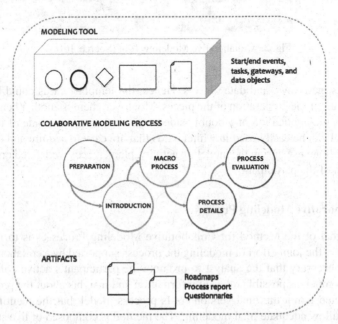

Fig. 1. Components of Meet2Map Framework.

5.1 Modeling Tool

The goal of the Collaborative Modeling Tool is to foster interaction among participants according to the design principles. So, we took the TBPM tool as a starting point. However, despite the benefits presented, the use of the tool in the exploratory case (commented in section "Research Methodology") pointed out some problems related to its design: building the acrylic pieces demands specific tools for the appropriate cutting and craftsmanship for the finishing of the pieces; and, applying the TBPM on a horizontal surface, preferably of glass, is required to allow the design of the swimlanes and flows. But since modeling activities usually take place at the customer's location, getting an environment with this type of table is not so simple. Thus, the solution was to replace the original material used in TBPM with the paper since it is low cost and accessible to everyone, as illustrated in Fig. 2.

To ensure a tactile experience, some resistance to handling and good appearance, several types of paper were assessed [7]. The BPMN elements are start events, end

Fig. 2. Collaborative Modeling Tool (Source: [6]).

events, tasks, gateways, and data objects. Hence, the template cards could be printed and easily cut for the preparation of the pieces. Moreover, there is another benefit of this design: with the application of a double-sided tape on the opposite side of the printed face, the tool can be easily used in whiteboards that are easy to acquire and commonly used. To facilitate use during the modeling activity, blank pieces were fixed on the board within reach of the participant.

5.2 Collaborative Modeling Process

The main goal of the Method for Collaborative Modeling Processes is to involve all participants in the joint effort of modeling the process supported by a collaborative tool. Thus, it is necessary that the analyst to promote the participant's active collaboration to extract as much as possible relevant information and insights about the process. The result expected is not the final version As-Is process model, but the identification of details that allow adequate final modeling. The method is composed of five steps.

The first step is the Preparation for modeling. This step is especially important since the team needs an appropriate environment. These are the instructions for this step: Select the environment: it is important to select a place free of distractions, favoring the attention of all involved; Guarantee full availability of participants: it is important to ensure that all key stakeholders are fully engaged in the activity during the proposed period, considering the human-centered, collaborative and multi-disciplinary characteristics of the design principles (DP1, DP2, and DP5 as shown in Table 1); Prepare the modeling tool: it is important to prepare the modeling tool so that it is ready for use and available for participants during the modeling activity encouraging experimentation (DP3).

The second step is the Introduction to essential concepts. The process analyst starts the workshop by introducing the team to the essential concepts to perform the task as listed below, stimulating process thinking (DP6): Understand the goals of the organization with the accomplishment of the task; Discuss the methodology that will be used for modeling; Introduce the basics of BPM, process modeling, and BPMN notation; Present the collaborative process modeling tool.

The third step is Modeling the macro-process. The main characteristics and macro tasks of the process must be identified (identification phase). It is recommended that the analyst conduct a semi-structured interview, driven by questions that help to understand the overall scope of the process. A proposal of questions is presented in the section "Artifacts". After the questions for macro tasks identification have been answered, providing new ways of thinking about the process (DP7), the participant should summarize the understanding of the macro process and its main activities, taking the time to do it (DP4).

The fourth step is Modeling of the process details. At this point, the macro tasks are already identified and, it is time to learn the details, such as responsibilities, interactions, transitions, rules, and events. First, it is necessary to present the modeling elements of the tool, recalling the basic concepts of BPMN.

To demonstrate the use of the tool to the participants, the analyst can represent the macro tasks at the top of the whiteboard. That way, while he demonstrates the use of the tool, he creates a reference guide. Additionally, the analyst can also draw the pool of the process with the swimlane of the first participant-role identified, and then place the start event. After this, participants should be encouraged to detail each macro task, according to the discovery phase. The top-down design technique can aid in the discovery of more details about the macro task, so each task is examined until the desired level of detail is reached. This human-centered activity (DP1) promotes reflexive thinking, process thinking, and new ways of thinking about the problem (DP4, DP6, and DP7).

The process analyst can intervene by asking questions about the process to guarantee the correct use of the modeling concepts. But it is important that he/she does not interfere by drawing conclusions about activities since this may decrease the participation of business experts and make them mere observers. Moreover, whenever necessary, the analyst can also introduce a more specific BPMN element to further enrich the process model. To do this, he/she must adapt the base element (e.g., transforming a basic task into a user task or manual task) by drawing the symbol corresponding to the most specialized task. However, the concept must always be presented to the participant, and his understanding must be confirmed, so that knowledge could be replicated in another situation by him. Additionally, throughout this step, critical points of the process and points of failure must be identified and pointed out in the model; however, it is important that process improvements are not addressed yet to prevent the participant from contaminating the as-is process model with insights of improvements that do not correspond to the actual process performed.

Finally, the last step is Process Evaluation. Once the process has been detailed, the analyst must validate it, corresponding to the as-is modeling phase (sketch of the model). Therefore, the process analyst should request participants to verbally summarize the process model from the start to the end to observe if any part of the process has not been represented. It is also important to check if the modeling goals were met.

5.3 Artifacts

To facilitate the application of Meet2Map, a script was defined with reference to the experiment designed by Luebbe and Weske [16], in which the following steps are defined: Step 1) Apply the preliminary evaluation questionnaire; Step 2) Conduct the interview to identify, discover and model the processes; Step 3) Apply the step evaluation questionnaire. About Step 1, the preliminary evaluation questionnaire is composed of four questions:

1. Have you participated in a business process mapping activity before?
2. Do you have any previous knowledge about business process management, by articles, books, or daily tasks?
3. Do you have knowledge of BPMN notation? Have you used this notation?
4. Do you have experience with any process or activity modeling language?

That questionnaire must be sent to the participant by email and must be completed before the discovery and modeling activity.

Considering Step 2, a roadmap is proposed, according to the following activities:

- *Activity 1* – Start the mapping session with a presentation to introduce main concepts to the team: Introducing the basic concepts of process methodology; What is the business process management and what is it for? What are primary, managerial, and support processes? What is an As-Is process? To present the basic concepts of BPMN notation, the objectives of the organization for mapping processes, the methodology that will be used for the mapping activity, and the mapping tool and collaborative modeling.
- *Activity 2* – After preparing the team, begin the interview for mapping the macro process as-is, following a script with pre-defined questions: What is the objective of the process? Why is the process important to the organization? Who is the client of the process? What is the customer's expectation of the process? Who are the suppliers of the process? How/when does the process start? How/when does the process end? What is the overall scope of the process and the most important activities?
- *Activity 3* – To demonstrate the modeling technique and summarize the understanding of the macro process and demonstrate the use of the modeling tool, the process analyst should model the macro process using the modeling tool.
- *Activity 4* – Start mapping to detail the process using the collaborative modeling tool. Conduct the participants in the as-is process model construction and perform questions to identify the parts of the process.
- *Activity 5* – Once the process is modeled, ask participants to flag existing problems in the current process. Signalize problems directly in the process using sticky notes that point to the point of the problem that the process has.
- *Activity 6* – Finally, the modeled process needs to be validated by participants. For this, the process analyst must: Ask the participant to verbally summarize the understanding of the modeled as-is process; Review whether the model made meets the reasons initially defined for mapping the as-is process; Question if there is any other information about the process that the participant would like to share.

Regarding Step 3, questionnaires were defined with the proposal to evaluate three main aspects: the information provided; the modeling activity; and the collaboration tool. These questions will be discussed in the results section of the case study. It is important to emphasize that the questionnaire was sent to the participant by email, right after the modeling activity was finished.

6   Evaluation

We conducted two cases studies in different contexts to evaluate the proposal according to the research question formulated. The first case study compared two approaches: this proposal and a traditional one. The second case study.

6.1   Case Study 1

The case study 1 was conducted within a company that provides services to the electric sector. This company had just acquired a new Enterprise Resource Planning (ERP) system and needed to start modeling and analyzing the current processes to align with the new system. The scope of the project encompassed the following areas of the company: Electrical Management; Accounting (tax and patrimony); Financial; Payroll and Point (Human Resources); Purchasing and Inventory (Supplies); Logistics. In total, 25 processes needed an analysis of conformance to the new system. It was necessary to map in detail how the processes were executed, observing activities supported by the legacy software, manual activities, rework due to the deficiency of the legacy software, and business rules related to the processes.

Process and Planning.   The internal analyst (first author of this paper) proposed to use the Meet2Map. It was selected four processes to be modeled following the proposed framework: Contract Billing (Electrical and Financial Management); Employee recruitment (HR); Maintenance of fleets (Logistics and Finance); Purchasing (Supplies and Finance).

The selection of the participants was made in two steps. The first selection was based on the departments involved in the processes to seek the support of the managers to identify all the key participants in the processes. The second one corresponded to the stage of identification of the participants for the modeling activities (Preparation for modeling). The identified participants engaged in the modeling activities of the selected processes, which were conducted by the researcher in the role of the process analyst. The number of participants for each process was: Kick-off Workshop (6); Process 1 (3); Process 2 (2); Process 3 (2); Process 4 (3). 80% of them had some coordination or management function, which benefits the modeling activity, considering their experience in the business. Regarding the participants' experience time, 90% had more than 2 years of experience in the area, with only one participant with less than one year in the company, who then participated in the modeling of the Employee recruitment process as a listener. None of the participants had performed process modeling before, and only 40% of the participants had participated as a collaborator in traditional modeling activities based on interviews. It is worth noting that a large part of the participants (80%) knew basic concepts of business process management but had never used these concepts in practice. This aspect reinforces the importance of an introduction to the theme for leveling the knowledge and the proposal. Finally, 90% of the participants already knew some language for control flow modeling and 40% of them had already used this technique to model processes, although none of the participants had previously used BPMN.

The study was conducted as follows: 1) two processes were modeled in a traditional way, without the support of a collaborative modeling tool; 2) two other processes were modeled using the tool. The case study started with a kick-off meeting, for which all the 6 leaders from the areas involved and the project sponsor were invited. During this meeting, the list of elected processes was presented together with the organization's reasons for choosing them, the methodology of work, and the structure of a form to identify roles and participants. Finally, the deadlines for returning the form and the agenda of the workshop were defined. After the kick-off, an email was sent to each leader, attaching the form, the presentation, and the preliminary calendar of the mapping activities. As far as the forms were being returned, the map of areas, functions, and participants had been defined. The next steps were the workshops for modeling each process. To exemplify how they were conducted, we will comment on two: first, using the traditional method; second, using the Meet2Map Framework.

The Contract Billing process was the first one to be modeled in a traditional method. The goal of this workshop was to identify the process, starting from contract measurement, billing approval by the contractor, registering in the Accounts Payable system, receiving of payment, and conciliation. The activity was performed as planned: all participants have actively provided information about their activities (2 leaders of the Electrical area, 1 supervisor of the Accounts Payable and Receivable area, all of them with 3–4 years of work in the company). The workshop spent 2:34 h. As main reports, the participants declared that they were unaware of the internal procedures of the other sector for the implementation of the process and the participants showed interest in knowing the work carried out by the other area of the company. Clearly, some weaknesses were evident in the first workshop: the dependence on an interview script, which needs to be detailed and precise to obtain quality answers; the low motivation in answering questions that require more dialogue; the difficulty in explaining processes from memories and experiences; the lack of interactivity between participants, which stimulates participation and reasoning through information sharing. At the end of each mapping workshop, a questionnaire was sent out to evaluate the respective stage. These results will be discussed in Sect. 5.3.

The second process mapped was the Purchasing process (one of the most complex), following Meet2Map Framework. The team identified all phases of the process, from the demand to purchase, quotation, approval, payment to delivery. The workshop spent 3:41 h. All participants (2 leaders of the Supplies area, one supervisor of the Accounts Payable and Receivable site, all of them with 2–4 years of work in the company) interacted actively providing information about their activities, this time generating an initial process design, as illustrated in Fig. 3. A clear picture of this process is available in [7].

All participants described the activities of their department, as well as their activities. Despite the longer duration, none of the participants felt unmotivated or disinterested. Before each modeling workshop, the researcher sent the team an email, attaching the preliminary questionnaire link.

Data Collection. A questionnaire composed of open and closed questions was used to collect data on the participants' perceptions about the framework application, organized into three groups: G1) information provided, G2) modeling activity, and G3) collaborative tool. Online questionnaires were applied to all participants in anonymous mode and preserving the anonymity of the participants. It is important to point out that the same

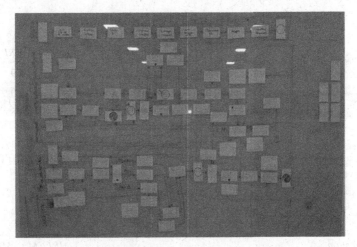

Fig. 3. Purchase process model (Source: [6]).

questionnaire was used for both situations, the processes modeled using the collaborative modeling tool and the traditional modeling through structured interviews.

The analysis approach was based on two ways of organizing results: Unified evaluation, in which all answers to the same question were grouped and evaluated uniquely; Distinct evaluation, in which the answers were separated into two groups, referring to collaborative modeling and traditional modeling. The goal was to assess if there is significant variation between the techniques. Additionally, the data is qualitative but uses a quantitative value scale to analyze the data obtained from the closed questions of the questionnaire. The closed questions verified the degree of influence of the factors involved in the modeling activity through the Likert scale. Thus, quantitative responses replaced the qualitative scale: I totally disagree (1), I partially disagree (2), Neither agree nor disagree (3), I partially agree (4) and, I totally agree (5). The Mean Ranking Index (MRI) was used to analyze the items answered in each question. MRI calculated the weighted average for each item, based on the frequency of responses as presented in the formula:

Mean Ranking Index (MRI) = WA/(NS), where WA (Weighted Average) = Σ(fi.Vi);
fi = Observed frequency of each response for each item; Vi = Value of each answer;
NS = Number of subjects

Results. We received the answers from 10 participants. Tables 2, 3, and 4 present the MRI for the questions of each group. For the first group (G1), as shown in Table 2, the results were considered positive, emphasizing the need for efficient communication proposed by the framework. The value of question Q1a (4.9) confirms that the information presented by the analyst was enough to guide the participants. That information consisted of the activity goals, introduction to BPM and BPMN. This reinforces the importance of the framework, which proposes the alignment of knowledge and goals to ensure the success of the modeling. The values of questions Q1b (5.0) and Q1c (4,8) emphasize the ability of the analyst to inform the concepts.

Table 2. MRI for Group of Questions 1 (Source: [6]).

ID	Question	MRI
Q1a	The information was sufficient to carry out the activity	4.9
Q1b	The information was presented in a straightforward way	5.0
Q1c	The information was presented objectively	4.8

The results in Table 3 compare the responses between the traditional (TM) and collaborative modeling (CM) teams, separately.

Table 3. MRI for Group of Questions 2 (Source: [6]).

ID	Question	MRI CM	MRI TM
Q2a	I felt motivated during the activity	5.0	4.0
Q2b	I would participate in the activity again for modeling another process	5.0	5.0
Q2c	This activity helped me to better understand the process	4.6	4.8
Q2d	This activity can really help me improve the current process	4.8	4.8

The second group of questions corresponds to the perception of the participants about the modeling activity. The results indirectly reflect the benefits of the framework in providing a good roadmap according to the following analysis. The values of questions Q2a (5.0/4.0) and Q2e (4.8/4.0) reinforce the benefits of the collaborative modeling tool since there is a significant difference between the values for the traditional and the collaborative modeling. This result is in line with the proposal of the tool and design principles. The values of questions Q2b (5.0/5.0), Q2c (4.6/4.8), and Q2d (4.8/4.8) support the perception that the activity was performed properly, showing that the participants in both cases were motivated and willing to participate in more activities of the type. The difference in the values of the Q2d was not significant, and thus, it is not possible to compare them.

The third group of questions (G3) corresponds to the perception of the participants about usability and utility aspects of the technique, as shown in Table 4.

The value of Q3a (5.0/4.4) presents the most significant result for this group, since it indicates, even with only a small difference, the participants' perception that the modeling technique with the collaborative modeling tool is better when compared with the traditional one. Although the value of Q3c (5.0/5.0) presents an equal result, it shows that all the participants agreed that people would learn to use the technique without problems. It was also a consensus that the BPMN notation is suitable for process modeling, this statement can be interpreted from the values of questions Q3d (5,0/5,0) and Q3e (1,2/1,2). The value of question Q3f (2.6/2.6) expresses those participants agreed that the presence of an experienced process analyst is required to perform the mapping activity. The values of questions Q3g (1,0/1,2), Q3h (4,8/4,6) and Q3i (5,0/4,8)

Table 4. MRI for Group of Questions 3 (Source: [6]).

ID	Question	MRI CM	MRI TM
Q3a	In my opinion, this technique is fully adequate for process modeling	5.0	4.4
Q3b	I felt confident in using the technique	4.0	4.8
Q3c	I agree that most people would learn to use this technique without problems	5.0	5.0
Q3d	The notation adopted is adequate for modeling processes	5.0	5.0
Q3e	The notation used is overly complicated to use	1.2	1.2
Q3f	I need the support of a technician to be able to use the technique	2.6	2.6
	The technique does not encourage the contribution of important observations	1.0	1.2
	I was enthusiastic about the technique used for process modeling	4.8	4.6
	This technique aided to extract my knowledge about the process	5.0	4.8

underpin the importance of the technique, and although they present better values for the collaborative modeling technique, the difference was exceedingly small.

This study confirmed the importance of preparing the execution of the modeling task, allowing the identification of critical success factors that the process analyst must know. In addition, strategies to promote the team's engagement during the activity helped the improvement of the collaborative modeling tool that transforms the domain analyst into the agent responsible for modeling the process.

6.2 Case Study 2

The second case study was conducted in a Brazilian state judiciary institution. The institution was implementing the BPM Office (BPMO) with the support of a consulting firm. We applied the framework proposed to support the modeling phase of two processes defined as the pilot project within the implementation of the BPMO.

Process and Planning. The pilot project of the BPMO were conducted by the consulting firm using two processes to judge incidents of repetitive demands, which are important instruments for the standardization and speed of decisions in the courts of justice. During the planning of the AS IS modeling phase, it was noted the need to use techniques that would support the collection of fundamental information so that process analysts could carry out their activities more efficiently, involving different process stakeholders. It was then suggested to use the Meet2Map Framework, whose process, tools, and artifacts were presented to the BPMO team.

After approval, discussions about who would be the participants for the workshop began. To select and divide participants into groups, the BPMO and consultancy team considered: (1) units involved; (2) phase of the process in which the unit is involved; (3) position held (magistrates, technicians occupying a managerial or operational role); and (4) ICT knowledge.

The chosen process permeates several roles and areas of the institution of justice, having identified the need for the presence, in addition to the stakeholders, of appellate judges, judges, representatives of senior management, the precedent management sector and the ICT area, in addition to the core of distribution and of the coordination responsible for the system that handles judicial processes.

To recognize the phases of the process, the activities previously identified through the study of legislation and interviews with stakeholders were taken into account, as well as the possible milestones reached. Thus, the following phases were defined: Application, Admission and Judgment.

Thus, the participants were divided into three groups, according to the definition of the phases, seeking a balance between the members of the judiciary and the technical professionals present, whether they held management positions. In addition, participants from the ICT area were distributed, leaving one representative in each group. Because the other participants had little or no knowledge in the area of BPM, the intention was that the ICT servers could, with a systemic view, strengthen the understanding of the activities to be performed during the workshop.

The strong hierarchy of the institution was a worrying factor, as it might have brought risks to the success of the event. Aiming to minimize this risk, it was defined that there would be a process analyst responsible for supervising each group. A magistrate in the role of sponsor of the project accepted the task of moving between the tables, aiding, and giving the necessary support. A process analyst at the consultancy was also given the same responsibility. Another action to minimize the high formality of the judiciary environment was to choose a place outside the institution, suitable for group work and the workshop with Mee2Map.

The workshop lasted approximately 2:45 h (two hours and forty-five minutes) and was divided into three moments: initial orientation, modeling in groups and consolidation of modeling. In the initial orientations, the participants received instructions about the pilot project, the reasons for being invited, the objectives of the event, what was not part of the scope of activities, basic information about BPM and process modeling, as well as explanations about its use. of Meet2Map materials and artifacts.

On site, in addition to the artifacts, sticky note pads, pens, markers and erasers were made available. The tops of the tables on which the groups worked were of the glass blackboard type, which facilitated the distribution of artifacts, the design of connectors and the writing of information. Participants had approximately one hour to model their respective phase of the process (Application, Admission and Judgment), as illustrated in Fig. 4. After that, the tables on which the participants worked were brought together, aligning the lanes of the models produced by the group, as depicted in Fig. 5. Each group had the opportunity to present their respective modeling, in order of occurrence. The other participants were able to contribute their perspectives. Process analysts recorded observations in workshop notes and on sticky notes so that they were visible to everyone in attendance.

Data Collection. The evaluation of the activities carried out, the thirteen (13) participants were asked to answer a questionnaire, divided into two types: closed questions (Likert scale) and open questions. In the open questions, participants were invited to indicate the positive points and points for improvement of the event. In turn, the closed

Fig. 4. Participants modeling the process during the workshop.

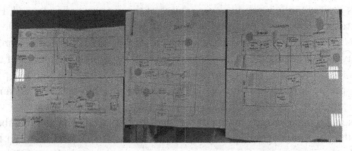

Fig. 5. Parts of the process brought together.

questions were divided into 5 groups of questions, each one intended to evaluate, respectively: (1) whether the information provided was sufficient to carry out the activities, (2) the chosen team (guests), (3) the activity performed (AS-IS modeling), (4) the adequacy of the adopted technique, and (5) the performance of the process analysts. The questionnaire was based on the work of Dutra [7], with the order of the groups inverted and only one question adjusted, regarding the chosen guests (Q2d).

The proposed answers to the closed questions, and their respective quantitative values, were: Totally agree = 1; Partially agree = 2; Neither agree nor disagree = 3; Partially Disagree = 4; and Strongly Disagree = 5. Thus, it was possible to measure the degree of agreement - or disagreement - of the participants with the statement. The Average Ranking was calculated using the following formula:

Mean Ranking Index (MRI) = WA/(NS), where

WA (Weighted Average) = Σ(fi.Vi);

fi = Observed frequency of each response for each item;

Vi = Value of each answer;

NS = Number of subjects.

Results. Twelve (12) responses were obtained. The results are presented in Tables 5, 6, 7, 8, 9 and 10. Regarding the closed questions, the average ranking presented in Table 5 shows that the information explained by the analysts in the initial orientations was considered by the guests as sufficient, clear, and objective. It is important to clarify to the participants the objectives of the workshop, the fundamentals of the BPM discipline and how to use the available materials.

Table 5. Average Ranking of Group 1 questions.

ID	Question	MRI
Q1a	The information was sufficient to conduct the activity	4.58
Q1b	The information was presented in a straightforward way	4.92
Q1c	The information was presented objectively	5.00

From the answers provided to question Q2a, as described in Table 6, we conclude that participants invited to the workshop were aware of the process to be mapped. However, five of the respondents only partially agreed with the statement. It is possible that this opinion was influenced by the fact that, until the present moment, the modeled process has low maturity, being conducted in an ad-hoc way, with no defined sequence of activities that should be carried out. The group responsible for modeling the Request phase, for example, found it difficult to define which event starts the sub-process, as can be seen in the representation recorded in Fig. 5.

We notice, by the average ranking of the questions about the chosen members (Q2b and Q2c), that there was good quality participation and interaction between the team members. In the responses to Q2d, ten participants agreed that all the people necessary for a complete understanding of the process were invited, with only one participant partially agreeing, and another remaining neutral (I neither agree nor disagree).

Table 6. Average Ranking of Group 2 questions.

ID	Question	MRI
Q2a	The participants knew the process	4.58
Q2b	The team was committed	4.92
Q2c	There were no conflicts among the team members	5.00
Q2d	Everyone who should have been in this event attended	4.75

As shown in Table 7, when evaluating the activities carried out, the respondents were unanimously satisfied with the workshop, having stated that they felt motivated and involved. The event was considered important for understanding AS-IS and enlightening for a future meeting (TO-BE modeling). They also stated that they would participate in another event with the same objective and tools used.

From the answers presented in Table 8, whose questions evaluated whether the technique used was adequate, we observe that Meet2Map Framework allowed the guests to feel confident, considering it as fully adequate for process mapping. The answers to questions Q4e and Q4f demonstrate that the BPM notation was considered simple by the participants, it can be understood and executed without a strong intervention of the process analysts. In Q4g and Q4i, we notice that, in addition to facilitating the externalization of important observations, the participants felt that the technique facilitated the

Table 7. Average Ranking of Group 3 questions.

ID	Question	MRI
Q3a	I felt engaged during the activity	5.00
Q3b	I would participate again in the activity to map another process	5.00
Q3c	This activity helped me to understand better the process	5.00
Q3d	This activity can really help to improve the process	5.00
Q3e	The activity was absorbing	5.00

extraction of their own knowledge about the process. Moreover, in the answers to Q4h, we see that everyone was keen on using Meet2Map.

Table 8. Average Ranking of Group 4 questions.

ID	Question	MRI
Q4a	In my opinion, this technique is adequate to mapping processes	4.92
Q4b	I felt confident to use the technique	4.83
Q4c	I believe most people would learn to use this technique easily	4.75
Q4d	The notation used is adequate to model processes	5.00
Q4e	The notation used is complicated to be used	2.42
Q4f	The support of a specialist is necessary to use the technique	3.83
Q4g	The technique does not foster important contributions	1.67
Q4h	I was enthusiastic about the technique used to map the process	5.00
Q4i	This technique allowed to extract my knowledge about the process	4.92

Finally, concerning the evaluation of process analysts' resourcefulness, Table 9 shows that they conducted the activity well, demonstrating experience in mapping processes and knowledge in the application of the technique.

Table 9. Average Ranking of Group 5 questions.

ID	Question	MRI
Q5a	I noticed the analysts' experience in process mapping	4.92
Q5b	I noticed that the analysts conducted well the activity	4.92
Q5c	I noticed that the analysts were knowledgeable of the technique	4.92

The results of the open questions, Table 10 summarizes the points raised by eight responses. Figure 6 depicts the main cited words (in Portuguese).

Table 10. Frequency of words used in citations of positive points.

Word	Frequency
Engagement	3
Improvement/Improve	3
Engaging/Involvement	3
Team	3
Innovative	2
Participative	2
Learning	2
Objectivity	2
Sectors	2
Efficient	1

Fig. 6. Positive points word cloud.

The statements of the participants confirmed that Meet2Map promoted the engagement, participation, and involvement. The event was considered interesting. The technique was understood as innovative, organized ("*it has a beginning, middle and end*"), objective, fast and efficient. It was also mentioned that there was a learning opportunity and that it helped to understand the points of improvement and how to achieve such improvements.

As for the points for improvement, six responses were received, but one of them did not indicate suggestions, she/he only considered the event as positive overall. Three participants understood that the workshop time should be extended, reinforcing the understanding that the technique was engaging and exciting. One guest suggested expanding the areas involved and another suggested using posters for process modeling, instead of tables, as she/he thought they would be a limiting factor.

Overall, based on the results, we could realize that Meet2Map was suitable for modeling processes in an institution with low maturity in BPM, having been considered by the participants as an innovative technique, which stimulates engagement, and which allows the "extraction" of knowledge.

7 Discussion

This study has limitations and threats common to all qualitative research: dependence on the application context and lack of repeatability. In addition, although action research has many advantages because it is applied, empowering, collaborative and democratic, it also has its flaws, especially in three important aspects: the subjectivity of the researcher, who may be over-involved or interpret results only under his perception; the influence of power or lack of it on the part of the researcher in the organizational hierarchy, who in the role of subordinate or leader may impact the results; and the time-consuming process, considering an authentic types of project presented and their complexity.

To mitigate these limitations and threats, this study first sought to define the research in two cycles. The first cycle sought to understand the problem better and define the research objectives based on actual experimentation with an existing proposal. In both cases, the multidisciplinary nature of the projects reduced the biases. As an IT manager, the researcher has a position that facilitates the articulation between other business area managers, but without power influence. Still, the survey involved several data collection tools, such as questionnaires and interviews, preserving the original data to avoid biased interpretations. To evaluate results, the proposal to use a qualitative questionnaire with the numerical value scale (Likert scale) sought to attenuate the subjectivity of the analysis. In addition, the collaborative process modeling included review and corroboration steps for all involved, avoiding misunderstandings and new biases. Finally, this work has some limitations that prevent an in-depth analysis of the framework's benefits. Nevertheless, even with these limitations, the results shown that Meet2Map can contribute to real-word BPM projects. We also highlight general lessons learned from both case studies, structured as Good Practices as following:

Phase 1 – Preparation

- *Select the environment*: It is important to select a distraction-free, attractive, comfortable, and stimulating environment for group work. The appropriate environment can help participants stay motivated with BPM project activities and provide greater focus on project discussions without the demands and distractions of the work environment.
- *Process analysts*: It is recommended to have at least two analysts, because while one conducts the mapping, the other takes note of all the details about the process and helps with questions. These functions do not need to be rigid, but it is important to be aligned in advance for the performance of each analyst.
- *Define information collection standards*: the standards must be defined to determine what information will be collected, by whom, how it will be validated, how it will be stored and organized, how it will be updated and how it will be used.

- *Identify areas, functions and participants in the process*: To choose the participants, it is important to: Assemble a heterogeneous team, with participants from the main areas and functions, to provide different points of view of the process; Choose participants who complement the knowledge about the process; Choose participants who are motivated and communicative; Teams of 2 to 4 participants are recommended; Invite participants with managerial and operational views; Prefer participants who work longer in the process; If the modeling involves several teams, ensure that at least one of the members is familiar with IT or that some IT specialists can support the teams.
- *Engagement of process participants*: The areas involved in the process need to be aware of the work to be done and the availability of time required. It is important to bear in mind that the chosen team must be willing to actively participate in the mapping and modeling activity. In this way, the chosen participants need to be people not only willing to verbally expose their knowledge about the process, but also motivated to collaborate directly in the construction of the model.

Phase 2: Introduction

- *Remember organizational goals*: Reinforce the business objectives of the BPM project and the importance and relevance of guest participation to the project's success.
- *Aligning objectives:* Align the organization's objectives with the accomplishment of the work.
- *Aligning knowledge*: Introduce participants to the essential concepts to perform the task of AS-IS process modeling: Introduce the basic concepts of BPM; Introduce the basic concepts of process modeling and BPMN.
- *Present the Meet2Map*: Present the methodology that will be used for the mapping activity, techniques and tools used;
- *Time control*: The suggestion is that this phase does not last more than 20 min, not compromising the objectivity of the mapping.

Phase 3: Macro Process Modeling

- *Identifying macro activities*: In this phase, the main characteristics and macro tasks of the process must be identified. This information can be anticipated via a form sent before the workshop, with key questions about the start and end events, customers, suppliers, systems, and artifacts linked to the process.
- *Presentation of the macro process*: With the information collected previously by the analyst, he can model the macro process at the top of the whiteboard. At this point, the macro process is mapped and based on it, the deployment of each macro task will begin, to find more detailed tasks, responsible parties, interactions, transitions, rules, and process events.

- *Initial kick-off:* Additionally, the analyst can also draw the process pool and the swim-lane of the first identified participant-function and, right after, position the process initiation event, stimulating the beginning of the process detailing.

Phase 4: Detailing of the AS-IS Process

- *Detailing the process:* After modeling the macro process, participants should be encouraged to conduct the process detailing activity from the macro process. For this, the top-down design technique can help discover more details about the macro task, because with it each task of the macro process is questioned until the desired level of detail is collected.
- *Stimulating participation:* It is necessary for the analyst to be able to promote the active collaboration of the participant to extract the maximum amount of relevant information and insights about the process. The process analyst can also intervene by asking questions about the process being modeled by the participants, as a way of guaranteeing the correct use of the modeling concepts.
- *Supporting the modeling:* Whenever necessary, the analyst can introduce a more specific element of BPMN, to further enrich the process. For this, he must adapt the base element (example: transforming a basic task into a user task or manual task), drawing the symbol corresponding to the more specialized task. It is important that the presentation of the concept must be made to the user, and their understanding must be confirmed, so that the knowledge can be replicated in another need.
- *Do not make assumptions:* As familiar as the process seems to the process analyst, it is particularly important that he does not make assumptions about the behavior of the process. This can cause bias in the participant's response and compromise the fidelity of the modeled process. One suggestion is to ask more open-ended questions, for example: Instead of asking "Upon receipt of the proposal, the user must submit it for management approval, right?" can be asked "What should be done after receiving the proposal?". There is a subtle difference between the two questions, but one that can completely compromise the respondent's response. In the first question, the user can be induced to answer only that it is correct, without thinking about other actions that may occur, while in the second form the user is led to answer the procedure following the receiving activity.
- *Register points of attention and improvements:* Throughout the activity, points of attention in the process and points of failure must be identified and pointed out in the process model, but it is very important that the process improvements are not addressed in this activity, to avoid that the participant contaminates the As Is process with insights about improvement that do not reflect the actual process performed. For this, problems can be signaled directly in the process using sticky notes that point to the point of the problem that has the process.

Phase 5: Evaluation of the Modeled Process

- *Validation of the process with the participants:* Once the process has been detailed, the analyst must validate the modeled process, for this, the process analyst can ask the participants to verbally summarize the modeled process, from beginning to end, with the objective to verify that no part of the process was left unmapped.

- *Validate the objectives*: It is also important to refer to the mapping objectives, to validate whether the details of the process meet the defined objectives.
- *Checking the completeness of information*: It is important to ask if there is any other information about the process that the participant would like to share.
- *Workshop Feedback*: After the workshop is over, the process analyst can submit a form to collect feedback from participants about the workshop.

8 Conclusion

BPM projects are constantly challenged by their characteristics and impacts on an organization's business processes, transforming the activities of people who are not always prepared for future changes. In this scenario, BPM can be used as an appropriate initiative to identify current processes, analyze bottlenecks, delays, and waste and based on the understanding of problems, enable the implementation of future improvements. However, in an organization with low BPM maturity, where few or no processes are standardized and documentation on processes is insufficient, the process mapping step becomes quite challenging. In addition, the improvement of complex projects involves the need to consolidate and integrate a lot of information, involving diverse perspectives of several process stakeholders, which, in individual interviews, is very difficult to analyze and integrate. As a result, people involved with business processes at various organizational levels (users, managers, customers) represent the primary source of information and, therefore, a critical factor in a BPM project.

With the motivation to support BPM projects in this scenario, this study defends the use of the Meet2Map framework. This framework comprises 1) an easy-to-use modeling tool, which includes some elements of BPMN notation; 2) a macro process based on design principles to support the collaborative modeling of processes; 3) suggestions of artifacts and templates to support the modeling activity. From two case studies in real BPM projects, it was possible to obtain evidence of the acceptability of the framework, with very positive feedback on its approach. The evaluations of the Meet2Map application in these studies also showed that the collaborative modeling of processes, involving the process analyst together with the process stakeholders, allowed a detailed view of the process and promoted the discussion, activities discovering, and the emergence of insights that, in a traditional process based on individual interviews, would hardly happen. The main contribution of this research is a prescriptive approach to process modeling in the context of BPM projects, which includes a proposal for discovering and modeling As-Is processes in a collaborative, engaging, and motivating way. This proposal stimulates participants in this activity to contribute to understanding processes and their improvement needs, keeping them, from the initial stages of the BPM cycle, aware and responsible for the organizational changes that will come with the improvements. From a scientific point of view, research advances in solutions for human-centered process modeling, encouraging that the As-Is modeling activity can obtain very positive results by involving the stakeholders in the process (users, process owners, and managers) together with the process analyst. From a practical point of view, it is expected that the application of the framework and its best practices can improve the process modeling task under diverse and rich perspectives, preparing stakeholders for future changes and supporting the following stages of the BPM lifecycle.

In future work, we intend to conduct case studies in more organizations and projects, extending the collaborative characteristic of Meet2Map to other steps of the BPM cycle. Moreover, we will extend the framework to incorporate the suggestions made by the participants. We will also study combining the user-centered techniques adopted with automated discovery to enrich the models even more.

References

1. Adamides, E.D., Karacapilidis, N.: A knowledge centred framework for collaborative business process modelling. Bus. Process. Manag. J. **12**(5), 557–575 (2006)
2. Antunes, P., Simões, D., Carriço, L., Pino, J.A.: An end-user approach to business process modeling. J. Netw. Comput. Appl. **36**(6), 1466–1479 (2013)
3. Baldam, R., Valle, R., Rozenfeld, H.: Gerenciamento de processos de negócio - BPM: uma referência para implantação prática. Elsevier, Rio de Janeiro (2014). (in Portuguese)
4. Brown, T.: Change by Design: How Design Thinking Transforms Organizations and Inspires Innovation. Harper Business, New York (2009)
5. Dumas, M., Rosa, M.L., Mendling, J., Reijers, H.: Fundamentals of Business Process Management. Springer, Heidelberg (2013). https://doi.org/10.1007/978-3-662-56509-4
6. Dutra, D.L., Santos, S.C., Santoro, F.M.: ERP projects in organizations with low maturity in BPM: a collaborative approach to understanding changes to come. In: Filipe, J., Smialek, M., Brodsky, A., Hammoudi, S. (eds.) Proceedings of the 24th International Conference on Enterprise Information Systems, ICEIS 2022, Online Streaming, 25–27 April 2022, vol. 2, pp. 385–396. SCITEPRESS (2022)
7. Dutra, D.L.: Um Framework para Mapeamento de Processos As-Is apoiado por Design Thinking (A framework for mapping As-Is processes supported by design thinking). Master Dissertation, Centro de Informática, UFPE, Brazil (2015). (in Portuguese)
8. Davenport, S.L.J., Beers, M.C.: Improving knowledge work processes. TH Sloan Manag. Rev. **37**(4), 53 (1996)
9. Davis, A., Dieste, O., Hickey, A., Juristo, N.: Effectiveness of requirements elicitation techniques: empirical results derived from a systematic review. In: 14th IEEE International Conference Requirements Engineering (2006)
10. Cereja, J.R., Santoro, F.M., Gorbacheva, E., Matzner, M.: Application of the design thinking approach to process redesign at an insurance company in Brazil. In: vom Brocke, J., Mendling, J. (eds.) BPM Cases (2017)
11. Forster, S., Pinggera, J., Weber, B.: Collaborative business process modeling. In: EMISA 2012, pp. 81–94 (2012)
12. Jeston, J., Nelis, J.: BPM practical guidelines to successful implementations (2006)
13. Van Looy, A., Poels, P.: A practitioners' point of view on how digital innovation will shape the future of business process management: towards a research agenda. In: 52nd Hawaii International Conference on System Sciences, Maui, pp. 6448–6457 (2019)
14. Johansson, B., Ruivo, P.: Exploring Factors For Adopting ERP as SaaS. Procedia Technol. **9**(2013), 94–99 (2013)
15. Lechesa, M., Seymour, L., Schuler, J.: ERP software as service (SaaS): factors affecting adoption in South Africa. In: Møller, C., Chaudhry, S. (eds.) CONFENIS 2011. LNBIP, vol. 105, pp. 152–167. Springer, Heidelberg (2012). https://doi.org/10.1007/978-3-642-28827-2_11

16. Luebbe, A., Weske, A.: The effect of tangible media on individuals in business process modeling. Technische Berichte Nr. 41. University of Postdam, Germany (2012)

17. Merriam, S.B., Tisdell, E.J.: Qualitative Research: A Guide to Design and Implementation, 4th edn. Jossey-Bass, Hoboken (2015)

18. Miron, E.T., Muck, C., Karagiannis, D.: Transforming haptic storyboards into diagrammatic models: the Scene2Model tool. In: 52nd Hawaii International Conference on System Sciences, Maui, pp. 541–550 (2019)

19. Recker, J.: From Product Innovation to Organizational Innovation – and what that has to do with Business process management. BPTrends **9**(6), 1–7 (2012)

20. Patton, M.Q.: Qualitative Research & Evaluation Methods, 3rd edn. Sage Publications, California (2002)

21. Persson, A.: Enterprise modelling in practice: situational factors and their influence on adopting a participative approach. Ph.D. thesis, Dept. of Computer and Systems Sciences, Stockholm University (2001)

22. Rittgen, P.: Collaborative modeling - a design science approach. In: Proceedings of the 42nd Hawaii International Conference on Systems Sciences (HICSS-42), 5–8 January, Wailoloa Big Island, Hawaii, USA (2009)

23. Rittgen, Peter: Success factors of e-collaboration in business process modeling. In: Pernici, Barbara (ed.) CAiSE 2010. LNCS, vol. 6051, pp. 24–37. Springer, Heidelberg (2010). https://doi.org/10.1007/978-3-642-13094-6_4

24. Sedera, D., Gable, G., Chan, T.: Knowledge management for ERP success. In: Proceedings of Seventh the Pacific Asia Conference on Information Systems, Adelaide, pp. 1405–1420 (2003)

25. Sedera, W., Gable, G., Rosemann, M., Smyth, R.: A success model for business process modeling: findings from a multiple study. In: Proceedings of the 8th Pacific-Asia Conference on Information Systems, Shangai, China (2004)

26. Seethamraju, R.: Adoption of Software as a Service (SaaS) Enterprise Resource Planning (ERP) Systems in Small and Medium Sized Enterprises (SMEs). Springer, New York (2014)

27. Taube, L.R., Gargeya, V.B.: An analysis of ERP system implementations: a methodology. Bus. Rev. Cambridge, **4**(1) (2005)

The Factors of Enterprise Business Architecture Readiness in Organisations

Tiko Iyamu[1](✉) and Irja Shaanika[2] (iD)

[1] Faculty of Informatics and Design, Cape Peninsula University of Technology, Cape Town,
South Africa
IYAMUT@cput.ac.za
[2] Department of Informatics, Namibia University of Science and Technology, Windhoek,
Namibia

Abstract. Before the deployment of enterprise business architecture (EBA), readiness assessment is often not conducted. Lack of readiness assessment results to fear and lack of know-how in the deployment and practice of the concept, and factors of influence are unknown. Consequently, the deployment and practices of the concept remain slow, despite the growing interest from both academics and enterprises. This chapter, through the case study approach, using five organisations, examines the factors that influence the readiness of EBA in organisations. The interpretivist approach was employed in the study and the hermeneutics approach is followed in the analysis.

Keywords: Business architecture · Readiness assessment

1 Introduction

Enterprise business architecture (EBA) is a domain of enterprise architecture (EA), which covers the non-technical activities of an organization [1]. Other domains of the EA are information, application and technology [2]. The focus of this study is to present factors of EBA readiness assessment in an organisation. AL-Ghamdi and Saleem explains how the concept of EBA deals more with business processes and modelling than with the technical and technological aspects [3]. The EBA is derived from the business strategy and it is mainly concerned with human resources, business processes and rules [4]. Business architecture can be described as a strategic tool that enables organisations, to drive business operations and determinants for information technology (IT), for competitiveness purposes. These facets, strategic, operationalisation and process model have attributed to the reasons why organisations (or enterprises) show interest and emphasis on the concept.

Furthermore, the deliverables of EBA are said to inform the design and development of other architectural domains, which include information, application and technology. AL-Ghamdi and Liu argued that even though EBA's focus is on business process, it eventually gets incorporated with the technical infrastructure, data architecture, hardware and software of the organisation [5]. Thus, EBA provides a roadmap for aligning

J. Filipe et al. (Eds.): ICEIS 2022, LNBIP 487, pp. 379–390, 2023.
https://doi.org/10.1007/978-3-031-39386-0_18

business needs with IT infrastructures. This aspect of EBA enacts the fact that business environments should not be studied in isolation but through a context [6]. The practice of EBA provides the context that allows for a better understanding, performance, and control of business operations [6]. Organisations that have implemented EBA are expected to reap benefits such as strategic alignment, customer-centric focus and faster speed to market [7].

However, many organisations have not been able to implement the concept of EBA. As a result, they lose out on the benefits, which would have fostered their competitiveness. The lack of implementation of the concept can be attributed to two main factors: (1) there are not many cases, which limit references and learning from practice. Consequently, it makes some organisations reluctant in embarking on the process [8]; and (2) many of the organisations that have implemented or attempted to implement the concept have failed in realising or articulating the benefits [9]. These factors are because of a lack of readiness assessment [10].

Given the strategic significance of EBA, there has always been a need for assessment, to determine an organisation's readiness for the implementation of the concept. Unfortunately, there seems to be no readiness assessment models tailored for EBA that can be used to guide this process. Many of the assessment models found in the literature focus on enterprise architecture (EA) as a whole and not on EBA as a domain [10]. A study by [11] developed an enterprise architecture implementation and priority assessment model comprising 27 assessment criteria. A model, based on an analysis of 9 factors and 34 indicators to assess organisations readiness when implementing EA is proposed [12]. Due to the lack of EBA assessment models, organisations find themselves deploying EBA even when the environment is not fit. The assessment models are critical as they enable organisations to determine the extent, they are ready before practising EA concepts and gaining a better understanding of the existing gaps [12].

This chapter is structured into six main sections, sequentially. The first section introduces the chapter. The section that follows presents a review of literature, which tries to unpack the gap that exists in the terrain that this study focuses upon. Next, the methodology that was applied in the study is discussed. Analysis and findings from the validation are presented in the fifth section, thereafter, a conclusion is drawn in the last section.

2 Literature Review

The development of EBA is a primary activity that needs not to be under-represented in the enterprise architecture program. As a management approach enterprise architecture aims to drive the alignment between business units and the supporting information technology in organisations [6, 13]. Without the business architecture, it becomes impossible to achieve the alignment between business and IT. Hence, EBA is the fundamental structure of the enterprise architecture responsible for defining the business strategy, governance and key business process [14]. The business architecture is a key component of the enterprise architecture development as it enables organisations to better articulate business objectives, goals and processes, which are - critical input for other architectural domains [15]. According to Caceres and Gómez, implementing EBA allows for a comprehensive vision of the organisation, and provides a tool that allows them to align strategic business objectives with applications and technological infrastructure [16]. Despite the benefits of EBA, there are persistent challenges.

Like other architectural domains, EBA is not immune to challenges. EBA encounters both technical and non-technical challenges, and how they manifest from the environment in which the domain is implemented and practised. When it comes to EBA practice strategic alignment remains a key challenge [17]. The strategic alignment aspect is mainly concerned with the fitting of business processes, objectives and organisational structures [15]. The challenges relating to the strategic fit are a result of dynamic markets that are often demanding innovation. Due to the constant changes in business environments, the business architecture is always under pressure to be responsive and ensure the changes are managed and controlled towards the organisation's strategic intent.

On the other hand, EBA is viewed to be abstract and lacking tangible delivery [18]. There is a report that some organisations believe that EBA is an abstract concept requiring significant investment but whose benefits cannot be proven [19]. As a result, the value of business architecture remains poorly understood in many organisations [20]. A clear understanding of the value and impact of EBA is critical for all different components of organisations as it drives their existence. The challenges of EBA impend organisations from exploiting the benefits of business architecture. Thus, the need for organisations to adopt a novel approach such as readiness assessments for EBA before its implementation.

EBA forms the foundation for other domains of architecture, namely information, application and technology, thus putting it in the most critical position in enterprise architecture development [21]. Due to its foundational role in enterprise architecture development, organisations need to conduct business architecture readiness assessments before its implementation. A readiness assessment is a benchmarking process that assists organisations in identifying potential blockages to the effectiveness of new system implementation [22]. Conducting a readiness assessment provide organisations with an opportunity to measure the amount of their readiness and if not ready, to better understand the issues and address them before implementation [12]. Implementing enterprise architecture is a complex and costly process and thus the need for readiness assessment for organisations to ascertain their resources before its implementation.

Consequently, when readiness assessment is carried out at the implementation level, it restricts product changes and generates higher costs [23]. Historically, unanticipated challenges contribute to implementation failure [24]. Therefore, by doing a readiness assessment organisations are taking stock of all relevant factors that have the potential to impact the successful adoption of an innovation in any way [25]. Broadly, business architecture is about the organisations operating model, strategy and human capabilities and by conducting a readiness assessment, the organisation reduces uncertainty and improves planning in the prioritization of the required resources [23].

3 Methodology

The qualitative method was employed in the study because of its focus on quality rather than quantity [26]. Thus, the method was most appropriate because, the aim is to understand the factors of readiness of EBA, which is beyond a 'yes' or 'no'; 'true' or 'false' type of event. The method is well documented, rationalised and is increasingly being

used in information systems (IS) research. It therefore, does not necessarily need introduction and explanation in the IS context – see [27, 28]. The qualitative method is applied
in this study primarily because the objective deters knowing insights in the rationalities
of the participants.

Given the aim of this study, the case study approach was suitable. A total of five
South African-based organisations partake in the study, to gain insight into the factors
that influence EBA reading. A preliminary question was used in selecting organisations
for the study. The organisations were selected according to a set of empirical criteria
thought to be most useful to the objective of the study [29]. These are: (1) thirteen organisations were invited, eight agreed to partake; and (2) of the eight, five have successfully
implemented the business architecture. The five organisations were assigned pseudo
names because the organisations strongly opposed to disclosure of identity. The organisations are South Bank South Africa (SBSA); (2) Ocean Bank South Africa (OBSA); (3)
Green Insurance South Africa (GISA); (4) Government Administration (GASA); and
(5) Essentials Retailing (ARSA).

The data was analysed through hermeneutics following the interpretivist approach.
This allowed a two-phase approach. Phase 1: the written feedbacks were repeatedly read
in conjunction with the interviews' transcripts to comprehend how the EBA is applied.
Eisenhardt suggests that this is crucial in the analysis of qualitative data; and (2) the
sets of data from each of the organisations were also repeatedly read with the EBA, to
gain an understanding of conclusions that were reached in the application (testing) of
the model [30].

The analysis focuses on 3 main standpoints, which are: (1) to broaden the logic of
the EBA, to the understanding of employees in the organisations; (2) its translation is
a means through which relationship between actors are established, and understandings
are connected; and (3) it helps actions to be coordinated, and meanings are transmitted.

4   Analysis of the Data

The key factors or areas are usefulness, value, design and automation, and ease of
use. The factors require translation in ascribing them into actors for implementation
purposes. Translation exposes how the interests of actors change in the implementation
of technology or processes [31].

4.1   Factors of Readiness

Thus, it is the primary role of a business architect to have a holistic understanding of
the business direction, context and strategies when developing EBA. There are multiple levels of translation in the process of testing the model in the organisations. In
the first level, the components of the model were translated to the participants. This
was to help them decide on their participation and provide useful responses. At the
second level, the participants translated the components in the context of their organisations, to ensure relevance and fit. Some organisations view business architecture as an
interlinked with organisational goal and objectives towards value creation and competitiveness [17]. Consequently, one of the participants concluded as follows: "The model

clearly presents the factors of readiness and also outlines the weight associated with these factors" (GASA_02).

Business model design and product design differ in several theoretically meaningful ways. Hence translation of the components was critical. According to one of the participants, "the main value is that it helps to improve the capabilities to achieve the goals and objectives of business architecture. Also, it helps with design, to capture and address all elements related to customers service such as digitisation of its services (OBSA_02)". Unswervingly, the managers (lead participants) established themselves as obligatory passage points by directly enforcing testing of the model. Such action ensures common understanding among participating employees, which helps with corroboration of responses from the employees. In addition, some of the managers used the study as an opportunity to test employees' theoretical know-how about the concept of business architecture which they have inscribed in them as the organisations' embarked on the route to developing the business. From the responses and actions, some of the managers translated employees' buy-in into indispensable interest.

4.2   Rationale of Readiness

The readiness assessment helps an organisation to understand capacity of its resources towards improvement of the mandatory requirements for successful implementation of EBA. Hussein considers readiness assessment as the first step for adoption as it can be useful in identifying gaps and risks [32]. In successfully implementing business architecture, a method that detects and traces means and ends at the domain's level was needed. One of the participants briefly explains: "The value stream allows an organisation to document its processes and procedures, create and improve business objectives" (OBSA_05). The business architecture involves the conceptualization of organisational boundaries and defines flows of processes [33]. Through translation, the assessment model is understood as "useful because it helps the organisation to deliver end-to-end business value to its customer" (ARSA_03).

Organisations must be able to assimilate change, for purposes of value add and realisation. "The model is useful in that it helps in the overall assessment of enterprise-wide business architecture model, more so in the identification key areas when gathering requirements" (SBSA_03). The factors used for testing make the model suitable for the assessment of an environment toward readiness of EBA. The factors are fundamental for both present and future including potential changes. Peppard and Ward argue that environment evolve to a point where change emerges, which therefore mechanisms for assessment [34]. The position of the participants was that the assessment "... Helps a great deal as a guiding plan when building the matrix to assess EBA maturity level" (GISA_01).

4.3   Process of Readiness

During the test, translation occurred at stages that further allow us to analyse the proliferation of related networks (groups within organisations), to gain explicit fathom on why and how the EBA RAM was evaluated and accepted as readiness assessment mechanism in organisations: "the EBA model designed to help organisations build a better visual representation of their business environment" (SBSA_02); "The model is useful because it helps an organisation to capture and futuristically assess how business activities fit together, to serve the end-to-end stakeholders' needs" (ARSA_02).

The readiness assessment helps an organisation to identify various factors that can impede the successful operationalisation of the business architecture. The factors are both technical and non-technical and can be unique for each organisation. "The model is well constructed, and it is easy to interpret and used" (GASA_04). As a result, "it further helped on establishing the gaps on the current departmental enterprise architecture effort" (OBSA_02). By conducting EBA readiness assessment, organisations do not only identify the risks and potential challenges but also opportunities that might arise when EBA is implemented [35].

Fundamentally, translation reduces cognitive biases and strengthens the proposition to gain an understanding of how the model can improve the stability, usefulness and value of business architecture. "In my view, the model is useful because it provides better business definition for every area of business architecture deployment which can lead to effective and efficient business processes and technology solutions within the environment" (SBSA_01). Assessment requires reconciling means with ends through translation in which change is ascribed in the actors within the environment. Consequently, actors prepare for known and unpredictable changes that relate to both ends and the means. "The model allows organisation to build business capability which can add value to the development and implementation of the business architecture" (OBSA_03).

5   Discussion of Results

The EBA has been better theorised in literature rather than practice [36]. This study provides empirically validated model for organisational practice. Testing of the model provides practicable evidence for implementing EBA in organisations. The test focused on four fundamental factors, which are usefulness, value add, design and automation, and ease of use as shown in Table 1. Typically, these factors are indicators for IT and business improvement, risk mitigation [33], and alignment (or co-existence) with existing IT solutions [37]. The factors enact structures, operations, governance, and alignment of the EBA with the current environment. The EBA implies realistic construction of structures and operationalisation of alignment and translation of strategies toward implementation of the EBA in an organisation. The factors are discussed below, and they should be read with the Table 1, to gain a deeper understanding of their criticality.

The notions of usefulness, value, design and automation and ease of use were prevalent in the conversation and written responses from the participants. These were because of translation of the business architecture goal and objectives. "The model also allows the measurement and monitoring of the key performance indicators within the environment" (SBSA_02).

5.1   Usefulness of the Assessment Model

An object or system is deemed useful when it enhances the performance of activities towards achieving the defined goals. Individuals accept and use systems to the extent

Table 1. Test result of the EBA-RAM [38].

#	Org	Factor	Weight				
			1	2	3	4	5
1	SBSA	Usefulness					X
		Value adds					X
		Design and automation					X
		Ease of use				X	
2	OBSA	Usefulness					X
		Value adds				X	
		Design and automation				X	
		Ease of use				X	
3	GISA	Usefulness				X	
		Value adds				X	
		Design and automation				X	
		Ease of use					X
4	GASA	Usefulness			X		
		Value adds			X		
		Design and automation		X			
		Ease of use				X	
5	ARSA	Usefulness					X
		Value adds				X	
		Design and automation				X	
		Ease of use					X

they are better at addressing their needs. The EBA was considered useful by enforcing practicality in assessing organisations readiness for business architecture implementation. Also, because it helps to fortify implicit decisions in business processes, towards achieving the goals and objectives. Thus, determining areas of an organisation for the EBA focuses, to improve performance. The absence of this type of model has made it difficult for many organisations to understand the extent of complexity and the readiness nature of their environments. In addition, the usefulness of the model also comes from its generalization because it is not designed for a specific organisation as it is flexible and can be applied by different organisations wishing to implement the concept of business architecture.

The model is useful in providing guidance to both business and technology managers in assessing the environment to detect factors of influence in the deployment of EBA. The realisation of strategic fit within the business architecture is an important challenge for many organisations, which has not been actualised [17]. The test conducted in the 5 organisations proceed the model from a theoretical antecedent into practice.

Significantly, the model illustrates how to carry out the assessment process. The model, through the weight associated with each cell provides a valid reflection about the current business environment, enabling the identification of the existing gaps and the analysis of the efforts towards each factor. Also, it enables detects potential risks in business architecture' multi-faceted view of the organisation's key components. The test validates the gap between an organisation's blueprint and the real-world readiness and capabilities required to deliver EBA.

5.2 Value Adds to EBA Development

Lack of understanding of factors that influence the deployment contributes to the inability to assess the value of business architecture in organisations [10]. Significantly, this is one of the contributions of the EBA to organisations in their pursuit of developing and implementing business architecture. The EBA brings a fresh perspective to organisations that enable management and employees to scrutinise their environments for readiness before committing to architecture activities. Thus, the business architecture is considered the genesis domain of the enterprise architecture because it is pivotal to value add. Also, the EBA can be viewed as a communication tool through which alignment of the various components necessary for successful operation of EA is achieved. This addresses concern that demonstrating the business value of architecture has proven elusive as many of the benefits are intangible [39].

The result from the test clearly shows that the EBA is resilient and adaptive to business architecture in organisations. There are some explanations on how business architecture resolves historical challenges in organisations and translates objectives into strategies, thereby aligning technology and operations [40]. This can hardly be achieved without an assessment, a significant value which the EBA presents. From value aspect, the EBA clearly addresses the gap in processes, designs, and communication within business units, which can be used to promote quality of business' functionalities and supports. The value is fortified through its provision of managerial approach to reveal reality about current state and guides processes toward performance improvement. Consequently, the approach removes the incessant of going in virtual circles, without valuable contribution.

5.3 Design and Automation of Processes

The implementation of EBA is influenced by various factors that are of technical and non-technical nature, which manifest from characteristics and categorisations. There are challenges of characteristics, constraints, and categorization of resources, which often hinder the implementation or practice of business architecture in organisations. Without initial assessment, it is difficult to detect some of these factors because of their uniqueness. The uniqueness of the factors requires a deep view, to gain better understanding and their impacts toward successful development and implementation of the business architecture. This is critical as it shapes the business process network and automation. Also, it enables management to develop a holistic view of an organisation's resources necessary for the design and development of business goal and objectives.

In theory, business architecture defines fundamental components such as transformation and strategy [40]. Through its design, workflow, and logical artefacts, enables alignment, an integrated bridge between business units and IT [36]. Therefore, its assessment should not be taken for granted in actualisation for the objectives. Also, the increasing complexity in business processes and operations require fixing and manageability, to promote cohesion and business-IT alignment [41]. The factors that influence these aspects can be detected at the readiness stage, to ensure stability and increase the chance of fulfilling the objectives for value purposes.

5.4   Ease of Use of the Assessment Model

In addition to other valuable components, the EBA is considered ease of use focus when assessing an environment. Davis argues that when a system is perceived as ease of use, there is high possibility that the users will continue to make use of the system [42]. This is important for the EBA, in assessing readiness of an environment and enhancing the model as technology and business evolve. The EBA's ease of use is attributed to it making complex environment look simply to understand design. The comprehensive description of each cell in the model enhances employees' understanding of factors.

Organisations in all industries operate in dynamic environments. The constantly changing environments affect the business and IT structures making some environment complex. Rakgoale and Mentz explained that IT-landscapes continues to be a challenge due to the constantly changing requirements and globalisation [43]. These add to environment complexity, pointing to why numerous research that have been conducted in the areas of business architecture do not aim to assess implementation gaps [9]. Therefore, implementation of the EBA has been slow primarily because many organisations do not have a clear understanding of how to transform from it being a concept to practice. Also, it is difficult to demonstrate and quantify the value of EBA changes without being able to detect the risks and the bridging mechanism. The EBA is an easy-to-use approach that supports business model- driven migration from a baseline to the deployment of EBA.

6   Conclusion

It is well documented that the deployment and practice of business architecture have been slow. The rationale for the slowness seems to remain a mystery, despite the many studies conducted concerning the subject. This is the main significance of this chapter, for empirically revealing the factors that can influence the readiness of business architecture in an enterprise. The findings are so important that they have bearing and influence from both academics and enterprises (business-organisations) perspectives. On one hand, the chapter demonstrates the criticality of readiness and assessment in the deployment and practice of business architecture. On the other hand, it creates awareness for the stakeholders of the concept in both academic and business domains.

From the academic front, the chapter theoretically positions the factors for further examination in the deployment and practices of business architecture. Also, the chapter adds to existing academic resources, towards the development of capacity and business architecture artifacts. Another significant aspect of the chapter is that it provides additional material for information systems (IS) and information management scholars or researchers that focus on the topic of business architecture.

In practice, the chapter reveals factors of readiness and assessment, which can be used as guidelines for practitioners, which include business managers, information technology specialists, and enterprise and business architects, in their quest, to deploy and practice business architecture. The empirical nature of the findings gives the stakeholders the confidence to use the factors in formulating policies, standards and principles, for business architecture. Thus, the factors contribute to bridging the gap that makes business architecture lags in organisations.

References

1. Kim, C., et al.: Ontology-based process model for business architecture of a virtual enterprise. Int. J. Comput. Integr. Manuf. **26**(7), 583–595 (2013)
2. Iyamu, T.: What are the implications of theorizing the enterprise architecture? J. Enterp. Transf. **8**(3–4), 143–164 (2019)
3. AL-Ghamdi, A.A.-M., Saleem, F.: The impact of ICT applications in the development of business architecture of enterprises. Int. J. Manag. Stud. Res. **4**(4), 22–28 (2016)
4. Kitsios, F., Kamariotou, M.: Business strategy modelling based on enterprise architecture: a state of the art review. Bus. Process. Manag. J. **25**(4), 606–624 (2018)
5. AL-Ghamdi, A.A.-M.S., Liu, S.: A proposed model to measure the impact of business architecture. Cogent Bus. Manag. **4**(1), 1405493 (2017)
6. Gonzalez-Lopez, F., Bustos, G.: Integration of business process architectures within enterprise architecture approaches: a literature review. Eng. Manag. J. **31**(2), 127–140 (2019)
7. Whittle, R., Myrick, C.B.: Enterprise Business Architecture: The Formal Link Between Strategy and Results, 1st edn. CRC Press, Boca Raton (2004)
8. Hadaya, P., et al.: Business Architecture: The Missing Link in Strategy Formulation. Implementation and Execution. ASATE Publishing, Cape Town (2017)
9. Gromoff, A., Bilinkis, Y., Kazantsev, N.: Business architecture flexibility as a result of knowledge-intensive process management. Glob. J. Flex. Syst. Manag. **18**(1), 73–86 (2017). https://doi.org/10.1007/s40171-016-0150-4
10. Zondani, T., Iyamu, T.: Towards an enterprise business architecture readiness assessment model. In: Iyamu, T. (ed.) Empowering Businesses With Collaborative Enterprise Architecture Frameworks, pp. 90–109. IGI Global, Hershey (2021)
11. Bakar, N.A.A., Harihodin, S., Kama, N.: Assessment of enterprise architecture implementation capability and priority in public sector agency. Procedia Comput. Sci. **100**, 198–206 (2016)
12. Jahani, B., Javadein, S.R.S., Jafari, H.A.: Measurement of enterprise architecture readiness within organizations. Bus. Strategy Ser. **11**(3), 177–191 (2010)
13. Niemi, E., Pekkola, S.: Using enterprise architecture artefacts in an organisation. Enterp. Inf. Syst. **11**(3), 313–338 (2017)
14. Costa, A.B., Brito, M.A.: Enterprise architecture management: constant maintenance and updating of the enterprise architecture. In: 2022 17th Iberian Conference on Information Systems and Technologies (CISTI) (2022)
15. Chen, J., et al.: Complex network controllability analysis on business architecture optimization. Math. Probl. Eng. **2021**, 1–7 (2021)
16. Caceres, G.J.L., Gómez, R.A.A.: Business architecture: a differentiating element in the growth of organizations. J. Phys. Conf. Ser. **1257**(1), 012007 (2019)
17. Roelens, B., Steenacker, W., Poels, G.: Realizing strategic fit within the business architecture: the design of a process-goal alignment modeling and analysis technique. Softw. Syst. Model. **18**(1), 631–662 (2017). https://doi.org/10.1007/s10270-016-0574-5

18. Whelan, J., Meaden, G.: Business Architecture: A Practical Guide, 2nd edn. Routledge, London (2016)
19. Lange, M., Mendling, J., Recker, J.: An empirical analysis of the factors and measures of enterprise architecture management success. Eur. J. Inf. Syst. 25(5), 411–431 (2016)
20. Gong, Y., Janssen, M.: The value of and myths about enterprise architecture. Int. J. Inf. Manage. 46, 1–9 (2019)
21. Musukutwa, S.C.: Developing business architecture using SAP enterprise architecture designer. In: Musukutwa, S.C. (ed.) SAP Enterprise Architecture, pp. 119–140. Apress, Berkeley (2022)
22. Waheduzzaman, W., Miah, S.J.: Readiness assessment of e-government: a developing country perspective. Transf. Gov. People Process Policy 9(4), 498–516 (2015)
23. González-Rojas, O., López, A., Correal, D.: Multilevel complexity measurement in enterprise architecture models. Int. J. Comput. Integr. Manuf. 30(12), 1280–1300 (2017)
24. Balasubramanian, S., et al.: A readiness assessment framework for Blockchain adoption: a healthcare case study. Technol. Forecast. Soc. Change 165, 120536 (2021)
25. Yusif, S., Hafeez-Baig, A., Soar, J.: e-Health readiness assessment factors and measuring tools: a systematic review. Int. J. Med. Inform. 107, 56–64 (2017)
26. Conboy, K., Fitzgerald, G., Mathiassen, L.: Qualitative methods research in information systems: motivations, themes, and contributions. Eur. J. Inf. Syst. 21(2), 113–118 (2012). https://doi.org/10.1057/ejis.2011.57
27. Markus, M.L., Lee, A.S.: Special issue on intensive research in information systems: using qualitative, interpretive, and case methods to study information technology - Foreword. MIS Q. 23(1), 35–38 (1999)
28. Gehman, J., et al.: Finding theory-method fit: a comparison of three qualitative approaches to theory building. J. Manag. Inq. 27(3), 284–300 (2017)
29. Yin, R.K.: Case Study Research and Applications: Design and Methods, 6th edn. SAGE Publications, Thousand Oaks (2017)
30. Eisenhardt, K.M.: Building theories from case study research. Acad. Manag. Rev. 14(4), 532–550 (1989)
31. Heeks, R., Stanforth, C.: Technological change in developing countries: opening the black box of process using actor–network theory. Dev. Stud. Res. 2(1), 33–50 (2015)
32. Hussein, S.S., et al.: Development and validation of Enterprise Architecture (EA) readiness assessment model. Int. J. Adv. Sci. Eng. Inf. Technol. 10(1) (2020)
33. Amit, R., Zott, C.: Crafting business architecture: the antecedents of business model design. Strateg. Entrep. J. 9(4), 331–350 (2015)
34. Peppard, J., Ward, J.: Beyond strategic information systems: towards an IS capability. J. Strateg. Inf. Syst. 13(2), 167–194 (2004)
35. Pirola, F., Cimini, C., Pinto, R.: Digital readiness assessment of Italian SMEs: a case-study research. J. Manuf. Technol. Manag. 31(5), 1045–1083 (2020)
36. Kotusev, S.: Enterprise architecture and enterprise architecture artifacts: questioning the old concept in light of new findings. J. Inf. Technol. 34(2), 102–128 (2019)
37. Ori, D.: Misalignment symptom analysis based on enterprise architecture model assessment. IADIS Int. J. Comput. Sci. Inf. Syst. 9(2), 146–158 (2014)
38. Iyamu, T., Shaanika, I.: Assessing Business Architecture Readiness in Organisations. SciTePress, Setúbal (2022)
39. Shanks, G., et al.: Achieving benefits with enterprise architecture. J. Strateg. Inf. Syst. 27(2), 139–156 (2018)
40. Hendrickx, H.H.M., et al.: Defining the business architecture profession. In: 2011 IEEE 13th Conference on Commerce and Enterprise Computing (2011)

41. Řepa, V., Svatoš, O.: Adaptive and resilient business architecture for the digital age. In: Zimmermann, A., Schmidt, R., Jain, L.C. (eds.) Architecting the Digital Transformation. ISRL, vol. 188, pp. 199–221. Springer, Cham (2021). https://doi.org/10.1007/978-3-030-49640-1_11

42. Davis, F.D.: Perceived usefulness, perceived ease of use, and user acceptance of information technology. MIS Q. **13**(3), 319–340 (1989)

43. Rakgoale, M.A., Mentz, J.C.: Proposing a measurement model to determine enterprise architecture success as a feasible mechanism to align business and IT. In: 2015 International Conference on Enterprise Systems (ES) (2015)

A Specification of How to Extract Relevant Process Details to Improve Process Models

Myriel Fichtner[✉][iD] and Stefan Jablonski

University of Bayreuth, Universitätsstrasse 30, 95447 Bayreuth, Germany
{myriel.fichtner,stefan.jablonski}@uni-bayreuth.de

Abstract. Defining process models can be quite challenging. Among other important aspects, a designer must decide how detailed a process should be represented by a model. The absence of relevant process details can result in reduced process success, e.g., lowered quality of the process outcome. Furthermore, in case a deficiency in process execution is detected, it is difficult to determine which part of a process model is insufficiently modelled and which kind of detail is missing.

In this work, we provide a conceptual approach that is able to extract relevant process details from data recorded during process executions. Then, identified process details are integrated into an existing process model as task annotations. Our method comprises phases for analyzing deficiencies in process executions, creating process annotations, and finally validating these modifications. A prototypical implementation of the approach using image data and explainable AI has been extensively evaluated in two experiments.

Keywords: Business process model enhancement · Business process improvement · Relevant process detail · Task annotation · Image data analysis · Model explanation

1 Introduction

Enterprises use process modelling languages like Business Process Model and Notation (BPMN), Event-Driven Process Chains (EPC) or Unified Modeling Language (UML) to represent internal and external processes. Process models provide transparency, standardization and potential for optimization. There are coarse guidelines which support process designers in their decisions on how to model a certain process.[4,20,22]. However, they do not prescribe the level of detail in which a process should be modelled. Finally, a process modeler decides whether to consider a process detail or not, i.e., to integrate it into a process model. As process detail we denote all information that describes a certain aspect of a process and typically supports the execution of it. When process executions - especially their results - vary extremely, it might be an indication that a process is specified too loosely and important process details are missing. Identifying and integrating these process details can lead to improved models with increased process success, i.e., higher qualitative process results. We identify three reasons why in general process details can be missing in process models:

J. Filipe et al. (Eds.): ICEIS 2022, LNBIP 487, pp. 391–414, 2023.
https://doi.org/10.1007/978-3-031-39386-0_19

Unknown Details. Usually, process designers are no process experts and therefore do not know all details of a process. Thereby, the process expert has to communicate all details to the designer. In such cases it may happen that some details are not explicitly communicated to the process designer since the process expert is not aware that they are relevant for process success. Input parameters or implicit sub-tasks are often taken as granted although they can have an impact on the process result. Furthermore, there is a high chance that even process experts do not know all process details [23].

Restricted Expressiveness. Existing process modeling languages are restricted in their expressiveness. In some cases, the provided modeling elements of a language are not sufficient to integrate a certain process detail appropriately in a model [24,36]. Therefore, a process detail can be missing in a process model, although the process detail is known.

Intentional Abstraction. Large and complex process models lead to overload while they should be avoided to maintain their readability. Therefore, abstraction mechanisms are applied on large models to reduce the number of modeling elements and details [5,26,29]. Thereby, the loss of detailed information is accepted in favor of model clarity and traceability. However, details may be lost by abstraction. Furthermore, to avoid overload from the beginning, rarely executed parts of the process or alternatives in the control flow are not modelled.

If a process detail is relevant, it has an impact on the overall process success and has to be represented in the process model to produce better process results. We face this issue in an industrial project from the manufacturing domain, where the arrangement of work pieces on a plate has an impact on the result of a subsequent process step. For example, the larger work pieces have to be placed in the rear of the plate because it is put into an oven with a non-uniform heat distribution. If the model contains only the instruction to place work pieces (i.e., *what should be done?*) on a work surface and no further information about where and how to place it (i.e., *how should it be done?*), it depends entirely on the operator how the work piece is placed. As a result, process results depend on human experience and decision-making. Arising thereby, relevant details regarding the positions of the work pieces are missing in the model due to one of the above described reasons.

In the area of process optimization, there are many methods to optimize control flow. Data from the process execution logs are considered. Naturally, a log only contains information that is defined in the process model. However, other approaches have shown that there is further potential for optimization when data is taken into account that is coming from outside of a process model, e.g., [23,25]. Inspired by previous work, we developed an overall concept to extend existing business process models by relevant process details that are gathered during process execution and that are not stemming from a process model.

In particular, the following research questions are considered:

RQ1. How can relevant process details be extracted by data which is recorded during process executions?

RQ2. How can an existing process model be extended by missing but relevant process details in order to increase process success for future model executions?

This article is an extension of our previous work [17] and augments it by the following aspects:

(i) the concept have been restructured and extended regarding several aspects, i.e., the approach is generalized, the "Annotation" step is added and the "Analysis" step is clarified and described in more detail.
(ii) types of missing process details and data sources are presented and discussed.
(iii) limitations of the overall concept are elaborated more clearly.
(iv) the implementation is described in more detail.

This paper is structured as follows. Section 2 gives an overview of related work. In Sect. 3 we describe the extended concept and discuss different types of missing process details, data sources and limitations. Section 4 describes and evaluates a concrete implementation approach to extract relevant process details from unstructured image data by using explanation models. Our work is discussed and concluded in Sect. 5.

2   Related Work

The modeling language *Business Process Model and Notation* (BPMN) is the standard to describe a series of steps that must be performed to achieve a certain business goal, e.g., the manufacturing of a specific product [8]. With BPMN a graphical notation of business processes can be created. Therefore, various modeling elements are provided which enable to map the procedure itself and involved entities like process participants.

To improve business processes, process models are essential, while the Business Process life cycle describes how they are used for this purpose. The life cycle includes six steps: identification, discovery, analysis, improvement, implementation, monitoring and controlling [12]. The repetition of these steps leads to a continuous improvement of the process and therefore to an increase of process success. In the process improvement step, a process model is redesigned and enriched by aspects coming from observations of process executions. An established technique to analyze data stemming from process executions, e.g., event logs, in order to improve process models is Process Mining [1]. Process Mining approaches are able to optimize control flows, i.e., the execution order of process steps. This information is then considered for future executions through redesigning the process model. However, event logs store the information which activities of the model were performed and therefore contain exclusively information that was previously modelled.

According to this criterion, we divide existing approaches into two classes. Approaches that analyze processes by restricting themselves to known information that is already contained in the process model use *intrinsic* parameters. Techniques that consider further data sources or external information that is not yet included in a process model work with *extrinsic* parameters.

A lot of approaches optimize processes from fundamentally different points of view using merely intrinsic parameters, e.g., [2,5,19,26,29,32]. We also classify techniques that consider multiple process perspectives in order to improve process analyzability as intrinsic, e.g., [18].

However, most work do not consider all relevant data sources or are restricted to a single data source [28], i.e., only a few approaches are based on extrinsic parameters. A (semi-) automated process optimization approach is suggested in [23], which integrates both, process and operational data [23]. Also techniques that extend the expressiveness of process modeling languages are extrinsic. In [36], an extension of BPMN is presented, which enables the attachment of media annotations to modeling elements. Therefore, more complex information stemming from different data sources can be included, e.g., image or audio data. Another approach uses extrinsic parameters to extract relevant process details by analyzing image data of process executions [16, 17].

3 Extending Process Models by Relevant Process Details

A first conceptual approach to optimize existing business process models by relevant process details has been presented in [15] and [16]. The concept describes a first overall approach to extract important information from data recorded during process execution event after the modeling phase has been completed. Thereby, it considers only image data as input and describes the challenging analysis step and the processing of the extracted process detail from a very high-level and non-technical perspective. For this reason, we aim on *i)* extending the concept to deal with different data sources and data types and *ii)* giving more insight into the analysis step and subsequent steps regarding process details which are necessary to extend the process model by useful information. These considerations motivate to split our extended concept into three phases:

1. The **Observation Phase:** The results of process executions are observed and differences in process results are recorded (cf. Figure 1a).
2. The **Improvement Phase:** The existing process model is enriched with relevant details (cf. Figure 1b).
3. The **Validation Phase:** The changes in the process model made in the previous improvement step are verified whether they lead to increased process success (cf. Figure 1c).

3.1 Extended Concept

Observation Phase. When domain experts recognize that the execution results of the same process model vary and are not satisfactory, they have to start observing and assessing the results of each performed process instance (cf. Figure 1a). Therefore, the daily processes are performed as usual, however, the outcomes (results) of each process execution are assessed. If the results collected in an observation period are not adequate, an Improvement phase is initiated.

Improvement Phase. This phase consists of 5 steps which are explained in detail in this Section. The prerequisite and input for this phase is an existing process model. Here, the term process model should be representative for any representation that describes a set of process steps. The motivation of the concept originates from the manufacturing domain where different actors (i.e., humans or robotic systems) participate to perform prescribed tasks. However, the application context of the approach is not

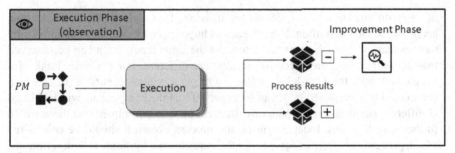

(a) Phase 1: Observation of varying process results.

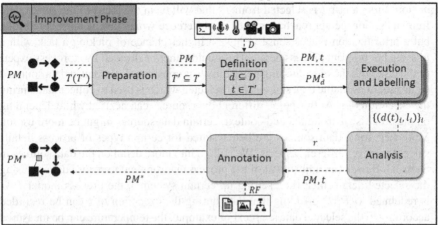

(b) Phase 2: Improvement of the existing process model.

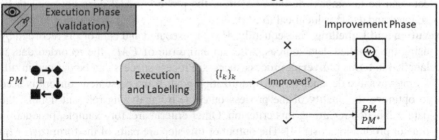

(c) Phase 3: Validation of the enriched process model.

Fig. 1. Extended concept to enrich process models with relevant process details based on the work of [15].

limited to that area. It is assumed that the single task descriptions in the existing model do not contain all the details. The instructions carry only the information what has to be done in a process step but lack of details describing how it is done. The set of all tasks defined in the model is summarized in T.

Preparation. In this step, T is reduced to $T' \subseteq T$ containing tasks which are modelled too unspecific and where process experts suspect that important details are missing, but they do not know which details are missing. This mainly includes tasks that, according to their definition, leave the actors huge space for decision-making during their execution. In such cases, all actors get the same input, but it can be observed that the output, i.e., the execution results, are different. For example, think of a simple task with the high-level instruction to place components on a pallet which is assigned to several actors. It can be expected that the components will be placed at different positions on the pallet by different process participants, as there are no further specifications. Independent of any process expert, it should be considered which process perspectives (cf. Section 3.2) are covered by single task descriptions, while missing perspectives are an indicator to include a task in T'.

Definition. First, a task t is selected from T' that will be in the focus of the subsequent four steps. For the approach, it makes no difference which task is considered first, but a prioritization makes sense to have a higher chance of picking a task with a missing but relevant process detail. This could either be decided by a process expert or the task with the most missing perspectives is considered. Second, a data source d from a set of available data sources D is selected which is used to collect data during the execution of t. At this point, different data sources can be used while depending on the process environment and context, certain data sources might be more useful. Moreover, some data sources are better suited for certain types of process details since they cover different aspects. We will go into more detail on the data sources in Sect. 3.3. Besides an adjustment of the process environment to record data by using the selected data source (i.e., setting up certain systems), the process model PM is redefined to PM_d^t enabling that data during the execution of t can be recorded according to the selected data source. For example, the temperature can be measured by an actor at a certain point in time during the execution of t. If the recording of data can be automated during execution, the process model does not need to be adapted and PM_d^t is identical to PM.

Execution and Labelling. Subsequently, PM_d^t is executed and data of t is recorded by using the installed data sources. After an entire run of PM_d^t, the recorded data is labelled in terms of **overall** process success. Process success can be defined in different ways and depends on the optimization goal. If the presented concept is used to optimize the quality of the process outcome by analyzing relevant details, the data is labelled regarding this criterion. Other criteria are, for example, production time or production costs [9]. The output of this step are pairs of the form $(d(t)_i, l_i)$ per execution i, where $d(t)_i$ is the data available for t from data source d and a label l referring to process success. Since a sufficient database is required for the subsequent analysis step, the process model must be run several times to collect enough data of t.

Analysis. In this step, the labelled data is analyzed while the goal is to find a general rule r or a comparable construct containing any information that explains how to distinguish between successful and less successful executions. This information is thus the decisive criterion that has an influence on process success, whereby r explains how the criterion must be present. Thus, r specifies which conditions must be met during the execution of t in order to achieve increased process success, while we define this information as relevant process detail. In this concept, the procedure to compute r is related to the common classification problem: The decision boundary that separates samples of different classes in a feature space is equivalent to r. The samples refer to the recorded execution data while their labels indicate the different classes. Finding a decision boundary can become very complex, especially, when the feature space is very large and data points from different classes are distributed over the whole feature space. These factors strongly depend on which features of the data are considered, which in turn strongly depends on the structure of data. Figure 2 shows that depending on the structure of data, different analysis approaches have to be implemented.

Structured data refers to data that has been formatted to a certain structure, e.g., a relational database. In contrast, unstructured data is stored in its native format and is not organized according to a predefined schema. Examples are audio data, text data or image data. To mine rules from structured data sets, techniques like association rule mining or decision trees are applicable. For unstructured data, in general, deep learning approaches with neural networks are preferred. Unstructured data can be transformed into structured data using various approaches [3, 21, 38] depending on the application domain. However, the core of all approaches is to extract a set of structurable information or attributes from the unstructured data.

Annotation. Depending on the structure of the data and the selected technique to find r, the output can have a very complex presentation. In general, rules describe logical relations and express them by a sequence of logical operations. Such representations are difficult to read for process actors, especially when it comes to large rules considering multiple items or features. For this reason, r must be post-processed to be integrated into the existing model in the most suitable way. This is done by creating a task annotation based on r with a readable presentation. While this is the challenge in the sub-step "creation", the sub-step "adding" deals with the technical implementation for integrating this detail into the process model (cf. Figure 2).

The creation of a task annotation depends on $i)$ the intended representation of r in the model, i.e., in which representation format RF the detail should be integrated and $ii)$ the current form of r, i.e., which analysis technique was chosen in the previous step so that r_{sd} or r_{ud} have to be processed.

For RF, three variants are distinguished: The process detail can be integrated textually, pictorially or diagrammatically into the process model. The goal is to extend the model by a task annotation that covers all information contained in r while preserving the readability of the model. Previous work evaluated the three representation formats on different task settings from manufacturing environments regarding intuitive comprehensibility [14]. Their study results shows that all three representation formats are usable for task annotations. However, pictorial and diagrammatic representations should be preferred when it comes to task settings having multiple dependencies between participating objects. The simplest approach to create a task annotation based on r and a specified representation format, is to do it manually. Depending on the form of r and the chosen representation format, an automation of this step is more or less challenging. We will present an example in Sect. 4.

Subsequently, the created annotation regarding t is integrated into the original process model PM. This can be done in different ways depending on the representation format. For all representation formats, a simple integration as an attachment is possible, as suggested by [36]. In this case, all representation formats are considered as

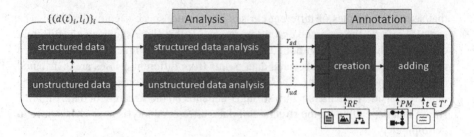

Fig. 2. Analysis and Annotation step depending on the structure of the input data.

a medium that is attached to a task and is available during task execution. Furthermore, in case of textual representations, the task description in a process model can be modified by changing the label of the analyzed task within the model. For this purpose, the underlying file of a model is modified directly. For example, BPMN models are based on XML files, which can be adapted programmatically in order to change model semantics. In case of diagrammatic representations, the structure of the model can be modified to include the detailed version. This implies that new modeling elements may be added that have a relation to the analyzed task and that the control flow is adapted accordingly. For example, in the case of BPMN diagrams, the underlying XML file can be adapted programmatically in the same way as for textual changes. A major difference here is that the number and hierarchy of elements can change and the relationships between elements must be adjusted. The output of the annotation step is the process model PM^* what corresponds to input process model PM but contains process details regarding t.

Validation Phase. In the third phase, it is examined whether the now enriched process model PM^* leads to increased process success. For this purpose, the same process success criterion as for the labelling step is used, e.g., the quality of the process outcome. The model PM^* is executed several times and the process result is labelled with respect to the selected criterion, resulting in $\{l_k\}_k$. If the modification leads to increased process success compared to the execution of PM, PM is discarded and PM^* is accepted as the new process model. This can for example be decided by comparing the positive-to-negative ratio from "Execution and Labelling" in Phase 2 and Phase 3. If the ratio shifts in favor of the positive labels, an improvement is confirmed and PM is replaced by PM^*. Otherwise, a new instance of the Improvement Phase is initiated. In this case it is assumed that the missing process detail refers to another task which is modelled insufficiently. Therefore, T' is either adapted to $T' \leftarrow T' \setminus \{t\}$ or a process expert redefines T' completely and Phase 2 restarts having PM and T' as input. In general, the process expert have to decide the procedure of the Validation step. A validation of the analyzed process detail that is based entirely by the assessment of the process expert is also possible.

To address the practical perspective, we want to explain how the three phases are integrated into the normal production flow of an enterprise. The Observation Phase evolves from ordinary process execution. Domain experts recognize that, for example, the production quality diminishes or varies. Thus, they start to observe the process in more detail, i.e., they continue the production flow, however, assess the outcome of process execution accurately. In case they detect changes in the production quality, they initiate the Improvement Phase. Nevertheless, the production flow continuous, merely the steps depicted in Fig. 1b are additionally performed. The outcome of this phase is an process model enriched with details, which contain information related to the process success. A consideration of these details should increase the quality of process outcomes. The production flows again continues, however, executing the improved process model. Thus, the Validation Phase is entered. Here, again process quality is assessed. Either the process quality is sufficient (i.e., the same or higher quality is achieved), then the production is continued. Or the process quality is not sufficient, then another instance of the Improvement Phase is initiated.

3.2 Types of Missing Process Details

Fundamental perspectives were identified in order to proceed systematically during process modeling [10]. The *functional* perspective described what happens in the process and what elements are part of it, e.g., tasks. The *behavioral* perspective refers to the control flow in a process model and describes when process elements are performed. The *organizational* perspective includes the actors, specifying who is responsible for the execution of a certain task. The *informational* perspective covers two sub perspectives: 1) The *data and dataflow* perspective which describes which data is necessary for the process or created during its execution and 2) The *operational* perspective which specifies the applications or tools needed within the process. According to these perspectives, missing process details can be divided into different types:

- *Type A* (data and dataflow): The input specification or parameters of a task is not modelled, e.g., instructions containing concrete placement information of objects are missing in a process task.
- *Type B* (functional): The task is not modelled in enough detail, i.e., the task contains sub-steps that are not modelled sufficiently. For example, a task having the instruction that two components have to be glued may implicitly contain pre- and post-processing steps that are not included in the model.
- *Type C* (behavioral): The execution order of tasks is not considered consciously by the modeler in the model. For example, the order in which different ingredients are mixed together is not prescribed.
- *Type D* (organizational): The assignment of tasks to the actors is not specified enough. Usually, assignments to roles are given, while assignments of single tasks to individual actors are missing.
- *Type E* (operational): The tools that should be used within a task are not described, e.g., the information that a particular tool should be preferred over another when executing the task is not modelled.

The occurrence of one or more types causes reduced process success. To illustrate this, we would give another example. Think of a task with the description "mix ingredients together" that is not further specified. The process participant will mix the ingredients together in a self-decided order during execution. It is observed that some sequences cause a delay in the process because the mixture starts to foam which which prevents to continue with the next step right away. In this example, Type B and Type C occur because relevant sub-steps are missing in the model and their order has an impact on the process success.

3.3   Types of Data Sources

In process environments, there are various data sources to record data which can be used to analyze missing process details using the described concept. In the context of this work, we define the term data source as a system that produces output. Therefore we distinguish between two types of data sources: *i)* sensors which record physical phenomenons in the environment and *ii)* software or digital tools which are used in context of the execution environments. In real work spaces, only a selection of data sources might be available while not all data coming from certain data sources is suitable for the analysis of a specific type of missing process detail. In this work, we focus on data recorded by sensors, while an overview of different types of sensors is given in [33]. We address an excerpt of sensors that are commonly used for monitoring process executions in manufacturing plants. Thereby, we do not describe the techniques to extract information from sensor data but focus on their potential to use them for analyzing certain types of missing process details.

Optical Sensors. Detectors that are able to sense light and to transform it in a digital signal are called optical sensors. The most common devices using that technology are camera systems. The output are images or sequences of images, i.e., videos. Camera systems are used in different domains for process monitoring or quality control, e.g., [6,37], while they are low in cost, do not require additional hardware and are easy to integrate into process environments. The authors of [31] further confirmed the benefits of image data and image mining techniques in the context of business process management.

Image data can contain a lot of process information and is therefore suitable for analyzing all types of missing process detail. An image which is taken during the execution of a task may contain missing process details of Type A, D and E. For Type B and C, a single image is not sufficient, so that videos showing the entire task execution can be used.

An image showing a certain scene after a task execution can contain relevant information about the participating objects that is not yet included in the process model (Type A). This can be, for example, the position or type of an object. However, only details that can be expressed by an image are recognizable. Depth information or other information like room temperature cannot be derived from a 2D image. Participating actors (Type D) or tools (Type E), can also be identified in such images.

Videos recorded during execution and showing a whole task, can be used to extract hidden sub-steps (Type B) and relevant execution orders (Type C). Type D and E can also be analyzed from video data, e.g., by using techniques to track actors and tools to identify their occurrence in certain activities.

Electromagnetic Sensors. Electromagnetic fields are used as sensing mechanism to capture information from the environment by electromagnetic sensors. A common used technology that works with electromagnetic fields for identification and tracking is RFID (Radio Frequency Identification). RFID tags are used in many manufacturing environments for activity tracking [34, 35]. They can be used to detect assembly steps (Type B) and their order (Type C). Furthermore, they can be attached to clothes and tools to detect which worker (Type D) is using which tools (Type E) for a certain task. Objects in the work space can also be augmented by RFID tags to identify further object-related information like their positions. One of the advantages of electromagnetic waves is that they penetrate objects. Accordingly, no line of sight is required to identify objects with RFID. In contrast, the communication between transmitter and receiver can be disturbed by liquids or metal which can be, depending on the domain, a major problem.

Acoustic, Temperature, Humidity, Force Sensors. All sensors that record a specific physical quantity are grouped in this category. Data from these sensors can mainly be used to analyze missing process details of Type A. Data which is recorded by those sensors during task execution can reveal that for example a certain temperature is relevant for the task to succeed. Although no further information in the context of the other types of missing process details can be extracted from such sensor data, a hint regarding the occurrence of Type B, C and E can be derived. For example, if a measured temperature changes differently within different execution instances, it can be assumed that the task contains non-modelled sub-steps (Type B), which have a certain order (Type C) or that the use of a certain tool causes the temperature change. However, the hypotheses have to be tested by considering data from other data sources.

3.4 Limitations

We identified three weaknesses in the overall concept.

Data Quantity. First, the concept needs a certain amount of labelled data for the "Analysis" step in Phase 2 to provide a reliable result. Although data can be recorded during daily operations in a company without disturbing the work flow, especially in small and medium-sized enterprises, the amount of data may not be sufficient. In such companies, only a small number of a product is produced or processes are executed significantly less frequently than in industrial settings.

Separability. To find a process detail in the form of a rule that has to be considered in order to produce positive process results, the recorded data related to different classes (i.e., labels) have to separable. This means that the feature space has to be chosen appropriately, which is quite challenging due to the unknown nature of the missing process details. Furthermore, all classes have to be sufficiently represented by the data.

Optimization Independence. The considered tasks in Phase 2 must be able to be optimized independently. The presented concept optimizes the process with regard to the global criteria process success. We assume that the considered tasks can be optimized regarding this global goal and that there are no optimization dependencies between the tasks. Thereby, an optimization dependency means that certain combinations of states of two tasks, i.e., multiple rules, lead to an optimization. If, on the other hand, two tasks are optimization-independent, there is one state per task that optimizes the global goal and is independent of the optimization of the respective other task. Thus, the order in which a task is optimized is irrelevant, enabling the incremental optimization approach.

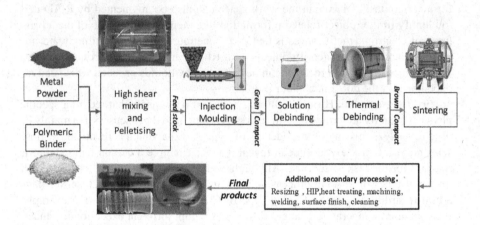

Fig. 3. The metal injection molding process [11].

4   Implementation of the Analysis and Annotation Step by Using Unstructured Image Data

As described in Sect. 3.3, image data contain a lot of context information thus there is a high chance to extract still unknown details from it. Among other advantages, we therefore focus in our work on using image data to extract relevant process details. Raw image data is classified as unstructured data, why the first prototype of our approach considers unstructured data analysis techniques to extract relevant details. Furthermore, motivated by a real process example, we focus on the identification of missing process details of Type A, i.e., insufficient modelled input parameters.

4.1   Motivating Example Process

The metal injection molding (or extrusion) process is a real-world example from the manufacturing domain. These processes are similar, only in the latter the molding has been replaced by extrusion. Figure 3 shows the whole process. The output of this process are small metal components used in many industrial settings and application. In a sub-step of this process, weakly bound work pieces (brown parts) are taken from a debinding oven and are placed on a charging plate. The charging plate then enters a sintering furnace, resulting in the solid and final products. Figure 4 shows the manual execution of these tasks in a real process environment[1].

---

[1]   https://www.youtube.com/watch?v=QaMdjKE7vT8 (last accessed 21 Nov 2022).

Fig. 4. Manual steps of the metal injection molding/extrusion process (See footnote 1).

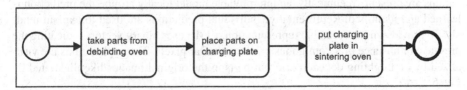

Fig. 5. BPMN model of a part of the metal injection molding/extrusion process [17].

In this process, the arrangement of the objects on the charging plate has an impact on process success. The heat distribution in an sintering furnace may not be completely uniform, why the position of parts on the plate, depending on their size and shape, is relevant for the quality of the final product. If binder remains in the parts, unacceptable cracks or blisters can occur. Therefore, it should be ensured that larger or thicker components should be located within an area where sufficient heat is present during sintering. For example, this can be the rear area of the furnace. However these areas may be unknown but are a relevant process detail that can be identified by using the presented concept. In our implementation approach we show how such a detail can be found prototypically. Therefore, we designed a simple BPMN model representing these steps as input for the Improvement Phase (cf. Figure 5). We assume that varying process results were recorded by process experts in the Observation Phase after executing this process model. The reasons for the partially unsatisfactory results are unknown, why the Improvement Phase is initiated.

4.2   Background and Technical Approach

The output of the Improvement Phase is a general rule which should be followed to achieve process success. This rule contains the information about a relevant process detail, where we define it as a feature in a feature space having a set of valid values. This feature space is spanned by the process context, which is implicitly contained in the data recorded during execution. In our approach, regions in that feature space are found that refer to process success and the criteria is extracted that separates them from those regions that refer to unsuccessful process outcomes. The separation of a feature space in two or more classes is the well researched issue of classification. In case the feature space is known, classical machine learning approaches can be used to determine the boundary of these classes. Otherwise, deep learning mechanisms are used to automatically generate a feature space [27]. Convolutional neural networks (CNN) have proven to be a successful technique especially in case of image data. Usually, CNNs are used to prediction a behavior about a particular input. Knowing the reason for a prediction is related to the topic of trust in the context of system decisions. In some applications it is essential to understand the reasons of a decision and to recognize wrong ones in order to avoid mistakes [13]. Different explanation techniques are summarized in [7]. One promising approach in this context are local interpretable model agnostic explanations (LIME) proposed by [30]. It explains predictions of any classifier in an interpretable and faithful manner [30]. Non-experts are enabled to identify irregularities when explanations are present. Technically, an interpretable model locally around the prediction is learned and identified. Representative individual predictions are used to explain models. To provide an interpretable representation, LIME uses binary vectors indicating the presence or absence of a contiguous patch of similar pixels. The explanations are visualized by highlighting decision-relevant parts in the original images like illustrated by Fig. 6.

The approach matches our requirements, because the search for a relevant process detail is strongly related to the explanation why bad process results are predicted for an execution. The non-consideration of both leads to reduced process success. By using LIME, we are able to identify *(i)* which features in an image are related to successful or unsuccessful process outcomes and *(ii)* which values of these features are required to increase process success.

| (a) Original input image | (b) Explanation of "electric guitar" | (c) Explanation of "acoustic guitar" | (d) Explanation of "labrador" |

Fig. 6. Highlighted parts explain the predicition of different classes [30].

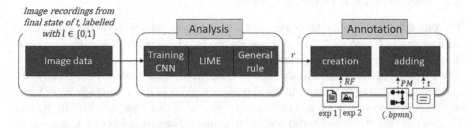

Fig. 7. The implementation steps embedded in the structure of the concept (cf. Figure 2).

A drawback of the LIME approach is, that is only provides local explanations. This means that the explanations are only valid for the input image. However, for our approach to work, we need a general conclusion or rule that contains an explanation across all input images. Therefore, we analyze each local explanation resulting after using LIME on an input image and derive a general rule. Giving a global understanding of image explanations is an open research problem [30], while we present an idea to tackle this issue for our experiments. Our implementation is further described in [17]. Here, we structure it in five parts, while we cover the "Analysis" and "Annotation" step defined in our conceptual approach. Figure 7 shows how the five parts are embedded into our overall concept.

1. A CNN is trained with sufficient labelled image data that refers to a certain task. In our prototype, we use images showing executions of the task to place parts on a charging plate described in the previous section. Multiple labels or classes are allowed, however, we use only two labels which refer to (0/*unsuccessful process executions* and 1/*successful process executions*). Our implementation is realized in Python, while we use TensorFlow[2] to create a basic net architecture following common patterns, i.e., a stack of convolutional and pooling layers. Some parameters are adapted due to available performance and settings, i.e., the expected image size, the batch size and epochs.

2. Then, LIME is used to explain the predictions of the trained CNN. Therefore we follow an open source implementation which presents the usage of LIME[3]. We have adopted most of the code and parameters. However, we restricted the number of labels for which an explanation is made to two classes ($top\_labels = 2$). Second, the number of similar pixel regions to be included in the explanation ($num\_features$) is adjusted for different experiments. In this step, we let the trained CNN predict the class of a set of images. Simultaneously, LIME explains each individual prediction. As a result, one output image per input image is obtained, in which the regions are highlighted that are relevant for the classification decision of the CNN. To be more precise, these images are copies of the input images but contain only those regions that explain the decision to the positive class. Irrelevant parts are colored black.

3. The images with local explanations are analyzed regarding different features and a general rule or conclusion is derived. Like stated in [30], a global explanation can be computed based on local explanations by finding a possibility to compare same content across different images. We will solve this requirement for our experiments. Per local explanation, we analyze the remaining visible regions in the image and check if they contain any objects that participated in the process (i.e., the placed parts). We recognize all visible objects and analyze them with respect to a set of features (color, shape[4], size and position (centroid)). These features will then be comparable across all local explanations. As output, there is one set for each feature containing all analyzed values. For example a color set contains all object colors that

---

[2] https://www.tensorflow.org/ (last accessed 22 Nov 2022).

[3] https://github.com/marcellusruben/All_things_medium/blob/main/Lime/LIME_image_class. ipynb (last accessed 22 Nov 2022).

[4] https://www.pyimagesearch.com/2016/02/08/opencv-shape-detection/ (last accessed 23 Nov 2022).

were analyzed across all local explanations. A comparison of those values with the values that can possibly occur for that feature in that task then provides the relevant process detail. If only a subset of all possible values occurs, it is relevant for the process to succeed. The allowed values for the corresponding feature are then the relevant process detail which has to be considered during task execution to enhance process success.

4. A representation format is selected in which the detail should be represented in the existing process model. The analyzed process detail is then converted into that representation. In our implementation, we considered the creation of textual and pictorial task annotations. For the latter, we create an image that shows the relevant feature and values. We will explain details in Sect. 4.3.

5. The details are then integrated into the existing business process model in the present representation format. In our implementation approach, we restrict to BPMN models created with BPMN.io[5]. This is an established toolkit to model BPMN 2.0 diagrams while they can easily be imported and exported via XML files. Each modeling element can be modified either by the interface or by editing the XML file directly. In our approach, we enrich an existing model with details by modifying the underlying XML structure of our process model programmatically. For textual modifications, the attribute *name* within the identified tag is extended by the process detail. In case of media annotations, an image that represents the process detail is created and attached to the task.

4.3   Experiments

In this section, we demonstrate the functionality of our implementation through two experiments. We recorded image data showing scenes regarding the task "place parts on charging plate" from the example process. To avoid problems related to insufficient data quantity, we recorded 1000 images per experiment. Each image is assigned to one label referring to process success, i.e., 0 or 1. For both experiments, 500 images showing successful process scenes and 500 images 500 negative examples were recorded. For both experiments, we use 800 images for the training and 200 for the validation of the CNN, while both sets contain equal distributions of both classes. For our prototype, we used simple objects with convex shapes and saturated colors, i.e., bricks, and a clearly distinguishable background. Furthermore, we post-processed our images by using object recognition techniques. We avoid disturbing factors, such as noise or uncontrollable lighting conditions. For each experiment, we know the criteria (i.e., the process detail) that impacts the assignment of an image to a class. In contrast, these criteria are not known in real applications. However, in our experiments, we exploit the knowledge of this criteria to evaluate our approach. We examine if the implementation of the "Analysis" step is able to output a rule that contains exactly this criteria.

Experiment with One Relevant Feature.   In a first experiment, the criterion that is relevant for the classification is contained in only one feature, i.e. the object color. Each image shows the top view of a scene, in which one rectangular object is placed on a

---

[5] https://bpmn.io/ (last accessed 15 Dec 2021).

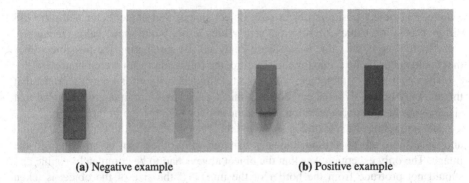

(a) Negative example   (b) Positive example

Fig. 8. Example image data for the first experiment [17]. Per class: original image (left) and post-processed version (right).

Fig. 9. Local explanations in the first experiment [17].

table. Images with blue objects are labelled with 1 (positive example) and images with green ones are labelled 0 (negative example). The background, the position of the object in the scene and other features of the object, e.g., shape or size are not relevant. The positions of the objects are determined randomly per scene. Figure 8 shows examples of both classes and their processed variants after using object recognition techniques. We computed local explanations of the images labelled with 1 by using LIME. Examples of results with $num\_features = 4$ can be seen in Fig. 9. All other images of local explanations look similar. From all resulting images, 22% exclusively highlight the object, while the rest of the image is blackened (cf. left images in Fig. 9). In the other cases, small parts of the background are highlighted as well (cf. right images in Fig. 9).

Subsequent, the remaining visible parts in the images of local explanations are analyzed. To restrict this analysis to objects only, pixels that do not have the background color are considered while this enables to process also images where small parts of the background can be seen. As described before, we derive a list for each analyzed feature, i.e., object position, object size, object shapes and object color. Allover, 498 different position values, 98 different size values, exactly one shape value (*rectangle*) and exactly one color value (*blue*) were analyzed. By comparing the possible values that could occur in the scene for each feature, the following result were obtained:

The origin size of the object were 46×118 pixels, while it is known a priori, that this the only object size that can occur in this scene. Therefore, it can be deduced that size can not be a decisive criterion. The same applies to the shape of the object.

The images have a size of 254×336 pixels. The table covers the whole background and since there are no further known restrictions, the object can be anywhere in the image. The only exception was that the object always had to be completely visible and should not protrude from the border of the image. If the size of the object is taken into account, it becomes apparent that the analyzed positions cover almost the entire tabletop. Therefore we conclude that also the position of the object is not a decisive criterion.

A comparison of the analyzed color value (*blue*) with the possible values occurring in the scene (*blue and green*) reveals that, the color is a decisive feature. In other words, the feature color with the value *blue* is the relevant process detail. It can be understood as a rule of the form "$(Color = blue) \Rightarrow (Label = 1)$" that must be followed. During the execution process participants have to make sure that only blue objects are placed

We integrate this information into the business process model by extending the considered task description. Therefore we edit the XML file describing the BPMN model and change the name of the task in the respective line. We show an excerpt of the modified file in Listing 1. Instead, this information can also be appended as text annotation like shown in Fig. 10.

Experiment with Two Relevant Features. In a second experiment we have two relevant features, i.e., the object shape and the object position in the scene. An image is labelled with 1, if at least one circular object is positioned in the upper seventh of the scene. Each image shows four objects with different shapes: Two rectangular and two circular objects. Their positions are again determined randomly. The background, the sizes of the objects and their color are no decisive features. We ensured that none of the

```
<bpmn:definitions [...]
 <bpmn:process [...]
  <bpmn:task id="Activity_1d7e3q8"
        name="place parts on charging plate;
        Color:{'blue'}">
   <bpmn:incoming>Flow_0k3x5p4</bpmn:incoming>
   <bpmn:outgoing>Flow_08y1hq5</bpmn:outgoing>
  </bpmn:task>
 </bpmn:process>
</bpmn:definitions>
```

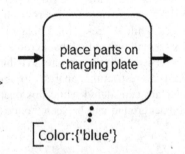

Listing 1. Extension of the task-related tag in the XML file by the relevant color information based on [17].

Fig. 10. Text annotation containing the color information according to [17].

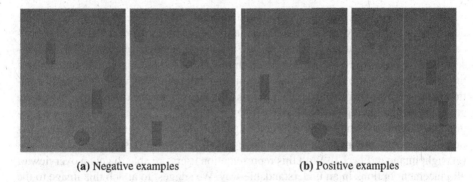

(a) Negative examples   **(b)** Positive examples

Fig. 11. Image data of the second experiment. At least one circular object has to be positioned at the top of the scene to be labelled as positive example [17].

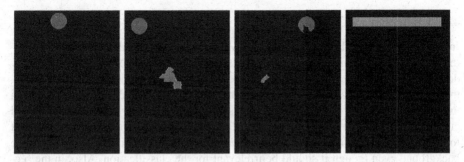

Fig. 12. Local explanations in the second experiment (left). Region of valid positions (right) [17].

objects intersects with each other and or protrudes beyond the image border. Figure 11 shows examples of the different classes. The local explanations resulting after applying LIME with parameter $num\_features = 2$ are presented in Fig. 12.

In all local explanations, one circular object is highlighted. Among them, 29% of the results are comparable to the leftmost image in Fig. 12. Others are comparable with the middle images and either highlight small areas of the background in addition to the object or do not contain the object completely.

Besides lists of positions and sizes, the analysis step outputs a single color value (*red*) and two shape values (*circle, pentagon*). We occurrence of the wrong shape value *pentagon* will be discussed in Sect. 4.4. From the comparison of these values with all possible values that could occur in the scene for each feature, we can deduce the following details:

The size and color are no relevant features.

The position values are a subset of all possible values and therefore describe a certain region in the scene. The position is therefore a relevant feature.

The shape *pentagon* is subsequently discarded because it cannot occur in the scene. The shape *circle* remains, which is a subset of all possible shapes indicating that the object shape is a relevant feature.

Both features can be integrated into the XML file of the process model like described in the first experiment. However, a list of positions is not readable and applicable for users in future process executions. For this reason, we select a pictorial representation for this process detail and create an image-based task annotation that contains the position information. We compute the minimal bounding box which covers all position values resulting from the analysis across all images of local explanations. We create an empty image having the same size as the input images (254×336). Then, we integrate the computed bounding box by coloring all pixels that are contained in the box. As output we receive an image that shows the region of valid positions for placing an object (cf. right image in Fig. 12). With this representation users are able to get an overview of all placement options in an understandable way. We suggest to attach this image to the respective process task in the model by using media annotation concepts proposed by [36]. The attachment will be shown to process participants during process execution.

4.4 Evaluation

In both experiments, the application of LIME outputs good results. In the first experiment, the local explanations reveal that the decision of the CNN for label 1 strongly depends on the object, while the background has no impact. In the second experiment, the LIME images represent the condition that one circular object has to be placed in the upper part of the scene. Irrelevant objects (i.e., the rectangular objects or the second circular object) and the background were blackened correctly. The only negative point to mention is that in both experiments, some local explanations highlighted parts of the background even though it should be irrelevant for the classification. Also, in the second experiment, not all parts of the object were always visible (cf. the third image in Fig. 12). The latter in turn led to an error in the subsequent analysis step, while the shape detection algorithm classified some objects as having the form *pentagon*. However, an adjustment of the segmentation step of LIME could increase these results.

The pre-filtering to relevant content by LIME is essential to derive a general conclusion. By analyzing the features of the highlighted image parts across all LIME results, we are able to compute a general rule or conclusion. In the context of Business Process Management, finding such a rule is an important issue regarding process success. A single positive example of a task execution is not sufficient as it restricts the scope of action of a process participant unnecessarily far. Nevertheless, we had to make many restrictions in the rule generation step to deduce the decisive criteria from local explanations.

Although the LIME images indicate which object is relevant, they can state which feature of the object is decisive. In the next step, we then analyze the relevant objects based on common features. For our prototype we determined a set of features. However, in real process environments, other features might be relevant and therefore a lot more features should be considered. In general, if a feature is usable depends on the domain and should be determined by a process expert. However, in this step we have to make assumptions regarding the feature space while we can no longer benefit from the fact that this was not necessary in the previous step.

In real process environments, scenes and process details can become much more complex. Although LIME has proven to be applicable for complex images [30], our rule generation step has to be adapted. The used image analysis techniques should be interchanged with more powerful ones to increase the robustness of out approach. If multiple objects are highlighted by LIME, the objects have to be analyzed separately. Creating combined list of values per feature would not provide useful results. Finally, a sufficient number of image data for training and validating the CNN may not be available in real process environments (cf. Section 3.4). We propose to integrate data augmentation techniques into the implementation to tackle this issue.

Concluding, the experiments confirm that our implementation approach is able to identify relevant process details from images recorded during task executions. Relevant parts of the images can be determined by using LIME without having knowledge about the feature space or any other assumptions. From a Business Process Management perspective, this is an essential aspect, since it is not known in advance which detail might be more or less important in a scene (cf. Section 1). With the subsequent analysis step, we are able to derive a general rule by extracting relevant features and values from highlighted regions per local explanation. The rule contains the information of relevant process details and has to be considered to increase process success.

5   Conclusions

We present a conceptual approach to extract relevant process details from data recorded during process executions. Thereby, data stemming from various sources are considered while we discuss their potential for analyzing certain types of process details. We describe in detail how existing process models can be extended by task annotations which reflect analyzed process details. Furthermore, we show how process improvement is ensured by validating the updated process model. Our comprehensive concept addresses both research questions, while we demonstrate applicability with image data by our implementation approach.

Our implementation is based on LIME to explain the predictions of a CNN. Therefore, we are able to explain which regions in a image were relevant for process success. By further analyzing these regions across all images, we show how to derive a general rule that contains the relevant process detail.

However, there is still space for improvement. Although our experiments show that LIME is able to correctly identify the relevant information, in some cases, regions in the image were highlighted that are not relevant, while in other cases, not all relevant regions were highlighted. This can led to errors in the subsequent step resulting in wrong rules. Further research is needed here, while we suggest that *i)* requirements regarding the quality of the data have to be investigated in detail, *ii)* other segmentation techniques should be tried in the explanation step, and *iii)* the approach to derive a general rule from local explanations has to be improved.

We aim on further evaluating our implementation approach with image data showing more complex scenes and objects from real process environments. Furthermore, we plan to investigate the applicability of structured data analysis techniques on images.

References

1. van der Aalst, W.M.P., van Dongen, B.F., Günther, C.W., Rozinat, A., Verbeek, E., Weijters, T.: Prom: the process mining toolkit, in bpm (demos). In: De Medeiros, A.K.A., Weber, B. (eds.) CEUR Workshop Proceedings, vol. 489 (2009)
2. Ahmadikatouli, A., Aboutalebi, M.: New evolutionary approach to business process model optimization. In: Proceedings of the International Multi Conference of Engineers and Computer Scientists. IMECS 2011, vol. 2 (2011)
3. Banziger, R.B., Basukoski, A., Chaussalet, T.: Discovering business processes in CRM systems by leveraging unstructured text data. In: 2018 IEEE 20th International Conference on High Performance Computing and Communications; IEEE 16th International Conference on Smart City; IEEE 4th International Conference on Data Science and Systems (HPCC/SmartCity/DSS), pp. 1571–1577 (2018). https://doi.org/10.1109/HPCC/SmartCity/DSS.2018.00257
4. Becker, J., Rosemann, M., von Uthmann, C.: Guidelines of Business Process Modeling, pp. 30–49. Springer, Berlin, Heidelberg (2000). https://doi.org/10.1007/3-540-45594-9_3
5. Bobrik, R., Reichert, M., Bauer, T.: View-based process visualization. In: Alonso, G., Dadam, P., Rosemann, M. (eds.) BPM 2007. LNCS, vol. 4714, pp. 88–95. Springer, Heidelberg (2007). https://doi.org/10.1007/978-3-540-75183-0_7
6. Brosnan, T., Sun, D.W.: Improving quality inspection of food products by computer vision–a review. J. Food Eng. **61**(1), 3–16 (2004)
7. Burkart, N., Huber, M.F.: A survey on the explainability of supervised machine learning. J. Artif. Intell. Res. **70**, 245–317 (2021)
8. Chinosi, M., Trombetta, A.: BPMN: an introduction to the standard. Comput. Stand. Interfaces **34**(1), 124–134 (2012)
9. Collins, A., Baccarini, D.: Project success-a survey. J. Constr. Res. **5**(02), 211–231 (2004)
10. Curtis, B., Kellner, M.I., Over, J.: Process modeling. Commun. ACM **35**(9), 75–90 (1992)
11. Dehghan-Manshadi, A., Yu, P., Dargusch, M., StJohn, D., Qian, M.: Metal injection moulding of surgical tools, biomaterials and medical devices: a review. Powder Technol. **364**, 189–204 (2020)
12. Dumas, M., la Rosa, M., Mendling, J., Reijers, H.A.: Introduction to business process management. In: Fundamentals of Business Process Management, pp. 1–31. Springer, Berlin, Heidelberg (2013). https://doi.org/10.1007/978-3-642-33143-5_1
13. Dzindolet, M.T., Peterson, S.A., Pomranky, R.A., Pierce, L.G., Beck, H.P.: The role of trust in automation reliance. Int. J. Hum.-Comput. Stud. **58**(6), 697–718 (2003)
14. Fichtner, M., Fichtner, U.A., Jablonski, S.: An experimental study of intuitive representations of process task annotations. In: Sellami, M., Ceravolo, P., Reijers, H.A., Gaaloul, W., Panetto, H. (eds.) Cooperative Information Systems. CoopIS 2022. LNCS, vol. 13591, pp. 311–321. Springer, Cham (2022). https://doi.org/10.1007/978-3-031-17834-4_19
15. Fichtner, M., Schönig, S., Jablonski, S.: Process management enhancement by using image mining techniques: a position paper. In: Proceedings of the 22nd International Conference on Enterprise Information Systems. ICEIS, vol. 1, pp. 249–255 (2020)
16. Fichtner, M., Schönig, S., Jablonski, S.: Using image mining techniques from a business process perspective. In: Filipe, J., Śmiałek, M., Brodsky, A., Hammoudi, S. (eds.) ICEIS 2020. LNBIP, vol. 417, pp. 62–83. Springer, Cham (2021). https://doi.org/10.1007/978-3-030-75418-1_4
17. Fichtner, M., Schönig, S., Jablonski, S.: How lime explanation models can be used to extend business process models by relevant process details. In: Proceedings of the 24th International Conference on Enterprise Information Systems - Volume 2: ICEIS, pp. 527–534. INSTICC, SciTePress (2022). https://doi.org/10.5220/0011067600003179

18. Front, A., Rieu, D., Santorum, M., Movahedian, F.: A participative end-user method for multi-perspective business process elicitation and improvement. Softw. Syst. Model. **16**(3), 691–714 (2017)

19. Gounaris, A.: Towards automated performance optimization of BPMN business processes. In: Ivanović, M., et al. (eds.) ADBIS 2016. CCIS, vol. 637, pp. 19–28. Springer, Cham (2016). https://doi.org/10.1007/978-3-319-44066-8_2

20. Kluza, K., Baran, M., Bobek, S., Nalepa, G.J.: Overview of recommendation techniques in business process modeling. In: Proceedings of 9th Workshop on Knowledge Engineering and Software Engineering (KESE9), pp. 46–57. Citeseer (2013)

21. Mansuri, I.R., Sarawagi, S.: Integrating unstructured data into relational databases. In: 22nd International Conference on Data Engineering (ICDE'06), pp. 29–29 (2006). https://doi.org/10.1109/ICDE.2006.83

22. Mendling, J., Reijers, H.A., van der Aalst, W.M.P.: Seven process modeling guidelines (7PMG). Inf. Softw. Technol. **52**(2), 127–136 (2010)

23. Niedermann, F., Radeschütz, S., Mitschang, B.: Deep business optimization: a platform for automated process optimization. In: INFORMATIK 2010-Business Process and Service Science-Proceedings of ISSS and BPSC (2010)

24. Patig, S., Casanova-Brito, V.: Requirements of process modeling languages-results from an empirical investigation (2011)

25. Petter, S., Fichtner, M., Schönig, S., Jablonski, S.: Content-based filtering for worklist reordering to improve user satisfaction: a position paper (2022)

26. Polyvyanyy, A., Smirnov, S., Weske, M.: Process model abstraction: a slider approach. In: 2008 12th International IEEE Enterprise Distributed Object Computing Conference, pp. 325–331. IEEE (2008)

27. Popescu, M.C., Lucian, M.S.: Feature extraction, feature selection and machine learning for image classification: a case study. In: 2014 International Conference on Optimization of Electrical and Electronic Equipment (OPTIM), pp. 968–973. IEEE (2014)

28. Radeschütz, S., Mitschang, B., Leymann, F.: Matching of process data and operational data for a deep business analysis. In: Mertins, K., Ruggaber, R., Popplewell, K., Xu, X. (eds.) Enterprise Interoperability III, pp. 171–182. Springer, London (2008). https://doi.org/10.1007/978-1-84800-221-0_14

29. Reichert, M., Kolb, J., Bobrik, R., Bauer, T.: Enabling personalized visualization of large business processes through parameterizable views. In: Proceedings of the 27th Annual ACM Symposium on Applied Computing, pp. 1653–1660 (2012)

30. Ribeiro, M.T., Singh, S., Guestrin, C.: Why should i trust you? Explaining the predictions of any classifier. In: Proceedings of the 22nd ACM SIGKDD International Conference on Knowledge Discovery and Data Mining, pp. 1135–1144 (2016)

31. Schmidt, R., Möhring, M., Zimmermann, A., Härting, R.-C., Keller, B.: Potentials of image mining for business process management. In: Czarnowski, I., Caballero, A.M., Howlett, R.J., Jain, L.C. (eds.) Intelligent Decision Technologies 2016. SIST, vol. 57, pp. 429–440. Springer, Cham (2016). https://doi.org/10.1007/978-3-319-39627-9_38

32. Schonenberg, H., Weber, B., van Dongen, B., van der Aalst, W.: Supporting flexible processes through recommendations based on history. In: Dumas, M., Reichert, M., Shan, M.-C. (eds.) BPM 2008. LNCS, vol. 5240, pp. 51–66. Springer, Heidelberg (2008). https://doi.org/10.1007/978-3-540-85758-7_7

33. Sehrawat, D., Gill, N.S.: Smart sensors: analysis of different types of IoT sensors. In: 2019 3rd International Conference on Trends in Electronics and Informatics (ICOEI), pp. 523–528. IEEE (2019)

34. Smith, J.R., et al.: RFID-based techniques for human-activity detection. Commun. ACM **48**(9), 39–44 (2005)

35. Stiefmeier, T., Roggen, D., Ogris, G., Lukowicz, P., Tröster, G.: Wearable activity tracking in car manufacturing. IEEE Pervasive Comput. **7**(2), 42–50 (2008). https://doi.org/10.1109/MPRV.2008.40

36. Wiedmann, P.C.K.: Agiles Geschäftsprozessmanagement auf Basis gebrauchssprachlicher Modellierung. Ph.D. thesis, University of Bayreuth, Germany (2017)

37. Yan, H., Paynabar, K., Shi, J.: Image-based process monitoring using low-rank tensor decomposition. IEEE Trans. Autom. Sci. Eng. **12**(1), 216–227 (2014)

38. Zappa, D., Clemente, G.P., Borrelli, M., Savelli, N.: Text mining in insurance: From unstructured data to meaning. Variance **14**, 1–15 (2019)

Considering User Preferences During Business Process Execution Using Content-Based Filtering

Sebastian Petter$^{(\boxtimes)}$, Myriel Fichtner, and Stefan Jablonski

University of Bayreuth, Bayreuth, Germany
{sebastian.petter,myriel.fichtner,
stefan.jablonski}@uni-bayreuth.de

Abstract. Shortening processing time and increasing the quality of process outcomes are two of the main goals of most companies regarding business processes. Business Process Management (BPM) is used to realize these company goals. One of the fundamentals of executing business processes often neglected during process optimization are human resources. Considering other user-centered domains, recommender systems often support users during decisions to increase user satisfaction. This inspired us to apply techniques established by recommender systems in the BPM domain since employee satisfaction significantly influences productivity. Thus, considering user preferences during business process execution leads employees to be more satisfied and productive. We proposed a recommendation service that exploits the information from event logs to guide users during process execution. In this paper, we refine this approach and make it more effective. In addition, as another major contribution of this paper, a proof of concept implementation is conducted to show the feasibility of our concept.

Keywords: Business process management · Recommender systems ·
Content-based filtering · User-centered process improvement · User
satisfaction · Worklist optimization · Process activity similarity

1 Introduction

Business Process Management (BPM) is a highly relevant research topic from a practical point of view [11,31]. The fundamentals of BPM are sequences of activities, namely business processes. Optimizing such business processes is one of the primary purposes of research related to BPM. Increasing the quality of process outcomes or shortening

The project is financed with funding provided by the Federal Ministry of Education and Research and the European Social Fund under the "Future of work" programme and managed by the Project Management Agency Karlsruhe (PTKA). The author is responsible for the content of this publication.

SPONSORED BY THE

   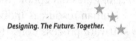

J. Filipe et al. (Eds.): ICEIS 2022, LNBIP 487, pp. 415–429, 2023.
https://doi.org/10.1007/978-3-031-39386-0_20

processing time are two of the main goals of most companies regarding BPM [17–19]. Companies need to deeply analyze the structure of their processes to reach optimization. This could be realized by visualizing them using process modeling languages like Business Process Model and Notation (BPMN) [23]. The resulting business process model is a formal representation of the sequence of required steps to execute in this process as well as the process entities involved, like process participants or data objects. To manage and execute business process models, Process-Aware Information Systems (PAIS) have been developed [28]. A PAIS determines which activities from various running process instances can be executed next by which process participant and provides them as items in a worklist. During process execution, the executed activity, as well as additional information like execution duration or resources (process participants) executing activities, are recorded in a so-called process event log [1]. Extracting knowledge from event logs through process mining allows process optimization [21]. Most approaches are activity centered and focus on process control flow [15,30]. Many processes executed in small and medium enterprises are human-driven. Since human resources are an essential component of business process executions in companies, they must be considered as a target for optimization.

The needs and expectations of human workers have to be regarded by companies since they are the centerpieces of a company. Increasing employee satisfaction and consequently decreasing fluctuation is a result of considering such aspects. Otherwise, the employees' productivity can be increased if the worker is satisfied with the assigned work. It is proven that "happy" workers are 13 % more productive [5,7]. In addition, when workers have more control over their work, i.e., they can widely decide what task to do in what sequence, human resources are more productive and motivated [29].

Regarding the user's preferences is one possibility to increase satisfaction and motivation of employees. Since BPM systems do not consider user preferences per se, we adopt the discovery of user preferences from the domain of recommender systems. Recommender systems take into account user preferences when assigning items, e.g., tasks of a process to be performed. They are mainly developed for e-commerce applications [3]. The use of such concepts has proven to be profitable for companies. These results inspired us to adopt the methods of recommender systems and apply them in the context of BPM for user-centered process improvement.

To the best of our knowledge in existing research, the primary purpose of worklist reordering or process optimization is to improve key performance indicators. However, user satisfaction has not yet been addressed adequately so far. To address the challenges mentioned earlier and to assist users during process execution, we previously presented an approach for intelligent user assistance in flexible PAISs [24]. In particular, we proposed a recommendation service that exploits the information from event logs to guide users during process execution. In this paper, we refine this approach and make it more effective. In addition, as another major contribution of this paper, a proof of concept implementation is conducted to show the feasibility of our concept. For our approach and the implementation, we need to enhance process activities and process event logs to adopt the concept of content-based filtering to BPM.

Human resources individually have specific preferences. As an example, consider a software process consisting of two activities "frontend development" and "backend development" besides others. Software developers must execute both. However, not all of them like, for example, JAVA as a development language for backend development; instead, they like HTML, which is suitable for frontend development. Due to missing knowledge about the preferences of software developers, the activities are randomly allocated to the available developers in a project. The developers must finally execute these activities independently of their and coworkers' preferences. This might diminish the motivation of the people to do their jobs.

The core idea of our research is to integrate recommender systems into BPM systems to enhance user satisfaction in process execution. Thus, the main research question we address in this paper is whether and - if possible - how we can derive recommendations during business process execution considering user preferences. Specifically, we want to apply content-based filtering methods known from recommender systems. Two research sub-questions can be identified:

1. Which information from BPM can be exploited by a recommender system for inferring user preferences?
2. How should these results be displayed to the users?

This paper is motivated through our cooperation within a joint project called PRIME (Process-based integration of human expectations in digitalized work environments), funded by the Federal Ministry of Education and Research and the European Social Fund. The project consists of an interdisciplinary consortium of three research partners, three application partners, and two implementation partners. In our work, we present an exemplary real-world process as running example. This business process is provided by one of the application partners to emphasize the relevance of this topic.

The remainder of the paper is structured as follows. The two main research areas, BPM and recommender systems in general, considered in this work are presented in Sect. 2. Furthermore, we show existing approaches combining these aspects. In Sect. 3, we describe the core idea of our recommending activities to the current user of a PAIS using content-based filtering. Section 4 presents the implementation of our approach as well as the application of two examples with different data sets and business processes. We conclude our work and present ideas for future work in Sect. 5.

2   Background

In this work, we use methods known from recommender systems in the domain of BPM to accomplish user-centered process improvement through worklist optimization. In this section, we first introduce Business Process Management and recommender systems, the fundamental research areas of our approach. We then present related work that addresses both topics.

Business Process Management. Increasing efficiency and reducing costs of business processes are two of the main goals of BPM. Therefore it is used in enterprises of all industries and sizes. Modeling, executing, and analyzing business processes are the fundamentals of BPM [11]. A business process model, a formal representation of the business process, is needed to execute a business process. This model is executed by a Process-Aware Information System, which handles the control flow and determines activities to be executed next. Those activities are placed into a worklist the users' interface to the PAIS. Each user has their own worklist containing all activities (s)he is eligible to perform. Usually, multiple users with the same roles are involved in the process execution. Consequently, the worklists of different users can overlap or even be identical. As long as there are no further restrictions, users can select any activities provided within the worklist, e.g., their preferred activity. Executing and completing these activities is recorded as an event in a so-called (process) event log. Besides information about the activity and involved resources, an event contains further information about the execution, like execution timestamp or data used during execution. Furthermore, other properties can arbitrarily extend this set of information [1].

Recommender Systems. One of the main attributes used to support users are preferences [13]. Recommender systems use these user preferences to predict results for alternative options [14]. In various domains like e-commerce and other user-centered web applications, recommender systems are used to show items to the user they may like [20,22]. As input data, recommender systems need information about users, items to be recommended, and user preferences, e.g., ratings. In generating potential ratings for an item list, four recommendation methods have been established: content-based filtering, knowledge-based filtering, collaborative filtering, and hybrid recommendation. These recommendation methods differ mainly in the input data they use to calculate recommendations. In our work, we focus on content-based filtering.

Content-based filtering methods learn user preferences from item ratings. The idea is the following. Items are attributed in order to cluster them. For example, books are classified according to their genres. Then, users rate items, i.e., they express their preferences. The rating can be simple (e.g., like/dislike) or more detailed (e.g., one to five stars). Considering all items available, we can split these into two basic lists: one list of rated items and one of unrated. Using the list of rated items, a content-based filtering method transfers user ratings to other items of a corresponding item class from the second list. For example, if a user rates books of class "drama" as very interesting, this rating is also assigned to books out of this class the user never read before. The latter are candidates to be recommended to the user afterward.

Content-based filtering needs two types of input data: (i) user feedback and (ii) item attributes. Then, the recommendation process is realized in three steps [3]:

1. Preprocessing and Feature Extraction. This is to extract features from considered items and convert them into a keyword-based vector space. These features are directly derived from the additional attributes provided before.
2. Learning User Profiles. A user-specific model (profile), based on the history of a user rating items, is created.
3. Filtering and Recommendation. The user model is used to derive recommendations on items for the corresponding user.

Recommendations are only possible if the user already has rated at least one item (cold-start problem). Another drawback of content-based filtering methods is that only apparent items are recommended because the algorithm recommends items with similar attributes. The advantage is that no information about ratings from other users is needed. Furthermore, new items can be recommended immediately since they are comparable to existing items due to the item attributes.

Related Work. Optimizing process results or shortening processing time are two of the main goals considering process improvement through recommendations. Recommending the next activity a user should execute is considered in [2, 10, 15, 30], and [16]. In [15, 30], and [2], the authors calculate recommendations based on the current execution trace and similar historical process instances. The goal of supporting the user by recommending next activities is to shorten processing time or to reduce risks (e.g., exceeding deadlines during process execution) [10, 16]. An optimized execution plan is generated using information about runtime estimation and resource availability, which were added to the business process model [4]. Deriving user preferences from event logs is presented in [6]. In [8], a priority-based optimal resource allocation at process level is presented using the Semantic Ontology of User Preferences [12]. The resource with the highest preference is allocated to the new activity. Furthermore, we consider two approaches using and measuring the effects of worklist reordering. In [25], the authors show that reordering the process participants' worklist does not negatively impact temporal parameters like throughput time. Another approach of reordering worklist items is presented in [26] to improve key performance indicators. We introduced an approach to combine content-based filtering with BPM to recommend worklist items to the user he possibly prefers [24]. This approach is extended in this paper. Summing up, recommending next activities and resource allocation for performance optimization are two research areas well considered, whereas user-centered process improvement still needs to be examined.

3 Conceptual Approach

To illustrate our concept, we first introduce a small running example of a real-world business process from one of our project partners. Figure 1 shows a simplified process model defined in BPMN [23]. The model visualizes a process where human resources with the role *Team Leader* must perform five different activities during process execution: "Accept order (ID1)", "Create roster (ID2)", "Dispatch roster (ID3)", "Preparation of order (ID4)", and "Receive special request (ID5)". The process starts with the customer sending an order by mail to the *Team Leader*. After accepting the order, the *Team Leader* creates the roster for the technician in Excel and mails it to the technician. Next, the *Team Leader* must prepare the order by obtaining detailed order information from the customer and mailing that information to the technician. As final step, the *Team Leader* must perform any special requests that the technician requests by mail.

We aim to use content-based filtering methods known from recommender systems in business process management to consider user preferences during process execution. To achieve this, we first need to identify which data from business process management can be exploited for the recommendation algorithm and whether we need to extend process descriptions. A content-based recommender system needs two kinds of input data to derive preferences: a list of items already rated by a specific user and a list of

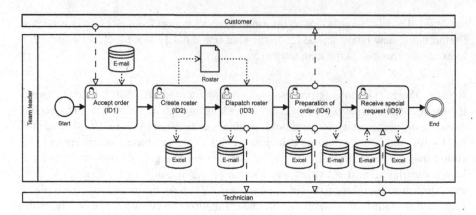

Fig. 1. Example: Technician service management process [24].

items not rated by this user so far. Now, we have to identify to what an item - from the realm of recommender systems - is corresponding in the BPM realm. Shortly, we assume that an item corresponds to a task. Again, there are items (tasks) already rated by users. This list of items can be found in the event log of a PAIS, together with their ratings. Besides, some tasks are still not rated. Many of them can be found in a user's worklist. A worklist contains all tasks a user can potentially execute next. Our goal is to reorder this list according to the user preferences.

Applying content-based filtering methods, tasks (items) have to be attributed. Thus, we extend task descriptions by tags that realize those attributes.

Figure 2 depicts our concept. The PAIS provides the worklists - including items to be reordered according to recommendations - and an event log storing all task executions together with their ratings. A component for attribute management and one for calculating recommendations still have to be added. In total, four steps need to be executed to recommend worklist items to users according to their preferences:

1. Activity Attribution (Tagging). This step must be executed once before the recommendation algorithms are able to start.
2. Content-based Filtering. This step is continuously executed during process execution.
3. Worklist Reordering. This step is executed before a user refreshes the worklist.
4. Activity Rating. This step must be executed for each step users perform.

Activity Attribution (Tagging). The input data for this step of our concept are the activities known from the process model. The goal is to enrich the given activities with more detailed semantic information since we want to make the activities comparable. There are two types of attributes: static and dynamic attributes. Static attributes are generated during modeling time and barely change. They can be derived from the business process model or extracted from process descriptions or interviews with the process owner. On the other hand, dynamic attributes are generated during process execution

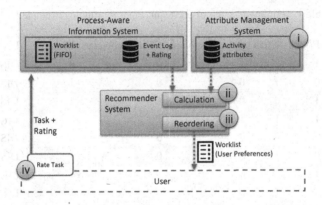

Fig. 2. Conceptual approach.

during runtime. During runtime, we can obtain more information about specific executed tasks with the help of the PAIS. For example, we can derive timestamps, which we can also treat as attributes. Furthermore, the user generates data during execution, like information about the customer, which can also be handled as attributes.

Since the attributes are handled the same during calculation, whether static or dynamic, we consider static attributes for our running example. Deriving attributes from the given business process model results in five attributes an activity can have in our running example: E-mail, Communication with customer, Excel, Roster, and Communication with technician. For each activity, it is determined whether the attribute is required or not during the execution of the task.

Table 1 shows a small cutout of the event log for our running example for a particular user. The event log consists of the activities, the attributes, and the user ratings, which are examined in more detail in step (iv). Typically, the event log contains more information about human resources executing the activity, process instances, duration, and more. Since this is not relevant to our approach, we have omitted this information.

Content-Based Filtering. To generate recommendations with a content-based filtering method known from recommender systems, we need the following input data: a set of rated activities as training set A_R and a set of unrated activities as A_U. We consider the event log enriched with attributes and user ratings introduced in step (i) as training data A_R to generate recommendations. The worklist provided by the PAIS containing all possible activities for the considered user, including attributes, is used as the set of unrated activities A_U. The goal is to predict the rating of each activity from A_U using the training data. Both A_R and A_U are user-specific sets of activities.

The problem we have to solve is similar to that of classification. To classify data sets, an established technique is the nearest neighbor classifier [9]. For our running example, we will use this classification. Note that this method is interchangeable and can be replaced by other classification methods.

In this work, we define a similarity function based on the nearest neighbor classifier where similarity or distance functions are used for multidimensional or structured data.

Table 1. Training data A_R of one team leader for running example [24].

ActivityId	E-mail	Excel	Com. Customer	Com. Technician	Roster		Like/Disklike
ID1	1	0	1	0	0	...	DISLIKE
ID2	0	1	0	0	1	...	DISLIKE
ID3	1	0	0	1	1	...	DISLIKE
ID4	1	1	1	1	0	...	LIKE
ID1	1	0	1	0	0	...	DISLIKE
ID2	0	1	0	0	1	...	DISLIKE
ID3	1	0	0	1	1	...	LIKE
ID2	0	1	0	0	1	...	LIKE
ID2	0	1	0	0	1	...	DISLIKE
ID3	1	0	0	1	1	...	LIKE
ID4	1	1	1	1	0		LIKE

The Euclidean distance is used to calculate the distance between two vectors. This is to compare the similarity of two activities. Due to better comparability, we transform the results into a similarity measure afterward.

The Euclidean distance is calculated as follows:

$$d(p,q) = \sqrt{\sum_{i=1}^{n}(q_i - p_i)^2} \qquad (1)$$

where p and q are a pair of vectors representing the n attributes of the two considered activities. Furthermore, a similarity metric is needed to improve comparability of distances:

$$sim(p,q) = \frac{1}{1 + d(p,q)} \qquad (2)$$

To identify the nearest neighbor, the similarity metric of each activity in A_U to each activity from A_R is calculated. The rating of the activity A_U can be deduced by considering the rating of the nearest neighbor activity.

Considering our running example, the training set A_R consists of the event log enriched with user ratings and the activity attributes created in the activity rating and attribution steps. In Table 1, the training set A_R is displayed. The current user has executed the activities with ID1 - ID4 several times and has rated them in contrast to activity ID5, which has never been executed by the current user. Furthermore, the table shows the attributes for every activity. The values 0 and 1 show whether an attribute is present for the activity. As an example, the vector representation of the attributes of activity ID4 is (1,1,1,1,0) since it is related to the attributes E-mail, Excel, communication with customer, and communication with technician.

The current worklist of the team leader is considered as set of unrated activities A_U. In our running example, the worklist the PAIS provides consists of five tasks in the following order, where every task can be executed independently of the others:

1. Create roster (ID2)
2. Dispatch roster (ID3)
3. Receive special request (ID5)
4. Accept order (ID1)
5. Preparation of order (ID4)

Since we only consider static attributes in our running example and all activities except Receive special request (ID5) have been executed at least once, we only need to calculate the nearest neighbor of activity ID5. As an example, we calculate the similarity between the activities ID5 (p) and ID4 (q). The attributes of the activities can be represented as vectors $v_p = (1,1,0,1,0)$ and $v_q = (1,1,1,1,0)$. Then $d(p,q) = 1$ and $sim(p,q) = 0,5$. Calculating all values for p of any activity in A_R, we notice that 0.5 is the highest value. A value of 1 means that the activities are equal regarding their attributes and a value of 0 means they are completely different. Considering the event log and the user rating, we determine that the user always liked activity ID4. Thus, the team leader possibly likes activity ID5, too. The team leader has already executed the remaining activities, following that the nearest neighbor is the activity itself, and the similarity equals 1. Deriving the average rating for activity ID2 from the event log, we determine that the activity execution has been disliked three out of four times. Thus, the team leader potentially dislikes this activity in the future, too. The potential rating of the remaining activities can be calculated in the same way since they have been executed at least once. Considering dynamic attributes, too, the calculations get more information about the activities, making predictions more accurate.

The result of this step of our concept is that the nearest neighbor of activity ID5 is activity ID4, with a calculated similarity of 0.5. Since activity ID4 has been liked by the user both times it has been executed, the potential rating for activity ID5 is $like$, too. Furthermore, the result contains the potential ratings of all remaining worklist activities derived from their past ratings.

Worklist Reordering. The main goal of our concept is to reorder the worklist the PAIS provides according to user preferences. Since we calculated the potential rating for all activities in the worklist in step (iii), we can now arrange the worklist according to the results and adapt it to the requirements of the current user. Since we do not omit any tasks nor add any new tasks, only the order of the worklist items is affected during the manipulation. The active user can still decide which activity to execute next.

Considering our running example, we use the results calculated in step (iii) and adapt the worklist according to these recommendations. Following reordered worklist results:

1. Receive special request (ID5)
2. Preparation of order (ID4)
3. Dispatch roster (ID3)
4. Create roster (ID2)
5. Accept order (ID1)

In the provided example, three activities ("Dispatch roster", "Receive special request", and "Preparation of order") are recommended to execute, because of the historical ratings of the user. The last two activities in the reordered worklist are outside the user's favorite tasks. Perhaps another team leader with opposite preferences performs the latter tasks, thus increasing that user's satisfaction.

Our running example considers a worklist in first-in-first-out order. However, approaches exist for reordering the worklist due to better process performance or other optimization goals. Regarding these preordered worklists, we do not want to falsify the results from previous calculations. One possibility of including our approach in preordered worklists is to reorder only tasks with the same priority.

Activity Rating. The last step for the user is to select one activity, complete it and afterward rate it. The last part of this step is the most important: rating the activity just executed. The activities executed must be rated by the process participant since a standard event log contains all information we need to calculate recommendations with content-based filtering methods except explicit user feedback. Every event log item is extended with the user rating, providing information about the user liked or disliked executing this activity. In our approach, we only consider the ratings *like* and *dislike*. More complex ratings are considered in future work.

4   Implementation and Feasibility

To validate our approach, we developed a proof-of-concept implementation for reordering worklists according to user preferences[1]. We then conducted the approach on two different datasets to validate its feasibility. First, we used the running example provided in this paper as a simple test for the application. After that, we took a freely available event log concerning the ticketing management process of the help desk of an Italian software company [27] with a sample worklist and calculated recommendations for the user.

The overall implementation to calculate recommendations is modularized into four parts: the main application, a recommendation engine, a rating engine, and an attribution engine. For the proof-of-concept implementation, we did not connect a real PAIS but replaced it with a PAIS mock providing a worklist and an event log. This demonstrates that we are independent of PAIS, which can be replaced by various systems already on the market. We only implement an adapter to retrieve the worklist and the event log. The main application is the central interface handling all data needed, combining the engines provided, and providing a user interface. The main task of the prototype is to provide a reordered worklist to the user according to his preferences. Furthermore, the user can create attributes and assign them to activities, and rate activities after executing them. In backend, the main application gets the worklist and the event log from the PAIS. The attribution engine provides functions to generate, read, and

---

[1] Code available at: https://gitlab.com/recpro/2022-iceis-extension.

delete attributes and assign them to activities (see conceptual approach, step (i)). Currently, we only consider whether an attribute is present within an activity. More complex attributes will be added in future implementation. The attribution of activities is a manual step a user must perform once before using the recommendation mechanism. To get user preferences, the rating engine is developed. It is responsible for user ratings after task execution (conceptual approach: step(iv)). Completing a task, the user can decide whether he liked or disliked the execution of the activity. The provided rating will be stored in a database with the event ID to identify the event in the event log. The rating engine can easily be extended by more complex ratings like five or ten star-ratings with more detailed gradations. Taking all data provided by the PAIS, the attribution engine, and the rating engine, the recommendation engine is used to calculate the user preferences (conceptual approach: step (ii)). The calculations are used in the main application to reorder the worklist according to the user's preferences (conceptual approach: step (iii)). Contemplating the schematic representation of the overall application architecture in Fig. 3, we notice that all engines are interchangeable and expandable. This allows us to extend our approach by different recommendation methods and more complex attribution and ratings without touching the fundamental structure. The single modules can also be integrated into an existing PAIS if it provides an interface to reorder the worklist.

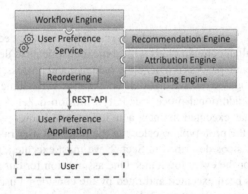

Fig. 3. Prototype architecture.

We demonstrate the feasibility of our approach by applying the presented example in our developed application. According to our steps presented in Sect. 3, we first have to create attributes for each task, which can be realized directly in the application. We can create new attributes manually and attach them to all available activities. The attributes should automatically be derived from the process model in a future implementation. The event log, as well as the worklist for our current application user, is provided by the PAIS. In contrast, the ratings for the single events are stored in our recommendation application since they are generated there. The worklist is shown in the default order in our application. Switching to the content-based recommendation order, our application provides, steps (ii) and (iii) of our approach are executed. The worklist, the enriched event log, and the activities' attributes are used to calculate each worklist item's potential ratings. The result is a worklist containing the same elements as the default worklist, ordered by potential user preferences (see Fig. 4). Furthermore, we display the information calculated by the recommendation engine. Besides information on whether an activity has already been executed and, if, how often, the worklist provides information about the predicted rating and the activity most similar.

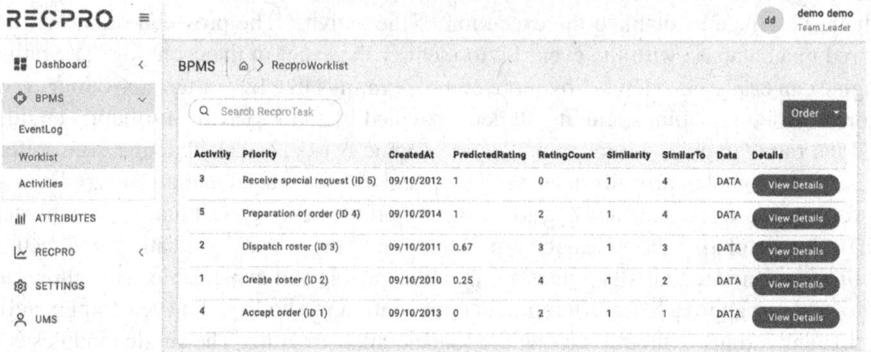

Fig. 4. Reordered worklist of running example displayed in prototype.

Since no existing approaches consider user ratings, we decided to use a real-life event log from an Italian company [27] with some adjustments to demonstrate the feasibility. The fundamental process is a ticketing management process. Deriving the activities from the event log, we receive eight different tasks. We assume that analyzing the process results in an additional process activity ("Inform ticket applicant") is necessary. Since no user has executed the new activity, no one knows whether he likes the activity. We can use the prototype to calculate the potential user rating of the activity. Before executing the steps described in Sect. 2, we must execute a preprocessing step to generate ratings for the event log items. One assumption for our approach is that at least one activity has been executed and rated by the considered user. To simulate this assumption, we can manually rate activities in the event log, but due to many events, we implemented a random rating mechanism, too. Next, we generated a worklist provided by the PAIS-mock application. The worklist contains the newly created activity "Inform ticket applicant" as well as four other tasks to execute ("Insert ticket", "Require upgrade", "Resolve ticket", and "Schedule intervention").

Now, having the event log and the worklist generated, we need to execute step (i) of our concept: activity attribution. For our example, we add a set out of eight different static attributes to the activities (see Table 2):

Considering the static attributions and calculating the similarities between the activities, we receive the following table displayed in the prototype (Fig. 5):

We can derive that the activity most similar to the newly introduced activity "Inform ticket applicant" is "Insert ticket". This information can be used to reorder the worklist according to step (iii) of our approach. Running the content-based filtering method for the worklist, we can observe that depending on the random rating, the only activity the user has never executed is consistently rated as its nearest neighbor.

Table 2. Activity attribution for help desk event log.

Activity	Ticket	Software	Priority	Timeline	Excel	Implementation	Ticket system	E-Mail
Assign seriousness	1	0	1	0	0	0	1	0
Create software anomaly	1	1	0	0	0	0	1	0
Insert ticket	1	0	0	0	0	0	1	1
Require upgrade	1	1	0	0	0	0	1	0
Resolve software anomaly	1	1	0	0	0	0	1	0
Resolve ticket	1	1	0	1	1	1	1	0
Schedule intervention	1	0	0	1	1	0	1	0
Take in charge ticket	1	0	1	1	1	0	1	0
Inform ticket applicant	1	0	0	0	0	0	1	1

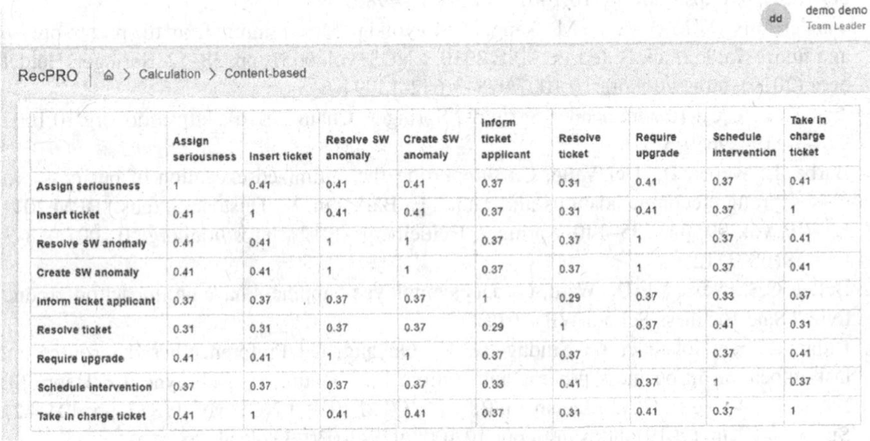

dd   demo demo
Team Leader

RecPRO | ⌂ > Calculation > Content-based

	Assign seriousness	Insert ticket	Resolve SW anomaly	Create SW anomaly	Inform ticket applicant	Resolve ticket	Require upgrade	Schedule intervention	Take in charge ticket
Assign seriousness	1	0.41	0.41	0.41	0.37	0.31	0.41	0.37	0.41
Insert ticket	0.41	1	0.41	0.41	0.37	0.31	0.41	0.37	1
Resolve SW anomaly	0.41	0.41	1	1	0.37	0.37	1	0.37	0.41
Create SW anomaly	0.41	0.41	1	1	0.37	0.37	1	0.37	0.41
Inform ticket applicant	0.37	0.37	0.37	0.37	1	0.29	0.37	0.33	0.37
Resolve ticket	0.31	0.31	0.37	0.37	0.29	1	0.37	0.41	0.31
Require upgrade	0.41	0.41	1	1	0.37	0.37	1	0.37	0.41
Schedule intervention	0.37	0.37	0.37	0.37	0.33	0.41	0.37	1	0.37
Take in charge ticket	0.41	1	0.41	0.41	0.37	0.31	0.41	0.37	1

Fig. 5. Content-based calculation for helpdesk event log.

5   Conclusion and Future Work

In this work, we addressed the problem of user-centered process improvement. We provide a novel approach to consider user preferences during business process execution and illustrate it with a running example. Using content-based filtering known from recommender systems allows us to reorder a given worklist according to the preferences of the current user of a PAIS. To use content-based filtering during business process execution, we introduced activity attributes and user ratings for executed tasks. In contrast to [24], we extended our concept with more details. Furthermore, we introduced an initial proof of concept implementation showing the feasibility of our approach. We outlined a prototype for generating recommendations on next activities according to the users' preferences. We show how established algorithms from recommender systems can be applied to the domain of BPM to discover similarities between process activities to predict the rating a specific user will give a specific task. Our concept shows that extended activities and enhanced event logs provided by PAISs can be used to infer user preferences. In our implementation, understanding recommendations is supported by displaying the activity that is most similar to the activity to be performed.

In future work, the concept will be extended to support complex ratings and runtime attributes. This allows for a more accurate prediction of user ratings. Further recommendation methods must be added to tackle drawbacks of content-based filtering, like the cold-start problem for new users. This includes adding collaborative filtering, knowledge-based filtering, and hybrid filtering methods. Finally, we plan to evaluate our approach in a real-world scenario.

References

1. van der Aalst, W.M.P.: Process Mining: Data Science in Action, 2nd edn. Springer, Heidelberg (2016). https://doi.org/10.1007/978-3-662-49851-4
2. van der Aalst, W.M.P., Pesic, M., Song, M.: Beyond process mining: from the past to present and future. In: Pernici, B. (ed.) CAiSE 2010. LNCS, vol. 6051, pp. 38–52. Springer, Heidelberg (2010). https://doi.org/10.1007/978-3-642-13094-6_5
3. Aggarwal, C.C.: Recommender Systems. Springer, Cham (2016). https://doi.org/10.1007/978-3-319-29659-3
4. Barba, I., Weber, B., Del Valle, C.: Supporting the optimized execution of business processes through recommendations. In: Daniel, F., Barkaoui, K., Dustdar, S. (eds.) BPM 2011. LNBIP, vol. 99, pp. 135–140. Springer, Heidelberg (2012). https://doi.org/10.1007/978-3-642-28108-2_12
5. Bellet, C.S., Neve, J.E.D., Ward, G.: Does employee happiness have an impact on productivity? Saïd Business School WP (2019)
6. Bidar, R., ter Hofstede, A., Sindhgatta, R., Ouyang, C.: Preference-based resource and task allocation in business process automation. In: Panetto, H., Debruyne, C., Hepp, M., Lewis, D., Ardagna, C.A., Meersman, R. (eds.) OTM 2019. LNCS, vol. 11877, pp. 404–421. Springer, Cham (2019). https://doi.org/10.1007/978-3-030-33246-4_26
7. Bryson, A., Forth, J., Stokes, L.: Does worker wellbeing affect workplace performance? Technical report, Department for Business, Innovation & Skills (2015)
8. Cabanillas, C., García, J.M., Resinas, M., Ruiz, D., Mendling, J., Ruiz-Cortés, A.: Priority-based human resource allocation in business processes. In: Basu, S., Pautasso, C., Zhang, L., Fu, X. (eds.) ICSOC 2013. LNCS, vol. 8274, pp. 374–388. Springer, Heidelberg (2013). https://doi.org/10.1007/978-3-642-45005-1_26
9. Chomboon, K., Chujai, P., Teerarassammee, P., Kerdprasop, K., Kerdprasop, N.: An empirical study of distance metrics for k-nearest neighbor algorithm. In: Proceedings of the 3rd International Conference on Industrial Application Engineering, pp. 280–285 (2015)
10. Conforti, R., de Leoni, M., Rosa, M.L., van der Aalst, W.M.P., ter Hofstede, A.H.M.: A recommendation system for predicting risks across multiple business process instances. Decis. Support Syst. **69**, 1–19 (2015)
11. Dumas, M., Rosa, M.L., Mendling, J., Reijers, H.A.: Fundamentals of Business Process Management, 2nd edn. Springer, Heidelberg (2018). https://doi.org/10.1007/978-3-662-56509-4
12. García, J.M., Ruiz, D., Ruiz-Cortés, A.: A model of user preferences for semantic services discovery and ranking. In: Aroyo, L., et al. (eds.) ESWC 2010. LNCS, vol. 6089, pp. 1–14. Springer, Heidelberg (2010). https://doi.org/10.1007/978-3-642-13489-0_1
13. Goldsmith, J., Junker, U.: Preference handling for artificial intelligence. AI Mag. **29**, 9–12 (2008)

14. Guo, Z., Zeng, W., Wang, H., Shen, Y.: An enhanced group recommender system by exploiting preference relation. IEEE Access **7**, 24852–24864 (2019)

15. Haisjackl, C., Weber, B.: User assistance during process execution - an experimental evaluation of recommendation strategies. In: Proceedings of the BPI Workshop, vol. 66, pp. 134–145 (2010)

16. Huber, S., Fietta, M., Hof, S.: Next step recommendation and prediction based on process mining in adaptive case management. In: 7th International Conference on Subject-Oriented Business Process Management (2015)

17. Jablonski, S., Bussler, C.: Workflow Management: Modeling Concepts, Architecture, and Implementation (1996)

18. Koulopolous, T.M.: The Workflow Imperative: Building Real World Business Solutions (1995)

19. Lawrence, P.: Workflow Handbook (1997)

20. Manouselis, N., Costopoulou, C.: Analysis and classification of multi-criteria recommender systems. World Wide Web **10**(4), 415–441 (2007)

21. Marin-Castro, H.M., Tello-Leal, E.: Event log preprocessing for process mining: a review. Appl. Sci. **11**, 10556 (2021)

22. Nguyen, H., Haddawy, P.: DIVA: applying decision theory to collaborative filtering. In: Proceedings of the AAAI Workshop on Recommender Systems (1998)

23. OMG: Business Process Model and Notation (BPMN), Version 2.0 (2011)

24. Petter, S., Fichtner, M., Schönig, S., Jablonski, S.: Content-based filtering for worklist reordering to improve user satisfaction : a position paper. In: Proceedings of the 24th International Conference on Enterprise Information Systems, vol. 2, pp. 589–596. SciTePress, Portugal (2022)

25. Pflug, J., Rinderle-Ma, S.: Analyzing the effects of reordering work list items for selected control flow patterns. In: IEEE 19th International Enterprise Distributed Object Computing Workshop, pp. 14–23 (2015)

26. Pichler, H., Edre, J.: Towards look-ahead strategies for work item selection. In: 2019 IEEE Jordan International Joint Conference on Electrical Engineering and Information Technology (JEEIT), pp. 752–757 (2019)

27. Polato, M.: Dataset belonging to the help desk log of an Italian Company (2017)

28. Reichert, M., Weber, B.: Enabling Flexibility in Process-Aware Information Systems: Challenges, Methods, Technologies. Springer, Heidelberg (2012). https://doi.org/10.1007/978-3-642-30409-5

29. Russell, N., van der Aalst, W.M.P., ter Hofstede, A.H.M.: Workflow Patterns: The Definitive Guide. MIT Press, Cambridge (2016)

30. Schonenberg, H., Weber, B., van Dongen, B., van der Aalst, W.: Supporting flexible processes through recommendations based on history. In: Dumas, M., Reichert, M., Shan, M.-C. (eds.) BPM 2008. LNCS, vol. 5240, pp. 51–66. Springer, Heidelberg (2008). https://doi.org/10.1007/978-3-540-85758-7_7

31. Weske, M.: Business Process Management - Concepts, Languages, Architectures. Springer, Heidelberg (2007). https://doi.org/10.1007/978-3-540-73522-9

Author Index

J. Filipe et al. (Eds.): ICEIS 2022, LNBIP 487, pp. 431–432, 2023.
https://doi.org/10.1007/978-3-031-39386-0

Printed in the United States
by Baker & Taylor Publisher Services